# Causality
# in
# Crisis?

Studies in Science and the
Humanities from the
Reilly Center for Science,
Technology, and Values

VOL. 4

# *Causality in Crisis?*

STATISTICAL METHODS
AND THE SEARCH FOR CAUSAL
KNOWLEDGE IN THE
SOCIAL SCIENCES

*Edited by*

VAUGHN R. McKIM
and
STEPHEN P. TURNER

UNIVERSITY OF NOTRE DAME PRESS
1997

Manufactured in the United States of America

Book design by Wendy McMillen
Set in 10.5/13 Times by Books International
Printed and bound by Braun-Brumfield, Inc.

**Library of Congress Cataloging-in-Publication Data**
Causality in crisis? : statistical methods and the search for causal
  knowledge in the social sciences / edited by Vaughn R. McKim and
  Stephen P. Turner.
    p. cm. — (Studies in science and the humanities from the
  Reilly Center for Science, Technology, and Values ; vol. 4)
    Includes bibliographical references.
    ISBN 0-268-00813-2 (alk. paper)
    1. Social sciences—Statistical methods—Congresses.   I. McKim,
  Vaughn R.   II. Turner, Stephen P., 1951– .   III. Series: Studies
  in science and the humanities from the Reilly Center for Science,
  Technology, and Values ; v. 4.
  HA29.C38 1997
  519.5—dc20                                                    95-43072
                                                                    CIP

∞ *The paper used in this publication meets the minimum requirements of the*
*American National Standard for Information Sciences—Permanence of Paper for*
*Printed Library Materials, ANSI Z39.48-1984*

# TABLE
# OF CONTENTS

# PREFACE

In October 1993 an interdisciplinary conference taking as its theme *Causality in Crisis? Statistical Methods and the Search for Causal Knowledge in the Social Sciences* was convened at the University of Notre Dame. Underwritten by grants from both the National Science Foundation and the National Endowment for the Humanities, the conference brought together for the first time a number of leading scholars from a variety of disciplines (notably statistics, social science methodology, and the philosophy of science) to debate the merits of recent claims and arguments that have been advanced on behalf of sharply differing answers to the question which provided the *leitmotiv* for the conference. The essays in this volume, all of which appear here for the first time, derive from presentations made at the conference.

Over the last half century, statisticians and social science methodologists have developed or refined a variety of closely related technical methods (multiple regression, path analysis, structural equation modeling, etc.) for analyzing data pertaining to social phenomena not amenable to experimental investigation. The value of these methods of statistical analysis for the purpose of data reduction, i.e., concise and illuminating description, and for prediction or estimation of unobserved values of particular variables, has never been in question. What has proven to be a matter of recurring concern and skepticism is whether such analytical tools can be appropriately used as a basis for drawing causal inferences when applied to observational rather than experimental data. Otherwise put, are there ever grounds for believing that the statistical manipulation of correlations present in nonexperimentally generated data can be used to identify the correct "causal model" of some complex social phenomenon? (A causal model is here to be understood as one that identifies not only factors causally responsible for a particular outcome of interest, but also the direct and indirect patterns of causally relevant relationships that exist among identifiable antecedent factors themselves.)

Applied statisticians and sophisticated social methodologists have long been critical of many of the ways in which causal modeling techniques have been utilized in specific studies appearing in the pages of social science journals.

Statistical assumptions required for the meaningful application of the methods are too frequently ignored, and alternative models that might have proven to be equally consistent with the data left unexamined. But in recent years far more serious objections to causal modeling have again been raised which, if sustainable, render the entire endeavor suspect. Multi-contributor exchanges addressing these "deep" criticisms of statistical analysis as an adequate basis on which to ground causal inference in nonexperimental research have begun to appear in the highly technical methodological literature of several of the social sciences. A central aim of the conference was to bring these controversies to the attention of the much broader audience of scholars who need to be aware of them, and to do so by seeking to focus on the philosophical and historical roots of the technical debates already underway.

At the same time, a novel response to recent attacks on the conceptual underpinnings of causal modeling has appeared in the work of several philosophers of science associated with the Artificial Intelligence community at Carnegie Mellon University. These scholars are engaged in a quite radical rethinking of the problems involved in seeking to determine the causal significance of statistical correlations in nonexperimental data. They too have been sharply critical of widely employed regression techniques. But their primary interest has been in seeking to reconceptualize the problems that must be overcome in seeking to identify causal structures that would best explain the correlations in large data sets. An early result of this work has been the development of computer algorithms which make it possible to comparatively estimate the "goodness of fit" of large numbers of potential models simultaneously, and in relation to one another. The ideas underlying this approach are certainly provocative as well as novel, but prior to the conference had not yet been widely discussed or evaluated. A second major aim of the conference thus was to provide a forum in which statisticians, methodologists, and philosophers of science could begin the important task of critically assessing this challenging new work.

While discussions of a number of germane technical problems have begun to appear in major "methodological" journals in several fields, e.g., *The Journal of Econometrics*, *The Journal of Educational Statistics*, *Journal of the American Statistical Society*, and *Sociological Methodology*, examination of the broader conceptual and philosophical dimensions of the fundamental issues at stake in these debates has scarcely begun. This volume, and the conference on which it is based, represent an attempt both to frame the issues in more accessible ways, and to emphasize the larger significance of the questions at stake in order to encourage critical and informed dialogue about their implications.

The fundamental premise on which this enterprise was conceived is that the incipient "crisis over causality" ramifies far beyond the relatively narrow confines of statistics and technical social scientific methodology. At stake is nothing less than the question of whether causal knowledge of many aspects of social life can in principle be achieved through available methods for conducting non-experimental research. In seeking an answer, one is thus inevitably led to address truly fundamental questions about what sort of enterprise "empirical social science" can intelligibly aspire to be.

We wish to express our appreciation to a number of organizations which offered indispensable assistance in bringing this project to fruition: to The Reilly Center for Science, Technology and Values at Notre Dame, and its founding director, Ernan McMullin; to COCTA (the International Committee on Conceptual and Terminological Analysis in the Social Sciences); and, to NSF and NEH for critical financial assistance. Above all, we are indebted to the scholars whose work appears here. All of the essays were written, or substantially revised, after the conference concluded in order to take into account perspectives and criticisms that arose in the course of the remarkably lively discussions which transpired at Notre Dame.

VAUGHN R. McKIM
*University of Notre Dame*

STEPHEN P. TURNER
*University of South Florida*

# Introduction

## *Vaughn R. McKim*

> Rather than characterizing the social sciences on the basis of
> their subject matter, it is at least as fruitful to start with the
> fact that most social phenomena are not amenable to investi-
> gation by laboratory experimentation.
>
> Paul W. Humphreys

> Nowhere is the gap between practice and its justification so
> large as in the social use of statistical methods.
>
> Christopher H. Aachen

In principle social scientists have access to a fairly broad range of distinctive
methodologies recognized as appropriate to the task of discovering, or seeking to
validate, causal relationships among phenomena of interest. At one extreme is
classical laboratory experimentation in which the experimenter actively intervenes to control factors that might interfere with the specific causal processes the
experiment is being utilized to identify or explore. A second type of experimental
methodology eschews laboratory control and experimental isolation in favor of
statistical controls, e.g., by assigning subjects randomly to treatment and control
groups. Here randomization is intended to shield the relationships to be measured
from contamination by factors (many of which may be unknown), the presence
of which could otherwise systematically alter or interfere with the hypothesized causal relationships being investigated. In contrast to these two explicitly
experimenter-controlled designs, quasi-experimentation is a more opportunistic
methodology. In the absence of laboratory or randomizing controls, researchers
may gather data pertaining to an observed intervention in a natural setting and
then, by means of statistical tests or further observations, seek to establish that
plausible confounding factors have not in fact biased causal inferences based on
the original data. Where biasing is suspected, the data may be reanalyzed in an
effort to assess the likely magnitudes of potentially distorting factors.

Each of these research designs has well-known strengths and limitations. No experimental paradigm can be implemented in a purely mechanical way to generate valid claims about causal relationships in either the physical or the social sciences. Substantial background knowledge, sensitivity to unanticipated features of an experiment as it proceeds, and much experience with the ways in which the idealized logic of experimentation may be vitiated in practice are always required to make causal inferences from experimental data truly persuasive. But at least the rationale for employing experimental designs to isolate causal factors is unimpeachable. Since the Scientific Revolution of the seventeenth century, experimental methods have been used effectively to advance our knowledge of causal relationships, to test the predictions of sophisticated theories in the physical sciences, and to establish robust, quantitative relationships among discrete types of properties and events in many fields of inquiry.

Experimental methods of one sort or another have become standard research tools in many areas of psychology and education and are frequently employed by sociologists focusing on small-group interactions and social psychology. But their utility across broad areas of social inquiry is sharply limited by virtue of the fact that the phenomena typically of interest to sociologists, political scientists, and economists are seldom amenable to laboratory or statistically controlled experiments. If one's concern is with, say, the relation between inflation and unemployment, the effect of a death penalty on crime rates, or differences in voting behavior in two *versus* multi-party democracies, to be told that the experimental method is the "royal road" to causal knowledge will scarcely be perceived as helpful. The critical question is rather whether nonexperimental research strategies can be devised that successfully emulate the acknowledged virtues of experimental methods.

Having posed this question let me quickly add that no attempt to answer it, let alone to explore fully the profound issues it raises, will be undertaken in this essay. The essays which follow address this question from a rich variety of perspectives, and the answers supplied are, as one might anticipate, sharply discordant ones. Fundamental disagreement about whether social scientific research *can* achieve causal knowledge on the basis of nonexperimental methodologies is precisely what gives the volume's title question its "bite." What this Introduction seeks to accomplish is twofold: first, to set the stage, both historically and conceptually, for the debates to follow by providing a broad overview of the sorts of empirical problems which led to the development of statistical methods of analysis by social researchers and of the basic insights on which such techniques have sought to capitalize. Second, it seeks to provide a "road map" of the conceptual terrain being explored in the volume by offering a preview both of the distinctive issues with which each author is concerned and of the varying perspectives from which these issues are approached.

## Toward a Scientific Understanding of Social Reality:
## Challenges and Responses

Students of social life became cognizant of the need to pursue novel ways of studying social phenomena early in the nineteenth century as they began to encounter for the first time what appeared to be some rather remarkable facts. As a result of the large-scale accumulation of statistical data initiated during this period by European governments and other public and private agencies, information about the frequency and distribution of such phenomena as crime, suicide, and poverty (along with much other demographic and economic data) first became available on a regular basis and immediately provoked new questions. On the one hand, some of these phenomena appeared to occur with remarkable regularity over time, in spite of obvious changes in population, wealth, political organization, the occurrence of war, disease, etc. How were the uniformities in such rates of occurrence to be made intelligible? On the other hand, it was also quickly discerned that the relative stability of such rates in a given population frequently coexisted with rather different rates of occurrence in nearby populations with otherwise similar demographic and economic profiles. What could explain such differences?

In neither case did it appear plausible to argue that the explanation of aggregate uniformities or differences could be given in terms of the unique constellations of reasons and circumstances influencing the behavior of particular individuals. No doubt every homicide or suicide, for example, has its own explanation. But appeal to reasons or circumstances operative in particular cases could throw little light on large-scale uniformities or changes in the rates of occurrence of such behaviors. Left unexplained would be why, say, intentional homicides or suicides attributed to insanity should themselves recur with remarkably stable frequencies in one population and with different frequencies in another. Clearly what was required was the identification of other phenomena occurring on a scale comparable to that of the behavior to be explained which might be shown to vary concomitantly (or inversely, in the case of negative causes) with the phenomenon of interest.

While it would have been admitted that the full range of causes operating to account for every instance of a widespread phenomenon might well be forever beyond our grasp, the seminal idea that seeking to identify correlations among variables representing phenomena differentially distributed in a population (e.g., disease, crime, poverty, suicides) might shed light on the causes and effects of such phenomena is one that was already being seriously explored by the middle decades of the nineteenth century. For example, there was a renewed interest in comparative methods of research in which efforts were made to identify situations representing "naturally occurring social experiments." Even more important were efforts directed toward using the newly developed tools of statistical analysis to attempt to isolate and identify non-accidental correlations in large sets of statistical data. In each case, the logic underlying these strategies was taken to be in conformity with the dictates of John Stuart Mill's *A System of Logic* (first edition, 1843), where Mill had contended, in his Fifth Canon of Inductive Reasoning, that "whatever phenomenon varies in any manner whenever another phenomenon varies in some particular manner, is either a cause or an effect of that phenomenon, or is connected with it through some fact of causation" (1965, 263). But the primary challenge faced by the early exponents of statistical analysis was how to disambiguate the many correlations with which nonexperimental data confronted them, so as to identify those which represented genuine cause-effect relationships. Given available techniques for representing statistical data (either large tables of correlations or graphs plotting changes in the values of variables against one another), convincing causal inferences proved very difficult to sustain. The techniques yielded many "relationships" between variables, but none exhibiting enough consistency between populations to be plausibly construed as invariant causal connections. In particular, the task of seeking to determine the "net effect" of one variable property on another, that is, the degree of correlation between them, absent the contingent and "local" influence of other variables correlated with each, proved to be enormously challenging.

Well into the early decades of this century, the struggle to overcome technical problems inherent in the available techniques of statistical analysis tended to divert attention from more substantive worries about such approaches. When doubts were raised about the ultimate relevance of correlational claims based solely on statistical analyses to inferences about causation, the tendency was to respond by emphasizing a "deflationary" interpretation of causal idioms themselves. This was, it will be recalled, a period in which causal concepts were under attack from many quarters. Ernst Mach in physics, Bertrand Russell in philosophy, and Karl Pearson, himself one of the founding fathers of statistics, each argued that ordinary causal discourse was intrinsically unscientific, representing a legacy of disreputable metaphysics if not of sheer mysticism. As Russell, in his inimitable way, was to put it, ". . . the reason why physics has ceased to look for causes is that, in fact, there are no such things. The law of causality, I believe, like much that passes muster among philosophers, is a relic of a bygone age, surviving, like the monarchy, only because it is erroneously supposed to do no harm" (1968, 174). In short, all that science requires, and all that it can hope to achieve by way of "causal" knowledge, is the discovery of relatively invariant functional relationships among measurable properties. From such a perspective it was tempting to infer that even the most fundamental laws of physics represented merely idealizations of descriptive correlational facts, and that the unraveling of basic correlations in data pertaining to social, economic, or political phenomena, of the sort to which social statisticians aspired, would mark a major step forward along the same path.

The subsequent history of philosophical work on causal concepts is too complex to be explored here. But its upshot may safely be said to involve a general repudiation of the facile reductionist or eliminativist approaches typified by the writings of the authors referred to above. Although there is still disagreement about whether the term 'causation' can be explicitly defined in an informative and non-circular way, few philosophers would be prepared to disagree with the view that causation is an indispensable notion in science as well as in everyday life, or that causal concepts share a number of well-understood features which distinguish them sharply from non-causal analogues. By contrast, social scientists have continued to maintain a much more equivocal attitude toward the notion of cause. On the one hand, it has proved difficult to dispense with, given the interests of researchers in providing not only genuine explanations of significant social phenomena, but knowledge relevant for policy-makers anxious to make effective interventions in the social world. Yet on the other hand, the exigencies of much social inquiry requires nonexperimental research designs. And the knowledge provided by such research is, at the most basic level, information about correlations in the data gathered, not about causes or causal relationships.

It is commonly agreed that the problem of inferring causes from correlational data proved to be not only challenging, but ultimately intractable, for nineteenth-century scholars armed with only the most elementary techniques of statistical analysis. But it is very widely held today that these problems have been largely overcome as a result of the enormous increase in sophistication of statistical methods pioneered by applied statisticians and quantitative methodologists over the last three-quarters of a century. However, this judgment has certainly not gone unchallenged, as the subsequent essays in this volume will quickly make evident. At stake in these essays is nothing less than the question of whether even the most advanced of statistical methodologies can be reliably employed to generate causal knowledge.

To set the stage for the controversies in which the contributors to *Causality in Crisis?* are engaged, a brief review of the fundamental features of the statistical techniques most frequently employed in contemporary social research may prove to be of some value. The observation (and measurement) of values of two or more properties distributed variably within a population is the raw material upon which various forms of statistical analysis are brought to bear by social scientists. In order to apply standard techniques for revealing the relationships that hold among properties whose values can vary, a procedure for measuring the distribution of the values of each variable must be selected. This will typically involve both a representation of central tendency, e.g., a variable's mean value, and a measure of the dispersion of its values, commonly represented by the average deviation of individual values from the mean, i.e., its variance.

Of course, as previously noted, associations between two or more variables can be established in more rudimentary ways, e.g., by constructing a table which displays for each value of each variable how many elements in a data set share just those values. For certain purposes, simple graphical representations can also yield relevant information about correlations, e.g., a scattergram (or scatterplot) constructed by plotting on a two or more dimensional graph individual data points representing specific values of each variable. But the critical breakthrough made by statisticians late in the nineteenth century involved capitalizing on the idea that the *degree* of association among variables represented in a scattergram could be represented algebraically.

A curve fitted to the data points in a scattergram by, e.g., calculating the values of an X-variable, measured along the Y axis, which minimize the square of the vertical distances between all X data points with a given Y value, provides an obvious way of summarizing the degree of association between two variables. And this curve can in turn be represented by an equation that will be linear when the curve chosen for fitting is a straight line. (In this connection it is important to be clear that the choice of a straight line rather than some more complex curve typically results from a decision to weight ease of subsequent computation at least as heavily as considerations pertaining to "best fit.") Given such a line (the linear regression line), the associated linear regression equation for two variables would take the form: $Y = a + bX$, where $a$ is the value of variable $Y$ when $X = 0$ (the $Y$-intercept) and $b$ represents the number of unit changes in $Y$ for every unit increase in the value of variable $X$ (the slope as measured by the so-called regression coefficient). In any case involving actual data and real measurement one will expect a spread of values of $Y$ for each fixed value of $X$, attributable to random errors in the measurement of $Y$ and possibly to the effects of unmeasured variables which have some influence on $Y$. Thus an error or disturbance term $u$ must be added to the equation to yield an expression of the following form:

$$Y_i = a + b_{YX} X_i + u_i \qquad \text{for } i = 1 \ldots n,$$

where $b_{YX}$ is the regression correlation of $Y$ on $X$, as estimated by the method of least squares and where $u_i$ is the error term. Note that while $a$ and $b$ are fixed coefficients for a given population or sample, $u$ is assumed to vary in value over different observations.

For present purposes it is important to recognize that regression equations as characterized summarize information about correlations that exist between measurable properties variably distributed across some population. Even so conceived, however, the advantages of regression techniques are obvious, and they have proven to be a virtually indispensable tool for many purposes. The discernment of significant correlations can and does provide an important stimulus for substantive theory construction, as can the discovery that variables previously assumed to be strongly related are not. Whether one's concern is with theory, policy making, social intervention, or program evaluation, the descriptive information provided by regression techniques is often invaluable. But in this connection it is critical to keep in mind that regression is providing new facts, not interpretations or explanations of facts, let alone unvarnished truths about the causal significance of the regularities it serves to characterize.

But do regression techniques also represent a tool with which more ambitious goals might be accomplished? In one sense, at least, the answer is clearly affirmative. In addition to describing correlations in some set of observational data, regression equations may be, and often are, put to use as predictive tools. Where relevant statistical assumptions are satisfied, correlations estimated from sample data may be generalized to unexamined portions of the population from which the sample was drawn. But when time enters into an extrapolation in an essential way, new risks of error are incurred. Estimating the value of one variable relative to another at some future time on the basis of correlations established at a previous time will always be hazardous unless there is independent reason to believe that the relevant population has not undergone changes that could affect one or both of the properties of interest. Similar caveats come into play when faced with extrapolating correlations from one population to another, where the degree of homogeneity required to make the inference plausible may be unknown. Of course, ignorance about such specific factors limiting reliable inference is not irremediable. By studying the predictive validity of actual forecasts made on the basis of regression studies, the credibility of such predictions can certainly be enhanced.

So far this account of regression techniques, along with their obvious virtues and limitations, has been predicated on the view that what their skillful employment produces is new information about what *de facto* correlations exist in some body of nonexperimentally derived data. While such information has many uses, it represents much less than we would like to know, namely, whether correlated variables are also causally related, and, more generally, what the underlying structure of causal relationships is that gives rise to a particular set of statistical dependence and independence relationships. Now regression methods can be used indifferently to analyze data drawn from experimental as well as non-experimental inquiries. In the former case, correlation between an independent variable under experimental manipulation and a dependent variable of interest may legitimately be treated as having causal significance, but precisely because steps have been taken either to randomize other potential causal factors, or to "hold them constant" through explicit laboratory controls. In such cases, correlations between values of a dependent variable and a manipulated independent variable do support inference to a causal relationship because (in the ideal case at least) other sources of variability in the dependent variable (with the exception of measurement errors) have been eliminated.

When we turn to the case of regression applied to nonexperimental data, the situation at least appears to be very different. Because none of the values of the variables in such data have been assigned or manipulated (as in the case of experimentation), but simply observed, we know in advance that they will be correlated in many ways. In the case of the simple bivariate equation model introduced above, we would expect there to be measurable correlations not only between values of the variables $X$ and $Y$, but between $u$ and $Y$, as well as $u$ and $X$. No causal inferences at all seem warranted in such a case, because the obtained correlations are compatible with a variety of different accounts of what the causes of the observed correlations are and of how such causes are themselves interrelated. But to grasp this difficulty is already to see how it might be surmounted. In order to learn whether a simple regression correlation between variables $X$ and $Y$ has any direct causal significance we need to take explicitly into account additional variables for which we possess relevant observational data, in effect "extracting" from the original residual term $u$ additional properties correlated with the observed variability in $Y$. Each new variable that is considered ($Z_1, \ldots, Z_K$) can be added to the original equation along with an associated error term to create a multiple linear regression equation, where the mean value of $Y$ is now represented as a function of the sum of the regression coefficients for each term on the right-hand side of the equation. The critical point is that multiple regression analysis involves calculating in a stepwise fashion a partial regression coefficient for each new term introduced. Each new coefficient is thus a measure of the residual correlation between a particular variable $X \ldots Z_K$ and $Y$ when the values of all other measured variables are held constant.

The investigator dealing with nonexperimental data cannot literally intervene to change the value of an independent variable of potential causal significance nor literally hold constant or randomize the values of other variables as the experimentalist always seeks to do. But multiple regression should be understood as attempting something analogous. Certainly the logic underlying the two activities appears to be the same: if C is a cause of E, then when the value of

E changes, the value of C must change as well if any other variables which are causally relevant to E do not change. Conversely, a change in C, all other relevant factors held constant, must be associated with a change in E. Though the non-experimentalist cannot intervene to "physically manipulate" the values of any variables, the calculation of the partial correlation between any of the variables $X$ or $Z_{1-K}$ and $Y$ yields a quantitative estimate of the value of the correlation free from the effect of other correlations in the data among any of these variables. Experimental controls are "replaced by" statistical controls by calculating the expected change in $Y$ associated with a specific increment of change in the $X$ or $Z$ variable under investigation, with the values of all other measured variables held constant in the calculations.

From one perspective, the computation of partial regression coefficients for each of the variables occurring in a multivariable linear equation can be thought of as providing merely a finer-grained description of the data, one which leads to a more perspicuous representation of the correlational facts than would simply determining the unanalyzed joint correlation between $X$ and $Z$ variables on the one hand, and $Y$ on the other. For example, the former enables one to identify the amount of $X–Y$ correlation in the observational data contributed by independent correlations between $X$ and $Z$ variables. And of course an apparent correlation between $X$ and $Y$ may even disappear once we control for the effects of $Z$ variables.

But having come this far it becomes increasingly difficult to resist the idea that statistical analysis of the data is revealing not only irreducible correlations, but important evidence about causal relatedness. Given the common underlying logic involved in experimental and statistical control, how can one avoid thinking that variables on the right-hand side of a regression equation, the parameters of which have been calculated as partial regression coefficients, as exogenous, or independent, i.e., as causes of $Y$ as well as predictors? And how to avoid thinking of $Y$ as an endogenous or dependent variable, i.e., as one whose values are caused by specific values of the $X–Z$ variables. Of course, even the controlled correlations do not in and of themselves determine the direction of causal influence. Here independent assumptions may have to be appealed to, e.g., the inferred temporal order of variables and "common knowledge" of what sorts of properties or events it is intelligible to think of as being related as cause to effect. But at least in some cases, these issues appear to be quite uncontroversial, e.g., parents' profession(s) and child's profession, or poverty, divorce, and juvenile delinquency.

It is still possible for the methodologically abstemious to consistently reject the temptation to make causal imputations based on regression analyses of non-experimental data for many reasons (some of the most significant of which receive careful examination in the essays gathered in this volume). But the situation appears to many to undergo a marked change with the introduction of more sophisticated forms of analysis of a kind which may be exemplified by Path Analytic Methods, a form of causal modeling. Path analysis, though based on regression techniques, is intended to provide a much more revealing picture of the relationships among numerous variables. Even when stepwise partial regressions are interpreted as elucidating the "net (causal) effect" of each of a number of independent variables on a designated variable of interest, the standard employment of such techniques typically leaves unresolved questions about whether particular variables are to be understood as direct or indirect causes of the effect variable, and does nothing to reveal what pathways of causal influence may exist among the independent variables themselves. In the conceptually simplest case one could seek to address such questions by sequential pair-wise calculations of partial correlation coefficients among all pairs of variables, in each case holding other variables constant. In fact it would make sense to pursue this strategy literally only if the number of relevant variables is very small due to "combinatorial explosion." As the number of variables increases, the number of possible relationships among them obviously increases exponentially. However, the rationale for thinking of the problem in these terms should be clear. Each positive or negative partial correlation coefficient would mark a path of potential direct causal influence between two variables. Similarly, a correlation coefficient arbitrarily close to zero would imply the absence of any direct causal relationship.

But it is critical to see that even such a correlational data base would not settle the question of how the variables in a proposed model are causally related. It is conventional to use the term 'path model' or 'causal model' to refer indifferently to graphical or algebraic representations of such relationships. A graphical path model might consist of nodes representing variables and arrows (directed edges) linking some, but not all nodes, with arrowheads used to indicate the direction of putative causal influence between two nodes. Each arrow between two nodes marks a "path" in the path model and is typically accompanied by a number (the path coefficient) calculated as a standardized partial regression coefficient indicating the strength of empirical association between the variables.

On this conception of a causal or path model it becomes clear that the number of distinct possible alternative causal models consistent with even a complete set of covariance data for a given set of variables is enormous, because correlations may fail to determine the direction of causal influence, or whether any causal pathways are cyclical, or whether some correlations are due to the effects of unmeasured variables, etc. Thus it is clear that, as in the case of regression analysis, substantive assumptions on the part of the investigator must play a critical role in specifying what will count as the correct (or, approximately correct) causal model of some phenomena. In fact, it is obvious that in real research contexts such assumptions will inevitably be brought into play very early on. In practice researchers are likely to be able to consider and test only a handful of models which they antecedently deem plausible. For each they will formulate equations, estimate parameters, and then run statistical tests to attempt to determine which best fits the data. Questions about whether many other models that have not been considered would also pass such tests, and would be as consistent with available background knowledge of the relevant domain, have neither been asked nor answered.

Thus in spite of the sophisticated conceptual rationale for causal or path modeling of nonexperimental data, it is scarcely surprising that the widespread use of such methods continues to generate controversy. Some of the recurring dissatisfaction voiced about these procedures results from the careless ways in which they have been employed by practitioners. Too frequently investigators appear to have selected variables for examination only because an available data set includes measures on them or because measures for some particular variable will be relatively easy to collect. A few models combining these variables are then articulated on the basis of intuitive judgments about what plausible causal linkages might hold among them. The researcher then selects the model which performs best on a relevant statistical test and concludes that it is at least a close approximation to a true representation of the pattern of direct and indirect causal relationships at work in the phenomenon under investigation.

This sketch, though intended as a caricature, would be endorsed by many thoughtful critics as far closer to the truth than many would like to admit. But given the range of truly formidable difficulties which the practitioner of causal modeling must attempt to overcome, the existence of some instances of questionable practice does not begin to settle the question of whether there is something fundamentally suspect about the very enterprise of causal modeling, construed as a methodology capable of generating causal knowledge from the analysis of nonexperimentally generated data. It may well be that there are well-founded criticisms of many published studies and of various practices commonly engaged in by causal modelers, but that none of these criticisms give reason to believe that the enterprise itself is essentially misguided. However, a reader who begins to explore the essays contained in this volume will quickly discern that the authors share quite profound misgivings about the current state of the art. While they disagree quite sharply about whether there are *envisionable* strategies that could to some degree circumvent the deep conceptual problems under discussion, none are sanguine about the prospect of providing anything like a straightforward "technical fix" for the statistical methodologies which currently represent best practice in social scientific research.

## Previews and Prospects

The remaining pages of this Introduction are devoted to a brief preview of the challenging arguments and counterarguments to be found in the essays which follow, emphasizing where interested readers without a strong background in statistical methodology might most easily gain a toehold on the issues at stake. The essays which comprise Part I (Cause and Correlation before Causal Modeling) each offer a historical perspective. Together they enable the reader to see how the problem of achieving causal understanding on the basis of large-scale correlational data first presented itself to social inquirers in the nineteenth century and to appreciate the sorts of limitations that quickly became evident with respect to the tabular and graphical techniques of analysis then available.

Stephen Turner's "Net Effects: A Short History" offers a good deal more than its title promises. It does indeed present an informative history of the vicissitudes encountered by early social statisticians in their attempts to employ tables of rates to distinguish spurious from genuine correlations and then to establish the causal significance of the latter by means of isolating the "net effect" of one variable of interest on another. However, Turner also argues that while the introduction of regression techniques in the 1890s radically altered the methodology of statistical analysis, it did little to close the gap between causal claims and the correlational evidence relied upon to support them. In a fascinating case study comparing John Snow's epidemiological approach to understanding the transmission of cholera in the 1850s with that of his contemporary, the prominent statistician William Farr, Turner argues that Farr's wholly statistical "net effects" approach not only yielded the wrong answer about cholera (that it was transmitted by "bad air") but that Farr's account could not have been definitively established to be wrong in the absence of the occurrence of several "natural experiments" to which Snow's, unlike Farr's, approach was highly sensitive.

The relevance of his case study to the issues mooted in this volume, Turner argues, is that social scientists who employ contemporary statistical methods of causal analysis are, like Farr, in no position to convert their causal hypotheses into strong causal knowledge of the sort we have come to expect from the natural sciences. In this sense, Turner lays down a provocative challenge to the received opinion among social methodologists that causal inquiry is not "in crisis."

Mary Morgan's complementary essay ("Searching for Causal Relations in Economic Statistics") offers a rich and detailed case study of attempts by some of the most prominent scholars of their day to unravel causal linkages between changing marriage rates and levels of trade over many decades in the nineteenth century. While this problem was, even then, of only modest substantive interest, it proved to be exemplary in at least two respects: first, because it embodied time-series data in a fundamental way, it served as a prototype of the distinctive sort of problem with which economists, and specifically, econometricians, quickly recognized they would have to deal. Second, (for historically quite fortuitous reasons) a number of the leading researchers of the day reverted to this problem both to illustrate and to test the ability of newly devised statistical correlation techniques to tackle issues that had proved intractable for graphical methods. Though Morgan's narrative does not directly seek to throw light on current methodological controversies, she does makes clear that the wave of optimism which accompanied the introduction into econometrics of modern causal modeling techniques has long since been muted. Difficulties facing the earlier methods she so carefully describes are beginning to be perceived to have troubling counterparts in the context of current modeling practices.

The two major essays in Part II (Causal Modeling and Causal Inference) take the reader quickly to the heart of the contemporary issues with which this volume is concerned, yet do so from quite different perspectives. The first, by Clifford Clogg (before his recent and untimely death, one of the outstanding sociological methodologists of the day) and Adamantios Haritou, opens with an extremely lucid review of the fundamental logic underlying what the authors call *the regression method of causal inference*. But it quickly emerges that the main focus of their essay is on the critical differences in assumptions required when using regression as a statistical tool for prediction and when employing it as a technique for justifying causal inferences (i.e., conditional *vs* unconditional regression models). They argue that a failure to clearly and unequivocally acknowledge these differences in much of the standard literature on statistical methods has created much confusion. For example, it frequently appears to be held that a regression model that predicts well is, *a fortiori*, a "good" model for estimating causal effects. But, argue Clogg and Haritou, such an inference cannot be justified because even very powerful predictive models may always be causally biased as a result of including non-causal variables and/or excluding causally relevant ones. And the "causal assumptions" required to rule out these possibilities cannot themselves be independently tested. As they soberly conclude, regression modeling as typically applied in social scientific research represent an inherently inconclusive method of causal inference.

David Freedman's very substantial essay ("From Association to Causation via Regression"), strongly complements Clogg and Haritou's contribution, but is written from a quite different point of view and focuses on a broader range of issues. As an applied statistician rather than a social scientist, Freedman launches a withering attack on what he contends is an indefensible use of formal statistical inference techniques by social scientists to establish cause and effect relationships. Readers who lack a strong grounding in statistical theory will inevitably find this essay "tougher going" than many of the other contributions to this volume. But Freedman's careful exposition of a wide variety of concrete examples repays careful study. At root, his worries are generated by a fundamental intuition about causality itself, to wit, that genuinely causal claims always involve a counterfactual component. If variable A really is a cause of another variable B, then an intervention to change the value of A, other things being equal, must alter the value of B. However, given that the circumstances in which statistical inference techniques are employed by social scientists permit only actually observed values of variables to be correlated, nothing can be inferred, he claims, about what would happen to the correlations were an intervention undertaken to modify or set new values for relevant variables. If in such circumstances we lack independent sources of causal knowledge about such variables (derived from experimental contexts or on the basis of independently confirmed theory), then statistical measures of association can yield virtually nothing of real interest about causes or effects.

When the critical thrust of Freedman's position is put in such stark terms, many will find incredulous the implication that social researchers have failed to take such issues into account in setting out the assumptions and proper criteria for employment of their sophisticated technical tools. But Freedman's response is always, "let's look and see." Thus the importance for his position of the, in many cases, quite detailed "case studies" which comprise the bulk of his essay here. One final point: while Clogg and Haritou's primary focus is on the general logic of what they refer to as the regression method of causal analysis, the scope of Freedman's essay is broader still. The latter half of his essay is devoted to critical discussion of the recently developed algorithmic approach to the identification of causal structures being developed by Clark Glymour and his colleagues at Carnegie Mellon. Since Part III of this volume is devoted to expositions and criticisms of this approach, readers for whom this recent work is still *terra incognita* may wish to explore the essays by Richard Scheines and Glymour himself before attempting to assess Freedman's interpretation of this project or to evaluate the strong interchange of views concerning it to be found in the Rejoinder to Freedman's essay by Peter Spirtes and Richard Scheines and Freedman's Response.

Readers previously unacquainted with the new approach to statistical causal inference pioneered by the Carnegie Mellon scholars, Clark Glymour, Peter Spirtes, and Richard Scheines, (hereafter SGS) will find Scheines' essay, "An Introduction to Causal Inference," essential reading. It represents by far the clearest and most straightforward introduction to their project yet to have appeared. In briefest terms, the SGS approach involves two related but distinguishable components. One is the development of powerful computer algorithms which operate on statistical data sets to produce directed graphs. The vertices of these graphs represent variables on which data has been gathered, and line segments connecting vertices (directed edges or arrows) are generated by calculations of conditional statistical dependence or independence among pairs of variables, i.e., while holding constant the values of other variables. The logic underlying this portion of the project is analogous to that embodied in standard multivariate regression analysis, but with some critical differences. Actual parameter values are of no interest here, only patterns of correlation and non-correlation are attended to. Second, the algorithm can generate all of the patterns of statistical dependence and independence among variables compatible with the data, outputting a set of distinct directed graphs that are equivalent, relative to stipulations in the program about what is to count as statistical independence and to an ordered set of rules for how graphs are to be constructed.

The technical features of this new approach to statistical analysis will themselves be of much interest, for the SGS algorithms hold out promise of providing investigators with critically important information that would be hard to duplicate either through piecemeal data dredging or on the basis of preconceptions about what sorts of linkages among variables are inherently plausible. However, the most striking (and controversial) dimension of the program is the claim that the graphs produced by the algorithms are not merely, as Scheines puts it, "compact and elegant representations of independence structures," but causal graphs. That is, it is claimed that the directed graphs generated by the algorithms may legitimately be understood to represent all the causal relations that hold among the specified variables included in the graph. Of course, when the algorithms generate a set of alternative graphs on the basis of the same data one may be in no position to decide, without additional research, which directed graph represents the real causal structure of the phenomenon being investigated. But the central claim is that directed edges linking vertices in the graphs produced by the algorithm are legitimately to be understood as identifying putative causal relationships in virtue of the conditions built into the program for graph construction.

The strategy SGS adopt to support these claims involves the identification of a number of formally specifiable conditions grounded in our intuitive understanding of causal relatedness such that if the graph for a set of data is constructed in a way which respects these conditions, the relations among the variables it identifies may be justifiably inferred to be causal. But if the strategy is clear, so are the potential difficulties confronting it. Is satisfaction of the conditions stipulated as relevant for underwriting a causal interpretation of the independence relations in a set of data really sufficient to warrant inference to conclusions about causal structure? In particular, is the notion of causality captured by the relevant conditions actually "strong enough" to underwrite counterfactual inferences about the consequences of ideal interventions? Could knowledge of the *de facto* relations of statistical dependence and independence in a body of data really provide strong support for the idea that if particular variables were to be directly manipulated, predictable changes in the values of other specific variables would occur?

A related issue is whether one can be sure that the causally laden assumptions used to specify the algorithm's procedures have really been satisfied when the latter have done their work. For example, can we know (or when is it reasonable to assume) that all of the statistical independencies among variables identified by the algorithm directly reflect causal structure rather than coincidence or chance? Can we be confident that the statistical patterns which exist in the data could only have been generated by one or another of the possible "causal structures" identified by means of the set of directed graphs produced by the algorithm?

Clark Glymour's long and detailed essay, "A Review of Recent Work on the Foundations of Causal Inference," provides the centerpiece of Part III. It offers a full-scale review and defense of the novel methodology for discovering causal structure he and his associates have advanced. He presents explicit arguments in support of the cogency and soundness of the fundamental principles on which their automated computational procedures are based and provides detailed examination of a number of empirical cases in which the performance of the algorithms can be directly compared with results actually achieved by researchers utilizing standard causal modeling techniques. In one of the most provocative sections of his essay, Glymour contends that there are actually many situations in which the statistical methods SGS endorse will yield information about causal structure superior to that which could, in principle, be attained by experimental methods. Needless to say, these are powerful and, indeed, quite remarkable claims. And in the essays which follow Glymour's, philosophers Paul Humphreys and James Woodward subject to critical examination a number of the central conceptual underpinnings of Glymour's provocative essay.

Humphreys' "Critical Appraisal of Causal Discovery Algorithms" consists of a detailed discussion of four fundamental features of the SGS strategy. He argues that each must, at the very least, be further clarified by Glymour before it will be possible to determine whether satisfaction of the formal principles intended to legitimate the employment of the research program's algorithm can guarantee that the latter will produce only representations of causal structure. "Causal Models, Probabilities, and Invariance" by James Woodward undertakes the challenging task of rethinking the rationale for the SGS research program. He devotes particular attention to a consideration of circumstances under which the assumptions of the program could be violated, whether such violations could be detected, and what the consequences of such violations might be. Attention is also directed to issues about underdetermination: Are the assumptions under which the algorithms operate powerful enough to ever uniquely identify the real causal structure underlying correlational data rather than some set of statistically equivalent but genuinely alternative representations? But at the heart of Woodward's essay is a sustained set of reflections on the critical difference between what he terms the "informational" and "control" conceptions of causations. In the latter sense causal relations are invariant, autonomous, support interventions, and, more generally, are understood to possess counterfactual import. By contrast, the former is an essentially epistemic notion, capable of underwriting predictions but not genuine explanation. Though SGS are clearly committed to identifying causes in the stronger, control sense, Woodward explores a number of reasons for being skeptical about whether the methodology they endorse actually does so.

Unlike previous essays, the essays on the fourth and final part of this volume (Putting Problems about Causal Inference in Perspective) are not directly concerned with the immediate assumptions involved in drawing valid causal inferences from statistical data. Yet each contributes importantly to those debates by calling attention to one or another relevant aspect of the larger context in which causal analyses are undertaken in the social sciences. In the essay which opens Part IV, "The Role of Construct Validity in Causal Generalization," Larry Hedges reminds us that the problems involved in seeking to justify "local" causal inferences about available data sets are not identical with those which must be confronted in justifying causal generalizations. Yet only when causal knowledge is generalized can it be put to work in theory construction or relied upon as a sound basis for policy intervention. Thus useful causal inference must be both locally valid and generalizable across a relevant domain of contexts. But as Hedges is at pains to make clear, errors in causal generalization may result from a variety of factors other than misspecification of a causal model. Conversely, misspecified models may nevertheless lead to consistent estimates of relevant relationships when generalized across contexts and measurements. That these "underdetermination" problems are wholly nontrivial is one of Hedges' key insights and provides the rationale for his attempt to identify and to propose independent ways of dealing with the diverse sources of error that plague attempts to make justified causal generalizations.

Nancy Cartwright's essay "What Is a Causal Structure?" can also be read as raising concerns about the generalizability of relationships among variables, whether established by traditional regression methods or by means of the algorithmic techniques championed by Spirtes, Glymour, and Scheines. But the perspective she brings to bear is very different from that of Hedges. As she notes, the argument of her essay represents "one stage in my more general Aristotelian campaign to replace the ontology of natural law with the ontology of capacities." However one views the promise of this larger program, its value in thinking about social phenomena seems undeniable. At root her thesis is that relationships among social variables, whether causal or merely associational, can be expected to exhibit stability only if they are grounded in what she calls "socioeconomic machines," i.e., relatively enduring systems of social structures, each with its own intrinsic capacities, configured in ways that are strongly resistant to externally induced change or evolutionary transformation.

The problem posed by this way of conceptualizing complex social phenomena is that nothing we can learn from the statistical correlations that happen to hold in a given data set will tell us whether the data were generated by a "socioeconomic machine" (a causal structure in Cartwright's sense), or if so, how stable the underlying structure actually is. In particular, because the relationship between a particular causal structure (in Cartwright's sense) and the statistical correlations to which such a structure may give rise *is not itself a causal relationship*, this difficulty cannot be solved merely by adding a new variable (one indexing a particular causal structure) to the equations used to formulate specific causal models. Thus Cartwright's essay poses a severe challenge to the idea that statistical techniques alone could ever suffice to establish causal relationships where such relationships are understood to be relatively invariant and to posses counterfactual import.

Stanley Lieberson's "The Big, Broad Issues in Society and Social History: Application of a Probabilistic Perspective," with which the volume concludes, contains challenging comments on a host of difficulties confronting the social scientist attempting to give causal explanations of complex social outcomes. But central to his essay are the claims that social structures frequently must be invoked to explain particular social outcomes and that there is much evidence pointing to the conclusion that such structures are in many cases hierarchically organized. That is, they may involve both a generic and fairly stable set of predisposing structures as well as more contingent local factors required to trigger the occurrence of certain outcomes. What transpires in such cases must be understood as the outcome of causal interactions among factors operating at different levels of analysis. There are a number of interesting connections between Lieberson's proposal to think of social phenomena as consisting of hierarchically organized, and frequently only probabilistically linked, structures of causal influence and Cartwright's emphasis on the need to think of statistical relationships among social variables as reflecting the operation of complex "socioeconomic machines." In both cases it seems clear that techniques for the "causal analysis" of statistical data of the sort that have been debated in this volume are ill-suited to reveal either hierarchically embedded relationships or boundary conditions determining the stability of high-level predisposing conditions, i.e., the conditions under which a particular "social machine" can be expected to operate in a reliable way. Yet in the absence of such knowledge, claims to have identified causal, let alone law-like, relationships on the basis of statistical dependencies and independencies in observational data will appear to many to have a hollow ring.

When this volume was first conceived, we could scarcely have imagined the richness of the extended dialogue among our contributors that was to ensue. But in retrospect it is clear that the scholars who agreed to participate in this project sensed, along with us, that the time was ripe to undertake a full-scale rethinking of the strengths and weaknesses of statistical techniques intended to generate causal knowledge of complex social phenomena in situations where experimental methods are unavailing. None of the contributors are prepared simply to endorse the statistical methods of causal analysis most widely taught and employed today in quantitative social research. But there are clearly deep differences among them both with respect to what issues most urgently need to be addressed and about whether the limitations inherent in present methods can be surmounted. Though technical matters figure prominently in a number of essays, none of the authors are inclined to argue that "technical fixes" alone will be sufficient to put the social methodologist's house in order. The real issues in every instance turn out to be conceptual rather than technical.

What is urgently needed, and what the essays gathered here certainly initiate, is a fundamental rethinking of the goals to which it makes sense to aspire by means of nonexperimental social research, and of how such goals properly understood should constrain and shape the articulation of appropriate methodologies. Among the scholars represented here, several argue with conviction that novel methodological approaches capable of generating causal knowledge in the strongest sense of the term are within our grasp. The reflections pursued by others have led them to embrace far more skeptical prognoses. And some have been more concerned to identify critical questions than to offer confident answers. But each of the essays makes an important contribution to the clarification of issues that must be addressed if we are to come to terms with the question of what sort of knowledge of social phenomena nonexperimental methods in the social sciences can achieve.

# I

# CAUSE
# AND
# CORRELATION
# BEFORE
# CAUSAL
# MODELING

# "NET EFFECTS"

## A Short History

## *Stephen Turner*

The essays in this volume deal, for the most part, with the contemporary use of statistical techniques (prototypically, regression analysis) for making causal inferences and with the problems that arise when the methods are used for this purpose. In the nineteenth century, before the invention of the theory of correlation and regression, similar sorts of inference, based on a wholly different technology of statistical presentation, gave rise to problems remarkably similar to those currently being debated. Indeed, the "crisis over causality" of the 1870s in some respects prefigures the present scene. Then, established modes of drawing causal inferences were yielding such widely variable results that the whole enterprise of employing statistics as a basis for causal analysis came to be seen by many as questionable, and the causal claims based on arguments about the proper interpretation of statistical evidence came to be regarded as "subjective," in contrast to the numbers themselves. The present issues are about causal inferences from regression analysis. But the concept of causation that is part of the dispute has roots in this earlier period, and the problems with the concept are not significantly different. In this essay I will identify these problems and show how they were carried over, in an obscured form, into the present century.

The work of Yule (1897) proved critical in giving renewed life to the idea that social statisticians might after all be able to identify the causes and effects of social phenomena on the basis of statistical data, in the absence of experiments or of strong knowledge of relevant causal laws or of specific causal mechanisms. Yule fully accepted the idea that statistical representations of data were not direct representations of causal relations. But he did not give up at this point. He reasoned that an empirical association must represent a mixture of effects of known and unknown causes even though the magnitudes of these various effects are unknown. If an observed association that we believe to be the result of a given cause (call this the "primary cause") could be "corrected" in a way that would identify the contribution of other causal influences that we suspect are contributing to the observed association, we would have a better estimate of the magnitude of the primary causal influence. And we can do some correcting of this sort on the basis of other observed associations. For example, one could take the observed association between a suspected causal contributor and the outcome of interest, and then subtract it from the total observed association to get a "corrected estimate" of the magnitude of the primary causal relationship.

The advantage of this procedure is that it makes use only of known empirical facts—the observed associations. The problem is that all such associations are themselves mixtures of known and unknown causes, none of whose magnitudes are known independently of the associations to which they contribute. One could, of course, repeat the "correction procedure" for a suspected third cause, and a fourth, and so on, until the resources of the available correlational data relevant to one or another possible causal influence have been exhausted. This idea of "correcting" associations suggests an important limit notion: that if one were able to fully correct a given association by taking into account all of the empirical associations relevant to assessing whether other specifiable causes were contributing to it, one would arrive at the true causal effect of the primary causal variable on another, that is to say, the true "net effect."

There are, however, two things that are both novel and odd about this reasoning. The first concerns warrants for causal claims. It is usually thought that to convert a causal intuition, that is to say a sense that something might be caused by something else, into genuine causal knowledge, it is necessary to identify either a causal mechanism or a law. It is usually thought that laws and mechanisms are subject to further vetting, on grounds other than statistical ones. Here, however, neither law nor mechanism is required, though it is not clear whether it is that they exist but are inaccessible or are simply dispensable. The actual underlying causal processes, if there are any, thus have a peculiarly equivocal relation to the claims. The second oddity is the supposition that the observed associations, rather than independent, non-statistical grounds for accepting claims about cause, can do the work of transforming causal intuitions, "plausible" causal hypothesis, into solid causal assertions. This supposition seems strange because all of the evidence used in this procedure, such as regression evidence, is thought to only imperfectly represent actual causal processes, and therefore to require correction. The evidence used to correct these imperfections, it seems, also possess imperfections, and of the same kind. How can an imperfect representation of causation become perfect by corrections using other imperfect representations?

In this essay I explore some of the history which lies behind the idea of correcting associations to identify net effects. I begin by examining some of the earliest discussions of how statistical data might be causally interpreted, because the debates that arose in this context about how to deal with the problem of unmeasured common causes are both accessible and illuminating. I then proceed by examining various methods developed in response to the problems of identifying and measuring the variety of "causes" that contribute to empirical associations in statitsical data. The aim of this historical *excursus* is to clarify the concepts of cause and causation present-day analysts have inherited from Yule. Genuine clarity, however, will prove to be elusive, because (or so I shall suggest) some of the equivocations implicit in the notion of net effects in the pre-correlational statistical literature were carried over intact into the era in which causal interpretation of multiple regression was being established, and from there into contemporary "causal modeling." I conclude that, with respect to the materials social scientists usually study, both the concept of net effects *and its successors* are highly problematic. First I explore in some detail how methods used to calculate net effects in the nineteenth century could fail to produce causal knowledge by examining a famous case of statistical reasoning undertaken in the wake of the London cholera epidemics at mid-century. And I then suggest reasons for believing that contemporary methods of statistical analysis are subject to the same liabilities. For present purposes, the virtue of the statistical work on cholera is that because we know the actual causal mechanisms at work in the epidemics we are in a favorable position to see why statistical techniques alone were unable to identify them. But the case also points to a larger moral. In a field such as epidemiology, there is a substantial amount of background causal knowledge derived from a direct or experimental study of causal mechanisms. In favorable circumstances, such knowledge can be used to provide relevant hypotheses for statistical investigations, and serve as an independent check on causal inferences from statistical evidence. Cholera was a case of just this kind. By contrast, in the typical research problems confronting social scientists, there is no independent source of such knowledge.

In her chapter in this volume, Mary Morgan describes some of the graphical methods early statisticians used to persuade themselves that the relationships they identified were not only real, i.e., not spurious, but genuinely causal. This essay thus complements hers, but I shall be focusing primarily on the problem of causality as it arose in what may be called the "tabular tradition" of social statistics, a tradition that pursued the causal understanding of large-scale social phenomena by means of the analysis of statistical tables of rates.

## Confounding, Common Causes, and Net Effects in the Tabular Tradition

A typical example of a problem about causation arose in 1835, when the Statistical Society of London proposed to answer several questions, among them the following:

> What has been the effect of the extension of education on the habits of the People? Have they become more orderly, abstemious, contented, or the reverse? (quoted in Porter 1986, 33)

This is a straightforward causal question: there is an apparent cause, education, and a set of possible effects. Both cause and effects have some more or less quantifiable aspects. Levels of education can be distinguished, as can degrees of "orderliness." Moreover, there was some relevant empirical evidence, which it was proposed be gathered. The evidence they had in mind was to bear on the following empirical question:

> What is the proportion of crimes to education? Are the educated found to be more exempt than the uneducated, or the reverse? (quoted in Porter 1986, 33)

The statisticians who formulated these questions were careful not to say what the inferential relationship between this question and the question of causation was. But they knew that inferences, *from* numerical evidence concerning proportions to answers *to* causal questions, about effects were fraught with difficulties. In the case of the relationship between crime and education the problems of interpretation were well known at the time the Statistical Society of London proposed its inquiry. The French statistician A. M. Guerry and his associates had reported in 1828 that districts with low levels of education were likely to have high levels of crime. But in the ensuing discussion of these results it became clear that other factors needed to be considered before the relationship could be properly understood. In this case, a variety of other possible causal factors were proposed, such as disparities of wealth, which were recognized to be potential confounders (Porter 1986, 173). The same districts that were high in crime and low in education were impoverished, and impoverishment was a possible cause of crime: the effects of impoverishment were likely to be mixed up with the effects of lack of education, and indeed impoverishment might even be the cause both of lack of education and crime, so that lack of education might have no separate effect on crime.

Statisticians in the tabular tradition became quite adept at thinking through problems of this kind concerning causality. Table 1.1, taken from Enrico Morselli's compilation and review of the statistical literature on suicide, is a typical example of the way in which data were presented, and of the problems of interpretation such tables presented. The table consists of three columns of figures, each representing the rate of occurrence of a particular phenomenon in one of three categories of province, those with high, moderate, and low suicide rates.

The grouping is thus by regions with different suicide rates. The first column presents the mean of the suicide rates in a particular grouping of provinces. The

**TABLE 1.1** Suicide Rates, Stature, and Military Exemptions for Height by Provinces Grouped by Suicide Rate

| | *Average suicide rate by province* | *Average stature in meters* | *Average percentage of exemptions for height* |
|---|---|---|---|
| Group I (under 20 per million) | 14.4 | 1.616 | 15.85 |
| Group II (from 21 million to 40 million) | 30.9 | 1.635 | 8.81 |
| Group III (over 40 million) | 46.5 | 1.643 | 7.34 |

(adapted from Morselli 1882, 103)

second column represents the height in meters of recruits from these districts; the third, the percentage of military exemptions on grounds of insufficient height. There is a clear parallelism here: suicide rates increase with stature and decrease with numbers of exemptions from military service on grounds of insufficient height.

But these relationships are not treated by Morselli as causal, on two grounds. One is that the relation between stature and suicide rate does not hold in the case of many other comparisons, such as the differences between England and Scotland—the taller Scots were less prone to suicide. Similarly for Slavs and Germans in Austro-Hungary: the Slavs are taller, but somewhat less prone to suicide. Morselli's conclusion from the fact that these cases do not conform to the rule is that stature itself cannot be a causal factor for suicide. But stature is associated with ethnicity, and in the case of Italy, ethnicity fits with the known regional variations in suicide rates. Italians of the German and Slavic racial type are found in different regions than those of the purely Latin type, and those regions have a higher suicide rate. Regions with higher proportions of Italians of German and Slavic descent had taller recruits and higher suicide rates because persons of German and Slavic descent were taller and the suicide rate for those of German and Slavic descent was higher. The original relationship between stature and suicide thus appears to have resulted from the mixing of populations with different average stature.

In tables 1.2 and 1.3, the argument that differences in "Public Morality" are the cause of differences in suicide rates is tested by comparing legitimacy rates with suicide rates by "nationalities and religions."

**TABLE 1.2**  Suicide and Illegitimacy Rates by Ethnicity

| Nations | Suicides per millions | Illegitimacy rate |
|---|---|---|
| Slavs | 36 | 18.3 |
| Latins | 60 | 16.3 |
| Germans | 147 | 8.6 |
| Germans proper | 165 | 6.5 |

**TABLE 1.3**  Suicide and Illegitimacy Rates by Religion

| Religions | Suicides per millions | Illegitimacy rate |
|---|---|---|
| Orthodox Greeks | 40 | 20.40 |
| Catholics | 58 | 11.15 |
| Protestants | 190 | 10.35 |

(Morselli 1882, 143)

Morselli uses these tables to indicate a basis for the hypothesis that Public Morality, which is not directly measured (and which, as he puts it, "is vain" to "define absolutely") is a common cause of suicide and illegitimacy. Morselli had grounds for rejecting this hypothesis too, however. The relationship between illegitimacy and suicide rates ought to hold *within* countries, that is to say for rates calculated at the level of administrative regions. But the relationship breaks down "if we confront countries and regions" (1882, 143): if we look at German provinces, for example, "the proportions . . . vary without any analogy" (1882, 144).[1] This failure was grounds for rejecting the causal claim: if a pattern or rule failed to hold when applied *within* a country, for example among administrative regions with differing rates, one could conclude that it did not represent a causal relationship. It might, instead, be the result of the mixing of populations, as the relation between stature and suicide in Italy was.

In these cases, tables serve the purpose of establishing that a hypothesis has some empirical basis, and then other tables serve to winnow the list of hypotheses empirically. Showing a parallelism between what are plausibly regarded as effects of the same cause, as in tables 1.2 and 1.3, establishes some unmeasured variable as a possible common cause. One can refute this possible common cause relationship by showing that other plausibly linked effects—in this case, an effect of "public morality"—are not parallel. So showing that a parallelism is merely local shows that it is not causal, or not unequivocally so. The equivocation is a result of the fact that in these cases we cannot decide among such interpretations as the following: that there is a causal mechanism which works locally but for some unknown reason does not work generally, that there is no causal mechanism that corresponds to the relationship that holds locally (so it may simply be an accident), or that some much more complex mechanism would account for both the appearance of the parallelism locally and its failure generally. Showing that statistical relationships hold only at a high level of categorization (e.g., for nations) or in regional groupings (e.g., in categories of provinces with high, moderate, and low suicide rates, as in table 1) yet do not hold at lower levels (as in the case of the illegitimacy rate–suicide rate relation in German provinces) shows that these are not unequivocally causal relations. But the strict application of these standards led to problems. Virtually no relationships, even very strong "local" parallelisms, were able to survive this empirical winnowing process. But looser standards let in too many relationships and placed the burden of choosing which of them were genuinely determinative, or causal, on considerations of plausibility, considerations that were "subjective" in contrast to the rates themselves. So it came to appear that causal analysis was itself subjective.

Because statisticians of the time often worked with maps, they were used to seeing large-scale patterns that did not hold within smaller geographical units. The large-scale relationships were undeniable and vivid, and usually could be given some highly plausible interpretation. So they were disinclined to ignore the relationships or think that some single mechanism produced the whole set of variations. It was far more plausible for them to suppose that the causal influences were simply additive—that local circumstances together with large-scale influences, such as religion and ethnicity, produced particular rates. Mary Morgan shows, in her chapter, how such additive effects could be handled graphically. The tabular tradition also had a means of dealing with additive effects—though it was less developed and rarely used.[2] Perhaps the classic example of this technique, cited by Morselli and also by Durkheim, who called it "an ingenious calculation" (1951, 172) was given by the elder Bertillon (1864–1869). Bertillon used the notion of net effects in trying to disentangle the effects of marriage from those of age on suicide. The basic tabular fact was that "the readiness towards suicide increase[s] with years." However, because single people have a lower average age than married people, and lower still than the widowed, "it is easy to understand how a part of the higher proportion" of suicides among the widowed and the married "may be attributed to the influence of age" (Morselli 1882,

234–35). He attempted to quantify this "part," which is the net effect of marriage with the effect of age eliminated. He called this the preservative effect of marriage. His technique was quite simple. He started with a table of age-specific suicide rates.

**TABLE 1.4**   Age-specific Suicide Rates for Men in the Principal States of Europe

| A. Men. | |
| --- | --- |
| From 16–20 years | 64 |
| "20–30" | 139 |
| "30–40" | 203 |
| "40–50" | 305 |
| "50–60" | 406 |
| "60–70" | 511 |
| "70–80" | 461 |

(Morselli 1882, 216–17)

As table 1.4 shows, the rate increases significantly until age seventy. Marriage and age are known to be correlated, so Bertillon estimated the effect of marriage by calculating the mean age of married men and looking up, on this table, the rate of suicide for this age. The rate on the table, which included the whole male population, was much higher than the rate for the subset of married men. The net effect was estimated by subtracting the rate for the subset from the rate for the whole population.[3] His result was this:

> the influence of age ought to have raised the average of 139 suicides a year per million of individuals, and about 250 of the married people; that is to say, that the probable loss by suicide, if it depended only on age, would be much larger among married people than among single, in about the relation 100: 55.5. But exactly the reverse happens as the proportion of suicides is less among the married, being as 100: 111.4. The influence of marriage as a preservative has, then neutralized that of age by diminishing the probable loss of the married by more than one-half. (Morselli 1882, 236–37)

In the case of widowers, the net effect is in the other direction: age and widowhood combine to increase the rate.

As I noted in the introduction, this simple notion of net effect represents a crucial conceptual step—or else a crucial bit of sleight of hand. It introduces causal language that would have seemed unwarranted in the case of mere associations. To call the numerical results a net *effect* implies that one is adding or subtracting the effects of some causal influences from those of other causal influences. And in some sense, the change in language seems justified. By using nothing more than tabular data to eliminate a known source of weakness in our causal knowledge of suicide, we have certainly, in some sense, strengthened it. But in general, the epistemic situation with respect to causation is unchanged. What one is adding and subtracting are numbers derived from tables. These numbers only represent "causal relations" by assumption. The assumption is that if there is a statistical difference of the sort represented by these numbers, there must be a reason for the difference, and this reason must be a cause of some kind. This is not too different from the argument that every association must be caused by something, but it is intuitively more compelling, perhaps because this is not just any association, but a "corrected" one. Of course it is subject to additional "correction." The relation between marriage and suicide may itself be explained by a common cause, for example, and other tables, with the form of tables 1.2 and 1.3, might be used to substantiate such a claim. But it does seem as though the process of correction, carried on long enough to eliminate possible confounding relationships, ought to lead to a fully "corrected" relationship—and that this relationship should be regarded as "causal."

This intuition is suggested by the language of "net effects" and its associated terminology, such as the notion of "residuals." But the intuition itself was not argued for explicitly, except in the somewhat backhanded manner to be discussed in the next section. The importance of the intuition is this: "weak" causal knowledge, of the kind that enables one to identify "possible" influences, can be transformed, through data-based means, into stronger causal knowledge, perhaps even into a stronger *kind* of causal knowledge, causal knowledge of the whole domain of relevant causes and their additive relations, and this may be done without the discovery of laws or "mechanisms" in a strict sense.

The idea that one can arrive at causes by correcting correlations has implications for the notion of cause itself. In the case of the influence of marriage on suicide there is no mechanism: there is only a numerical difference between a rate that would be expected on the basis of age alone and an observed rate for the subset of married persons. A difference, to use the language of the time, needed to be "referred" to a cause. This particular difference is being called a net effect *of marriage* on the grounds that the fact of possessing the status "married" is the defining property of the subset, together with the vague notion that marriage might plausibly be said to have an "influence" on suicide. One suspects that here we are talking about a qualitatively different notion of "cause" than that which occurs in such expressions as "causal law" or "causal mechanism." But the differences are hard to state. One might think that the differences are merely that social "causes" are just as real and operate in a manner similar to those described by physics, but that our knowledge of the details of their operation is much weaker. The notion of "details of their operation," however, seems largely irrelevant here. We think of the relations between marriage, age, and suicide not in a way that can or need be undergirded by "mechanisms," or sharpened up into "laws," but rather as a way of making sense of the manifestations of complex phenomena that we will never have access to in any other way than through the statistics that represent them, and which, if reduced to actual causal mechanisms, would perhaps need to be described in terms that would be irrelevant to the questions we wish to ask about suicide and marriage. The notion of cause that we use when we try to make such an "influence" seem plausible as a cause is at best analogical, and perhaps irretrievably so.[4] And this suggests an alternative interpretation of the difference: that these hypotheses simply do not refer to causal mechanisms of the kind that are familiar from the natural sciences. I will defer discussion of these two possible interpretations—the "different in epistemic accessibility" and the "different in kind" interpretations—until later in the chapter, at which point we may test them against some other historical examples.

## Yule and Partialling

The regression/correlation revolution of the 1890s that marked the effective end of the tabular tradition had an impact on these issues, but only an indirect one. Even the relevant terminology did not immediately change. Yule used the terms "net regression" and "net correlation" to describe the results of partial regression and correlation analyses (1897, 833). This language was subsequently taken up by social and economic statisticians.[5] The term "net" meant for them something closely akin to what it had meant for the elder Bertillon. The statistical technology did change. Tables were replaced by regression equations fit by least squares. The visual model of adjustments or correction also changed, gradually to be replaced by what is now called "partialling."[6]

The term "partialling" itself was not taken from correlation analysis, but from a procedure used to deal with associations of attributes. "Partial associations," as Yule explained in his influential textbook, are associations between two attributes in subuniverses, and are called partial "to distinguish them from the **total** associations between A and B in the universe at large" (Yule and Kendall 1937, 51). The reasoning behind partialling is best understood in terms of examples. The two Yule gives in his textbook date from the early part of his career and represent work from the late nineteenth century. The first example involves school children and the relation between mental dullness, developmental defects, and nerve defects. Other analysts had concluded, on the basis of a large association between developmental defects and dullness and the association of both with "nerve defects," that "nerve defects" was the connecting link between the other two. This is a good example of plausible causal reasoning based on what I have been calling "weak" causal knowledge: the output variable is dullness; nerve defects seem like a plausible direct cause of dullness; and developmental defects seem to be plausible causes of, or surrogates for, background weaknesses which might be thought to cause nerve defects. Yet, as Yule points out, plausible as this reasoning is, it does not fit with what one finds when the cases are divided into those with nerve defects and those without. The association is *not* high for the children with nerve defects. So nerve defects cannot be the connecting link. Partialling serves here to *exclude* a causal interpretation, by identifying relationships that failed the test of holding in the subuniverses. This use of subuniverses to eliminate a causal claim is familiar from Morselli and his predecessors: showing that parallelisms in rates that were impressive on the national or international level failed to hold on the level of regions or administrative districts represents an instance of the same type of reasoning.

The second example Yule gives is the inheritability of traits from various generations of ancestors. Here the issue is whether the trait of light-colored eyes is passed solely from parent to child or is subject to the genetic influence of grandparents as well. He takes the subuniverse of associations between grandparents and grandchildren (measured simply by the proportions of light-eyed children with light-eyed grandparents), and divides it into two sub-subuniverses: children with light-eyed parents and children with dark-eyed parents, and examines the proportion of light-eyed children in each. If grandparental eye-color made no difference, there would be no differences between the proportion of light-eyed children with light-eyed grandparents and the proportion without light-eyed grandparents in either subuniverse. In fact, there are clear differences in both (see table 1.5). His conclusion is that:

> In both cases the partial association [i.e., the difference in proportions] is quite well-marked and positive; the total association between grandparents and grandchildren cannot, then, be due wholly to the total associations between grandparents, parents, and children respectively. There is an ancestral heredity, as it is termed, as well as a parental heredity. (1937, 54)

The reasoning is this: if the net effect, the difference in the proportions one was interested in, such as the effect of grandparental eye-color, did not disappear within the relevant sub-subuniverse, it could not be accounted for entirely by the fact of parental eye-color. Showing that the relationship does not disappear in the relevant subuniverse establishes the causal significance of ancestral heredity. Here, in short, we prove the existence and the strength of a plausibly hypothesized causal relationship by showing there is a net effect.

**TABLE 1.5**

*Grandparents and Grandchildren: Parents light-eyed*

Proportion of light-eyed amongst the grandchildren of light-eyed grandparents $\left.\right\} = \dfrac{(ABC)}{(BC)} = \dfrac{1928}{2231} = 86.4$ percent

Proportion of light-eyed amongst the grandchildren of not light-eyed grandparents $\left.\right\} = \dfrac{(AB\gamma)}{(B\gamma)} = \dfrac{596}{821} = 72.6$ percent

*Grandparents and Grandchildren: Parents not light-eyed*

Proportion of light-eyed amongst the grandchildren of light-eyed grandparents $\left.\right\} = \dfrac{(A\beta C)}{(\beta C)} = \dfrac{552}{947} = 58.3$ percent

Proportion of light-eyed amongst the grandchildren of not light-eyed grandparents $\left.\right\} = \dfrac{(A\beta\gamma)}{(\beta\gamma)} = \dfrac{508}{1009} = 50.3$ percent

This reasoning is not entirely novel. Bertillon's use of the notion of net effect serves essentially the same purpose of establishing rather than undermining a causal claim. But Yule is more explicit about one aspect of the problem that Bertillon did not directly face, the problem of completeness, which is to say the problem of whether there might be unincluded variables which confound, or could independently account for, the putatively causal relationship. In Yule's use of the notion of partial correlation and regression to account for changes in pauperism he did face this issue directly. His procedure is paradigmatic of the tradition in which causal inference is grounded in regression analysis. His first step was to list "the various causes that one may conceive to effect changes in the rate of pauperism" (1899, 249). He then observed that "although I have classified . . . all the causes of pauperism into separate and distinct groups, it is not to be supposed that they are independent" (1899, 250). He then listed the ways in which changes in one class of causes "may effect" changes in others, as "for example, changes in trade will affect the internal migration of the country" (1899, 250). Of course, he did not have data on all of these things, but he could deal with some of them. And he could measure the total changes in the rate of pauperism. Administration was a plausible direct cause of change, but some of the total change could not be ascribed to administration. He hoped, "if possible, to ascribe such residual changes not due to administration to their respective causes or cause groups, social and moral" and to see whether changes in social and economic factors had any effect on administration (1899, 251–52). The data did not permit him to do so, but the idea is clear.

Yule knew two basic things about the consequences of partialling for new, potentially relevant variables. He knew that the slope of the regression line and the degree of correlation—and therefore the estimate of the net effect—could be significantly affected by partialling for new unincluded variables. He also knew that after most of the variables he thought might be relevant were partialled for, partialling for additional variables had little effect. He reasoned that when partialling for new variables made little difference, one's understanding of the relevant causal processes could be taken to be substantially correct and complete. He conceded "that there is still a certain chance of error depending on the number of factors correlated both with pauperism and with proportion of out-relief which have been omitted, but obviously the chance of error will be much smaller than before" (1899, 251). The phrasing is suggestive. The quantitative idea that the chance of erroneous interpretation could be diminished to something "much smaller than before" makes sense only if one thinks that one can produce valid causal results by partialling for all the variables that represent possible causes, and eliminating all that are not efficacious on empirical grounds.

This reasoning shifts the burden of proof for establishing a causal interpretation in a new direction. We do not know all the possible causes. Only if we had comprehensive knowledge of the relevant causal mechanisms could we claim to be able to identify all possible causes. In the absence of this knowledge, we can only test as many alternative "causal" hypotheses as possible, even if their credentials are very weak, and even if the supposed causal processes are poorly understood or not understood at all, but merely conceivable. Put simply, since the point of the exercise would be to eliminate or correct for possible causal factors whose behavior is very poorly understood, the standards for consideration of particular phenomena as possible causes could not be high. The advantage of the method was that it appeared to avoid the problem of subjective judgments about the plausibility of causal interpretations. Corrections of rates of the sort undertaken by Bertillon would have helped with this problem, but not much, because he did not have a criterion for when to stop—for when to say that enough "corrections" had been made. Yule provided such a criterion, though in a somewhat backhanded fashion: you were through when you couldn't make additional corrections that made any difference. Yet with this criterion in hand, statistical facts could apparently be delegated the task of deciding what causal inferences are valid—no "subjective" judgments about the plausibility of causal hypotheses were needed, other than the minimal judgment that an influence was possible.

## Net Effects as Causes

What could go wrong with such a procedure? David Freedman has repeatedly referred to the example of John Snow's epidemiological work on cholera as a model of epidemiological inquiry, and as a case which shows the limitations of causal modeling (1991; this volume). The resistance to Snow's results mounted by statisticians provides a clear example of what can go wrong with the notion of net effects that bears specifically on the problem of using statistical evidence to "correct" estimates of net effects. The dominant figure in biostatistics in Britain during Snow's career was William Farr, Superintendent of the Statistical Department of the Registrar General's Office. Farr studied the distribution of cholera and the correlates of the occurrence of the disease and supplied much of the data

**TABLE 1.6**

| Elevation of Districts in feet | Number of Terrace from bottom | Deaths from Cholera in 10,000 Inhabitants | Calculated series (1.) | | |
|---|---|---|---|---|---|
| FEET | | | | | |
| Under 20 | 1 | 102 | $\dfrac{102}{1}$ | = | 102 |
| 20–40 | 2 | 65 | $\dfrac{102}{2}$ | = | 51 |
| 40–60 | 3 | 34 | $\dfrac{102}{3}$ | = | 34 |
| 60–80 | 4 | 27 | $\dfrac{102}{4}$ | = | 26 |
| 80–100 | 5 | 22 | $\dfrac{102}{5}$ | = | 20 |
| 100–120 | 6 | 17 | $\dfrac{102}{6}$ | = | 17 |
| 340–360 | 18 | 7 | $\dfrac{102}{18}$ | = | 6 |

available to Snow. Unlike Snow, Farr believed that cholera was transmitted primarily through air, not water, and he had a substantial statistical basis for this idea. He found strong associations with altitude, which he took as a surrogate for the presence of miasmata in the air. He proposed an elevation law, based on a curve fitted to the levels of various terraces in London, that predicted observed mortality very nicely, at least for the data on one major epidemic (1885, 343–47). Farr noted that this table suggested a "law" with impressive numerical properties, since "the mortality at 100 feet is 17, at 50 feet 34 in 10,000; consequently *at half the elevation the mortality is doubled"* though he conceded that the relationship did not hold strictly at the lowest levels of elevation—the law would have predicted much higher levels of mortality at very low elevations. We know now that the account of the causal mechanism Farr based on this impressive table was wrong. But would he have discovered it was wrong through the device of correcting for possible causal influences in the manner of Bertillon or Yule?

The history of Farr's response to Snow's argument that cholera was waterborne (and in London was associated with particular water-supplying companies) is intriguing, for it suggests that he would not have, and that mistaken "net effects" hypotheses may be very difficult to eliminate by partialling for conceivable causes. More important, it shows how the methods produced false positives.

Snow did not attempt to explain away Farr's results statistically—by determining, for example, whether elevation had an independent effect by adding variables to the model to show that the relation might have been spurious. Nor would this have been likely to work, for Farr, like other statisticians of the period, was quite careful to consider such variables as weather, income, population density, age, and sex, the variables that would, typically, diminish a net effect. Instead, Snow attempted to determine the manner of transmission. He hypothesized that cholera was conveyed by minute quantities of material and tried to identify the means by which it was conveyed. This led him to examine many non-statistical, qualitative facts, such as stories about how a number of people were stricken after a meal of cow-hooves from a cholera-bearing region (1936, 21), and to consider the possible relevance of a great many details that ordinarily would not have presented themselves even in the most elaborate listing of background variables. He observed, for example, that the poor often kept food under the bed, near the chamber pot (1936, 22).

The associations studied by Farr might have led him to notice these things: the poor, for example, were known to have higher cholera rates. But the most impressive statistical relationships, such as the law of elevation, pointed to the wrong hypothesis. The "law" held strongly only for the 1849 epidemic. Snow concluded that the relationship was a result of unknown coincidental circumstances specific to that epidemic and ignored it. He also did not attempt to explain the many associations that Farr had discovered by examining health statistics. The evidence that struck him as most interesting, indeed, took the form of reports about the spread of the disease and about the many situations in which a ship or group or soliders escaped the disease entirely while others in virtually identical circumstances succumbed in large numbers. His conclusion was that cholera was transmitted in the evacuations of victims, that this explained all the cases, and that no other account explained some of the cases. He claimed, for example, that "whatever hypothesis of the native cause of the malady be adopted," his was the only one that "affords an exact explanation" of the outbreak associated with the famous Broad Street pump (1936, 54).

Presented with Snow's hypothesis about the source of cholera, Farr modified his own theory to allow for the possibility that impure water had a role in its transmission. He first tried to accommodate this hypothesis to his elevation law by suggesting that the cholera-producing material was particularly deadly when it entered the air by evaporation from water, and was then transmitted by air to the victim. He treated Snow's mechanism, in short, as one contributory causal variable among others. Indeed, he provided some compelling statistical support for Snow.[7] Farr reasoned in a now-familiar way: by the calculation of net effects. In discussing a table of Mortality by Cholera by Districts supplied by different Water Companies in order of elevation, he remarks that

> After correcting the above Table and the tables of cholera 1848–49, for the effects of elevation, it is found that a large residual mortality remains, which is fairly referable to the impurity of the water; for it is least where the water is known to be sweetest, and greatest where the water is known to be most impure. (1885: 359)

The impurity of water was a net effect, established by subtracting the supposed main effect of elevation. This was precisely the procedure later to be used by Bertillon. Farr's conclusion, however, was dead wrong: the causal process was not an additive one, not a combination of bad water and something else.

Eventually Farr himself reached this conclusion—but not before a great deal of bad water had gone under the bridge. He was persuaded in a step-by-step way that can have few analogues in standard present-day uses of regression analysis as a basis for causal inference. The first step involved an experiment of sorts. Changes in the way in which water was taken from the Thames, motivated by public health measures based on the partial acceptance of Snow's ideas, provided circumstances yielding new statistical evidence which undermined the elevation law. But this evidence was not conclusive. Another epidemic in 1866 provided more evidence, of a very special kind. The customers of the East London Waterworks Company suffered a disproportionate level of mortality during the epidemic. It was then discovered that the company "had been illegally supplying

**TABLE 1.7** Mortality from Cholera in Districts Supplied by Different Water Companies

| Water Companies | Sources of Supply | Aggregate of Districts supplied chiefly by the respective Water Companies | | | |
| --- | --- | --- | --- | --- | --- |
| | | Elevation in feet above Trinity High-water Mark | Population enumerated 1851 | Deaths from Cholera in 14 weeks ending Nov. 26, 1853 | Deaths to 10,000 Inhabitants |
| London | | 39 | 2,362,236 | 744 | 32 |
| *(1) Hampstead and (2) New River | Springs at Hampstead and Kenwood, two artesian wells, and New River | 80 | 166,936 | 8 | 5 |
| New River | At Chadwell Springs in Hertfordshire, from river Lee, and four wells in Middlesex and Herts. | 76 | 634,468 | 56 | 9 |
| Grand Junction | The Thames, 360 yards above Kew Bridge | 38 | 109,636 | 16 | 15 |
| Chelsea | The Thames, at Battersea | 7 | 122,147 | 22 | 18 |
| Kent | The Ravenbourne in Kent | 18 | 134,200 | 31 | 23 |
| West Middlesex | The Thames, at Barnes | 72 | 277,700 | 89 | 32 |
| East London | The river Lee at Lee Bridge | 26 | 434,694 | 162 | 37 |
| *(1) Lambeth and (2) Southwark | The Thames, at Thames Ditton and at Battersea | 1 | 346,363 | 220 | 64 |
| Southwark | The Thames, at Battersea | 8 | 118,267 | 121 | 102 |
| *(1) Southwark and (2) Kent | The Thames, at Battersea, the Ravensbourne in Kent, and ditches and wells | — | 17,805 | 19 | 107 |

* In three cases (marked with an asterisk) the same districts are supplied by two companies.

water from its reservoir at Old Ford which had been contaminated by the discharge from the recently completed sewage system of West Ham" (Eyler 1979, 119). Farr himself performed the analysis that established this. The actions of the water company again provided circumstances closely akin to those of an experiment. Soon other developments confirmed the implications of this experiment-like event.

Consider, however, what the situation would have been had this series of events not transpired. Farr's initial response to Snow—his attempt to combine the effects of elevation with the effects of water—would have to have been corrected on the basis of statistical evidence alone. Indeed, in terms of the model of causality later employed by Yule, Farr's ultimate embrace of Snow's mechanism for the transmission of cholera was unjustified. The possibility that some cholera was transmitted by air had *not* been eliminated, on statistical grounds, as a possible cause. On the contrary, if we invert Farr's reasoning with respect to table 1.7, it was the case that, to put it in Farr's language, there was a *significant* residual mortality not "accounted for" by impure water that was "fairly referable" to elevation. In 1859, this was in fact what Farr was reporting: that "a part of the excessive mortality is referable to the depression of the ground" in certain districts supplied by suspect water companies (1885, 360). Absent the later "natural experiments," there would have been no justification for eliminating this variable. One might well have concluded that elevation accounted for less of the total mortality in the epidemics after 1849, but, for reasons that will become apparent, one would not have been able to eliminate this variable entirely.

When Farr changed his mind in 1866, he did so not on the basis of reasoning about net effects, but on the basis on reasoning about the actual mechanism of transmission, the phenomenon of diarrhea, and the complexities of the physical and biological properties of the cholera flux (1885, 365–70). Experiments had been performed that were designed to see whether there was a mechanism for transmission by air. The experimenters attempted to produce a vapor from cholera flux. They were, essentially, unsuccessful, and it was concluded "that the amount of zymotic matter evaporated from cholera flux, and entering the system through air, must be inconsiderable as compared with the amount that may enter through a water supply contaminated with sewage" (1885, 370).

Other experiments, as it happened, supplied Farr with a way of explaining why the elevation law held for the early epidemic, but not as strongly for the later ones. Experiments showed that cholera flux, when put in a tube of distilled water, tended to settle. Farr reasoned that the elevation law held in the 1848/9 epidemic precisely because London's water was highly and universally contaminated at the time of this epidemic. The water pumping technology of the time was such that water had an opportunity to settle when it was held after pumping, and thus the cholera flux settled—and went disproportionately to taps at lower elevations. In the later epidemics, changes in the quality of some companies' water diminished this effect: some companies were pumping water free of cholera flux, and others were not.

The statistics for the 1849 epidemic suggested that elevation made a difference, and this was correct. But the interpretation was wrong—*no* mortality should have been "referred" to elevation as an index of the presence of cholera vapors. Elevation provided a clue, which if properly interpreted, would have pointed to the actual mechanism. But it pointed more directly to a mechanism that was in fact not operating. Only physical or experimental evidence could have shown that the interpretation was simply wrong, and that nothing should have been "referred" to miasmatic causes. Of course, it might be argued that, in a sufficiently extended analysis of cholera data, elevation would have been shown to be a bogus variable, and in a sense this is what happened. The law fit ever less well the cholera epidemics that occurred after the water companies began to change their procedures. But the results of regression analysis are well known to vary when subjected to a variety of differing conditions, and there are quite standard responses to problems like Farr's in the statistical literature. If we accept the notion, defended by Glymour (1983), that causal relations in social science are always likely to be "local" we would not expect the relations to hold up under extended analyses. By analogy, one might have concluded, in Farr's case, that the epidemic of 1849 represented a somewhat different disease, one following a different "law" than did the later epidemics, in which the variable of elevation was less influential. In any case, the statistical "net effect" of elevation would not have entirely disappeared. Some settling in the course of holding pumped water would have occurred, so there would have been some effect of elevation until the technology was radically changed.

## Causal Mechanisms and Net Effects

Is there a principled way of understanding why "net effects" reasoning misled Farr that also has implications for contemporary uses of regression analysis? Perhaps there is a problem with the concept of "net effects" itself, as it is employed in statistical social science research. If we begin with some ideal uses of the notion of net effects, it is possible to see where the typical statistical use of the concept goes wrong. The important historical uses of the concept of net effects in physics involve attempts to arrive deductively at quantitative conclusions about effects that cannot be directly measured. Determining a net effect requires a complete set of relevant laws, adequate measurement, and accurate estimates of the effects of all other relevant causes. Under these conditions, the notion of net effects is not only legitimate but indispensable. But the history of physics is rife with errors that arose through failures to identify relevant variables, thus mistaking the significance of experimental or observational results (cf. Conant 1951, 107). Net effects are not, after all, primary facts, like the fact of the death of a person from cholera, or brute statistical facts, such as the death rate from cholera in a specific population. They are facts that have been derived from a set of premises. The concept of "net" requires derivation. If one wishes to determine the net weight of a substance in a container, one must know the weight of the container, the weight of any other substance in the container, and whether any other things affect our weight measurements before one can subtract all these weights to derive net weight. If one wants to determine net profits, one needs a complete and accurate accounting based on properly categorized entries in the books. Anything less means that the results will be erroneous.

Yule and his successors employed the term "net effects," but they had no "laws." They had only observed regression lines and correlations, together with a sense of what factors might be causally relevant to an outcome. So "net regressions" and "net correlation" represented net *effects* only on the assumption that regression surfaces represented causes. The regression analyst does not know the actual "effect" of any putative cause, but can only begin with an empirically observed regression surface which is assigned a causal interpretation. Moreover, the regression analyst does not have a definitive list of the things that need to be subtracted from the total to generate the "net" results. And the categories employed may themselves be suspect—they do not necessarily correspond to real causal mechanisms. The promise of partialling is that it establishes causal relatedness by eliminating non-causes leaving only genuine causes, and accomplishes this job of winnowing on empirical grounds. Essential to this use of partialling is the claim that if a category or measure truly represents a causal influence of some sort, it should survive the winnowing process of partialling.

There are various reasons why this winnowing process might fail. One is the unavailability of the kind of data that differentiates between two categories, one of which represents a genuine causal influence and one of which does not. If there are no cases to partial with—if the relevant intensional categories are extensionally equivalent—then there is no way to decide between them on the basis of data. Another reason the winnowing process might fail is that a relevant variable is missing. This would allow the analysis to produce false positives—relationships that would have been eliminated had the absent variable been included. Spirtes, Glymour, and Scheines's (1993) technique for detecting latent variables may be understood as a technical solution to this problem. But the case of cholera points to the possibility of a failure of quite a different kind. The "results" Farr produced were impressive. The problem was that his categories did not directly represent the actual causal mechanism. Many of them *were* connected to the real mechanism, but each in an imperfect way. And the connection could be understood only once the mechanism was.

Yet no strategy for directly examining the actual causal mechanisms of complex social phenomena is open to the social scientist. There is no analogue, for most social scientists, of the independently accessible world of physical fact that could serve as a source of the correct categories for causal analysis, as exemplified by the experiments on the cholera flux discussed above. Typically, social scientists invent analogies to "explain" net effects such as the effect of marriage on the suicide rate. Durkheim, for example, offers something like this: "think of marriage as something that binds peoples' desires and thus limits their sense of dissatisfaction." This is not a causal mechanism that could be separated, so to speak, from the statistical relationship it "explains," and independently tested. Social scientists, to put it simply, are stuck with the kind of reasoning represented by Farr's first, and erroneous, efforts. In the case of the effect of marriage on suicide, there is no way to imitate Snow, for there is no unique causal mechanism to be found.

It is more plausible to think of the distinctive sort of statistical relationships that concern social scientists as abstractions from a quite heterogeneous set of situations in which there is no common element, such as the cholera bacterium, shared by all these situations. Snow had, in the vast body of evidence he accumulated about unusual cases of transmission (e.g., 1936; 3–9, 25, 34–38), a considerable amount of information that was wholly independent of the statistical relationships studied by Farr. And Snow could, and did, employ this information as a check on Farr's interpretations. The only additional forms of analysis available in the typical social science case, however, remain within the framework of the notion of "net effect," seeking through an extension of the method of partialling to adjust for potential interfering effects. This means that there are no non-statistical controls on results of the sort that Farr produced. On the other hand, and for the same reason, there is no danger that a given correlational model will be undermined by the discovery of a mechanism. Farr's elevation law and his mistaken interpretation of it would have been quite safe from refutation, had it been handled strictly in the fashion recommended by present methods.

This line of reasoning suggests a different way of thinking about the practical problems of social scientists. Social scientists are a bit like Farr before his complete conversion to Snow's account of cholera. They have no causal mechanisms of the sort that Snow had, but they have lists of possible causes and a means of eliminating those that do not fit the linear regression model for a given data set. This is an exercise not unlike fitting pegs of various shapes into a square hole: if the true cause is correctly represented by the categorization or measures used by the analyst, and if the actual relationship is of the right (i.e., linear) form, and if all the relevant variables have been included (are properly represented) and have the right form, and if the data is sufficient to distinguish between the available alternative hypotheses, the exercise works—the wrong causal hypotheses are eliminated and the right one survives. But nothing in this exercise excludes the possibility that the *mechanisms* are not correctly represented by the model.

In Farr's case, the variables did not correctly represent the true cause, and the model that fit the data, at least for the 1849 epidemic, was wrong, though not entirely wrong: there was a connection between elevation and mortality involving a mechanism that Farr had not thought of. Nor did "correcting" the elevation law for the effects of another categorization that was imperfectly representative of the cause—impure water—turn the results into a correct analysis. This suggests that, in order to employ these techniques, one must already possess strong causal knowledge. There may be cases in which, by some miracle, the representations employed in the model and the actual causal mechanisms do perfectly correspond, and in these cases the quantitative estimates of net effects would be meaningful. But the methods produce "net effects" results whether or not the model and the actual mechanisms correspond. The winnowing procedure does then, in principle, under certain conditions, convert weak causal knowledge into strong: but the analyst cannot know it to have done so—or know the conditions to be met—without having the kind of strong causal knowledge that would assure that the categorizations correctly represent mechanisms. And this of course is precisely what social scientists, and most other users of these methods for that matter, lack.

The methods are, in the absence of strong causal knowledge, highly imperfect means of eliminating imperfections. But these analyses clearly provide something—an empirical assessment of the possible causal relations (at least those that fit the framework) that currently strike us as intelligible and plausible. Historically, there are principled arguments—notably Weber's—for the claim that this is all we should want in social science: description in terms of the real mechanisms, Weber thought, would constitute gibberish that failed to answer the questions we want to answer.[8] Similar arguments can be given to the effect that making practical decisions requires that we test our best causal beliefs against available data as well as we can, and then decide accordingly, even if we know that whatever real causal mechanisms may be present are misrepresented, perhaps radically, by our models. The decisions made about how to prevent cholera prior to Snow's work are defensible in just these terms. Farr was able to improve on his original ideas; but by using the methodology of "net effects" on the categories with which he began he could not have come to the right answer.

**NOTES**

I would like to thank Ted Porter, who read an early draft of this essay, and David Freedman, who read two drafts, for their very helpful comments, as well as Vaughn McKim for his editorial advice. The final essay is much altered from the versions they read, and the interpretations—and of course any errors—are my own.

1. Morselli understood perfectly well that there were problems with illegitimacy data and looked for a better measure of public morality. A high level of public morality ought, for example, to mean a low level of crime. But crime rates and suicide rates were negatively associated in such places as the south of Italy. The rule seems to be that crime substitutes for suicide in the south, but as one goes farther north in Europe, this relationship gradually disappears (1882, 148).

2. As well as graphically, as Mary Morgan points out in her essay.

3. A somewhat odd procedure: it would be better to compare the suicide rates of the subsets of the married and the unmarried.

4. Of course it is possible to argue against every form of the attempt to make sense of the distinction between types of causality by collapsing one type into the other—by claiming that all causal explanation is analogical in character, for example. But it should be noted that similar reasoning about probabilistic causality was not only current in the late nineteenth century, it was influential in several contexts, including legal causality, and through legal writing on causality was imported into social science by Max Weber.

5. A typical example in a sociological study of the 1920s is Brunner, Hughes, and Patten's *American Agricultural Villages*, in which the notion of net correlation is used extensively (1927, 265 and passim). Sociologists worked with measures that were not quantitatively significant in themselves and were therefore more interested in establishing "net correlations" (thereby establishing the non-spuriousness of the relationships) than in numerically estimating the causal effects.

6. The image given in Yule and Kendall's textbook is of a physical model consisting of a box with data points represented by knobs on steel wires which could be tilted until the plane that best fit the data points could be seen as a straight line (1937, 276).

7. A brief discussion of Farr's research and his response to Snow is to be found in Eyler (1979, 114–22).

8. An explication of Weber's views on the subject may be found in Turner and Factor (1994, 119–71). A similar general defense of causal models might be mounted: it could proceed by arguing that what we want out of these models is the answer to causal questions that we ask in terms of a pre-given set of interests and ideas about how the world works; how the world "really works" is, in these domains, impossible to know, but also irrelevant to the actual decisions we want to make and the things we want to know about how we can hope to intervene. Glymour (1983) offers a defense of this kind of knowledge.

# SEARCHING FOR CAUSAL RELATIONS IN ECONOMIC STATISTICS

## Reflections from History

### *Mary S. Morgan*

### I. Introducing Correlation to Economics

Since the development of Quetelet's *homme moyen* in the early nineteenth century, social scientists have employed statistical concepts and methods to help them understand their subject matter. At times, these statistical procedures have been explicitly linked with causal arguments, and perhaps no statistical concept has been more prone to such interpretation than correlation and, in particular, the correlation coefficient in its various forms. My focus here is on the detail of a particular case chosen to enable us to explore (a) how economists and statisticians adopted and justified the employment of methods of correlation in the early twentieth century, and (b) to what extent the method of correlation changed the focus and form of their causal arguments. This historical case reflects not only their interests in searching for explanations of particular historical events as well as general causes, but also the difficulties they had in accessing multiple causes which combined in complex ways and remained obstinately unstable. In comparing the way in which causal relations were established on the basis of economic data before and after the development of correlation, we shall see that the new-found statistical measure did not entirely drive out the alternative method (itself relatively new) of arguing from representations of statistical data described in graphs; rather they remained complementary technologies, at least until the advent of structural modeling methods in the 1930s.

The usual case chosen to illustrate the development of correlation on social data is that of Yule's work (1895 and 1896) on the relationship between poverty and relief (discussed by Stigler [1986], Desrosières [1993], and Freedman in this volume), which was one of the first social science problems in which data sets were subjected to the new method. I discuss here another contemporary case which is more relevant for economics, namely, the controversy about the relationship between the marriage rate and trade, which wound its way through the statistical journals between 1890 and 1910. This question emerged from an older problem in economics—namely the possibility of a Malthusian population crisis growing out of the relationship between marriage (and birth) rates and wage levels. I discuss this case not because of any current inherent interest in the relationship between marriage rates and trade, but because it reveals both how late–nineteenth-century investigators used "the graphic method of statistics" to search for causal relationships and how correlation was introduced into this argument. The case allows me to describe and analyze one of the earlier attempts to use a new statistical technology to support causal inferences on the basis of statistical data, and it has the added advantage that it concerns time-based relations in time-series data, a topic of critical concern in much modern econometric work that deals with causality.

In order to see how the method and notion of correlation were incorporated into economics, the first part of the case study details how economists and statisticians searched for causal economic relations in statistical data before correlation entered the field as a way of measuring causal relations. As we shall see, the then-current ways of arguing about causes were very informal—investigators argued quite happily from graphs and tables about causes in a variety of ways. They made inferences, told stories, speculated from the data, and made use of a variety of different notions of causation. We might call these methods commonsensical, pre-scientific, or informal, to give an idea both of the lack of agreed inference rules and of the variety of ideas about causal relations that were in play at the time. How well would correlation fit into this environment? We have a number of clues about the impact of correlation from recent work in the history of statistics.

The biometric background to the development of correlation is well known. Ted Porter in *The Rise of Statistical Thinking* (1986) describes the key development in the late nineteenth century as the conversion of "error" into "variability"; the law of error became the normal distribution, and natural variability replaced the earlier focus on the average or mean. Statistical notions of association such as correlation allowed a more flexible concept of causal relatedness to replace older deterministic ideas of causality. Donald Mackenzie's *Statistics in Britain, 1865–1930* (1981) details the contribution of eugenics to these developments, while Ian Hacking in his book *The Taming of Chance* (1990) describes how correlation and regression (understood as representing law-like relationships as well as measures of co-variation) tamed randomness and chance, leading to the development of the concept of autonomous statistical laws. An appeal to correlation or regression could now be conceived as an appeal to a law-like relationship, yet one which did not require for its intelligibility any further—or deeper—causal stories about, e.g., underlying "mechanical" processes.

There is, of course, some reason to expect that the notion of statistical correlation would be put to use somewhat differently in economics than it was in the field of biometrics, where it originated. We know that the "probabilistic revolution" was not a uniform one—different disciplines saw it operate at any one or more of several distinct levels: measurement, data description, inference, theory and concepts (see Krüger et al. 1987, vol. II, Introduction). Thus those key changes documented and described by Porter, Hacking, and others are not necessarily the most relevant ones for economics. Why not? First because much socioeconomic data was not thought to be subject to the statistical laws obeyed by biometric data. Such good authorities as Bowley argued that economic laws did not depend on random or chance occurrences (like sampling balls from an urn) in the same way that population laws did, and so the applicability to social and economic phenomena of statistical methods based on probability theory was a matter for empirical investigation (see Bowley 1901, 297–98). Thus Hacking's arguments for the autonomy of the statistical laws which bound together data from known probability distributions in the biometric case would not necessarily be applicable to economics. Indeed, as Stephen Stigler discusses in his *The History of Statistics* (1986), for just these reasons there were initial differences in the sorts of correlation techniques suggested for social statistics by comparison with those utilized in biometrics. Second, economic laws did not deal with variability within a population, thus differing again from the situation described by Porter. Rather, socioeconomic data often consisted of single observations (each from a different point in time) connected together to form a series which exhibited variability over time. Available correlation techniques were not designed with such time-series data in mind, nor with the desire to solve the problems of historical or temporal explanation. This all suggests that the concept of correlation, along with appropriate measures of it, would have to be adapted to fit the distinctive features of socioeconomic data. Economics was not alone in this position. Judy Klein's new book (forthcoming) suggests that the problems of variability and of explanation over time provided the dynamic for developing new statistical methods and thinking in geology and meteorology as well as in economics.

## II. Marriage Rates and Trade Using the Graphic Method

Social investigators around the turn of the century wanted to explain what lay behind the variations over time in the marriage rate, one of the basic demographic data series commonly available and, as such, of interest to any social statistician. Their difficulty was to understand what had happened over specific periods covered by available data, and why: the data had to be made to speak—or rather to show—the relations between marriage rates and the price of wheat, unemployment, trade, and other variables which they thought might be causally associated.

This first part of the case study demonstrates how arguments about the causal significance of various factors were constructed by using "the graphic method" of analysis. This method needs some introduction. It was Bowley who offered the most professional treatment of "the graphic method."[1] One of the preeminent applied statisticians of the social sciences in the early twentieth century, Bowley lectured in statistics at the LSE, both close to and yet remote from the hotbed of statistical theory and eugenics at University College, London, where in the late nineteenth century, correlation and regression analysis were first developed. His *Elements of Statistics*, first published in 1901, was designed as a practical manual on how to do statistics, rather than a tract in mathematical statistical theory (though it included a considerable treatment of the latest results in that field).

Bowley described the graphic method as one of the "two main methods of elementary statistics," the other being the method of averages. The graphic method was similar to the method of averages in that its purpose was to yield a grasp of a whole series easily. But with time-related data, diagrams proved to be more instructive. Bowley paid particular attention to diagrams showing the relationship between two or more variables which might be used to support causal claims:

> Series of figures are often compared graphically with a view to discovering or illustrating causal relations. In such cases we do not only study relative growth . . . but look throughout the period for any signs of resemblance in rates of growth, dates of maxima and minima, or synchronism in any changes. The methods by which such comparisons are made are difficult, and need careful analysis. . . . In order to show direct connection, we shall try to make one line lie as nearly as possible over the other. (Bowley 1901, 172)

The argument being that if one line lies exactly over the other, it must replicate the same variations, and thus the statistical data on the two variables must be as closely related as it is possible to be. Nevertheless, before 1900, the number of graphs in which two variables were plotted with a view to establishing causal relationships, correlations, or associations between them was very small.[2] The example of the marriage rate and trade was one such case.

In this case study, I focus on three particular users of the graphic method: Ogle, Hooker, and Bowley. Their arguments involved causes in a number of ways. First, the pattern of variations was perceived as complex. Evidently, for the statisticians of the day, the primary problem was not to explain the average level of the marriage rate, but rather the variations in it over time, both long-term "trends" and short-term oscillations, and these were associated with different causes. Secondly, it is evident that they were happy to argue that different causes were at work at different times and in different places. There was no assumed stability in the operation of causes, nor in the types of explanations they might afford. Thirdly, although there might be a "main determining cause," there were also other causes which combined in unpredictable ways, and sometimes canceled each other. Finally, note that the graphic method of analysis did not, of course, provide any measures of the inferred relationships.

To a late–twentieth-century audience, these graphic techniques may well seem naïve, and the approach "alchemical." But these men were not stupid, and nor were the audiences they addressed; they were amongst the brightest of their generation. They were genuinely trying to investigate the causes of change in the marriage rate and attacked the problem with scientific spirit and with what (at the time) passed for scientific method: namely the study of tables, graphs, and finally, correlation coefficients. Stephen Turner's essay in this volume shows how tables were used in the nineteenth century to investigate partial causes in spatially recurring patterns of events. We have parallel examples in the attempts to unravel causes in temporal patterns. For example, William Farr, writing in the Registrar-General's annual report for 1878 provided a short discussion of a four-page table (his Table 69) with the title "Fluctuations in the marriage rate of England and some of the causes which had contributed thereto. 1839–78."

Although it became standard to include short tables in the Registrar-General's Annual Reports of the 1880s showing the marriage rate, trade, and other information in columns, little analysis of such tables was undertaken until William Ogle read his paper "On marriage-rates and marriage-ages with special reference to the growth of population" before the Royal Statistical Society in 1890. His starting point was the solidly Malthusian thesis,[3] expressed by the political economist Henry Fawcett as follows:

> There are few statistical facts better substantiated than that the marriages among the labouring class increase with the fall in the price of bread. (Fawcett, quoted by Ogle 1890, 256)

So, Ogle began his study with a conventional table showing prices of wheat and the marriage rate from 1820 to 1888. Judging from the number of years in which the elements in the two data series changed either in the same or in opposite directions, Ogle claimed that such an inverse relationship could not be supported by data from England and Wales for the mid-century (though it might be the case for earlier times and for other countries). He suggested that, if anything, a direct relationship was more plausible: "the marriage-rate varies not inversely but directly with the price of wheat" (Ogle 1890, 258).

In order to justify this retreat from the classic Malthusian story, Ogle developed the following argument:

> When the standard of living is extremely low among the working classes, the question of marrying or not marrying is decided simply by the possibility of getting sufficient food, and consequently the price of food is then the main factor in determining the marriage-rate; but, when the standard of living has been considerably raised, mere sufficiency of food is no longer held to be an adequate justification of marriage, and consequently the price of wheat ceases to be the one determining factor, and becomes quite subordinate to the amount of wages. This however only explains why marriages do not increase in this country when food is cheap; it does not explain why they increase when food, or rather when wheat, is dear. . . . If then the changes in the cost of food will not explain the fluctuations in the marriage-rate, what better explanation can we find? (Ogle 1890, 258)

In sliding neatly from the price of wheat (representing the price of food) to the more general question of the standard of living, represented by the wage rate or other measures of national prosperity, Ogle moved away from the agrarian basis of the Malthusian theory. By the 1890s, the accepted available measures of national prosperity were the value of trade (imports and exports) and the bills/checks cleared at banks.

His next step was to focus on the relationship between the marriage rate and exports, for the good reason that they seemed to be more closely related than the other series. In what sense were these closely related? What judgment was used? The sense employed was visual: the inspection was done by looking at and manipulating graphs of the two series, and judging their similarity by eye. His original graph of the data (his Diagram A, our figure 2.1) did not convince Ogle, so he moved quickly away from the trend elements to focus on the short-term oscillations, by re-graphing the series in such a way as to focus on their short-term similarities. From this he inferred that:

> The marriage-rate then goes up and down synchronously with the value of exports. This can clearly only be because the changes in these values are an indirect indication of corresponding changes in the employment and the wages of the labouring classes. (Ogle 1890, 260)

In other words, exports are a proxy for the well-being of the working classes, whose behavior dominates the demographic data.

**FIGURE 2.1** Ogle's Diagram A, showing the marriage-rate per 1000 for England and Wales and the value of exports per head of the British population.

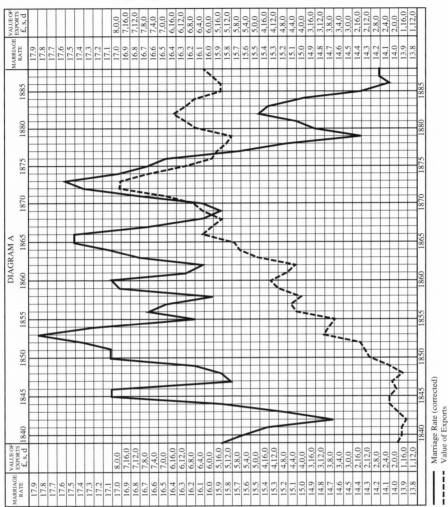

Ogle then tells us that "hunting about for such a measure" of these changes in employment and wages, he "lighted upon" the figures kept by trades unions of their unemployed members (Ogle 1890, 260). The total number of members represented was around 125,000, of which the number unemployed varied from 718 (in 1872) to 13,851 (in 1886). Again, with the use of a table, Ogle compared these data with the marriage rate and found a considerable similarity in movements between the series. Going further, and taking a "still smaller sample" by focusing on a particular union: the "Glass Bottle Makers of Yorkshire United Trade Protection Society" with a membership of 1600, proved equally efficacious! Ogle argued as follows:

> that the fluctuations in the marriage-rate follow the fluctuations in the amount of industrial employment; and secondly, that the various industries are so solidly bound together, that the changes in even a single and a small one afford a very close index of the changes in the whole of them. (Ogle 1890, 262)

Note how cautiously causal claims are being broached here. There is no claim that the fluctuations in marriage rates are other than "following" the fluctuations in employment, for which the little trade union offers an index. Nevertheless, these careful claims are gradually building up into a causal story.

Ogle had still to account for the fact that though the relationship between unemployment and marriage rates was inverse, there was nevertheless a positive relationship between the price of wheat and marriage rates.

> Such a coincidence as this cannot possibly be only casual, and the apparent paradox of increased marriages with dearer food, and diminished marriages with cheaper food, requires explanation. (Ogle 1890, 262)

His explanation depended on a third variable, the cost of transport: as trade expands, freight rates rise, increasing the price of wheat in a classic causal story which served to link all three of his original variables:

> In conclusion then the marriage-rate rises and falls with the amount of industrial employment, which in its turn is determined by the briskness of trade, as measured by the values of exports, which also rise and fall concomitantly, and produce by their effect upon freights a simultaneous rise or fall in the price of wheat. (Ogle 1890, 263)

Ogle seemed to have sorted out the short-term fluctuations, but there still remained the problem of the trend divergence of the two series of marriage-rates and exports (evident in his Diagram A). To solve this, Ogle picked out the value of exports per head in 1888 and in 1866 as being equal, and focused on the difference in the marriage rate between those dates.

Clearly then if the marriage-rate were determined simply by the export values, the rates should have been practically identical in these two years. (Ogle 1890, 264)

The actual decline of marriage rates by 3.3 per 1000 over the twenty-year period "must have been due to something else than changes in export values, which remained unaltered" (Ogle 1890, 264).[4] He assumed that this "unknown influence" had worked throughout the whole period from 1857–1888, and, extrapolating this trend decline backwards, calculated a drop of 4.65 per 1000 over the thirty-one years since 1857. "Correcting" for this long-term decline, on his Diagram C (our figure 2.2), he found

> the two curves tally in the most remarkable degree . . . irresistible evidence that the hypothesis by which the correction was made was a true one. It appears then that for many years past there has been some cause in operation, which, in spite of the increasing wealth of the country, in spite of the increased value and the still more increased bulk of our exports, in spite of higher wages and cheaper commodities, has tended to make men more and more unwilling to venture upon marriage, and it would be highly interesting to know what this influence may have been. Possibly, and probably, it has been no one single cause, but an aggregate of causes that has had this depressing effect. (Ogle 1890, 265)

Ogle went on to suggest two possibilities for this missing cause: the agricultural depression which had reduced marriage in that sector and the ever-increasing standard of living which had raised expectations elsewhere, so that marriages were delayed until higher levels of income could be obtained.

Ogle then went on to a cross-sectional analysis of his data before attacking the policy agenda: the attainment of a stationary population. But the remedies proposed for reaching this end did not draw on the analysis presented in the first part of the study. Rather, Ogle concentrated on how the change in the age of marriage (by retarding marriage) might (or more probably might not) solve the problem. As was usual at meetings of the Royal Statistical Society, a discussion followed the presentation of his paper, during which Ogle's data and hypothesis were criticized on matters of detail. But no criticisms were raised concerning the essential methods and ways of arguing he had employed, or the inferences about causal relations he had drawn on the basis of his tables and graphs.

R. H. Hooker took up this question of marriage rates in a paper read to the Manchester Statistical Society on January 12, 1898. He started from Ogle's data, but disagreed with his inferences, arguing for direct causal links between the marriage rate and trade. Whereas Ogle had argued that some "regularly increasing influence" had caused a "continuous fall in the marriage-rate between

**FIGURE 2.2**  Ogle's Diagram C, showing his "corrected" marriage-rate per 1000 for England and Wales and the value of exports per head of the British population.

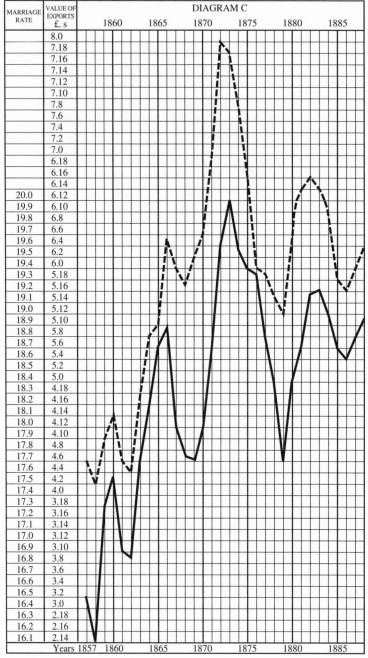

————— Marriage Rate (corrected)

- - - - - Value of Exports

(*Source:* Reset from Ogle, 1890, opposite page 265.)

1857 and 1888" (Hooker 1898, 103), Hooker focused on the sudden fall in the marriage rate in the 1870s (which can be seen clearly on Ogle's Diagram A, our figure 2.1) and argued that the data showed that a

> deep fall in trade prosperity might account for the . . . prolonged fall in the marriage rate. (Hooker 1898, 102–3)

Why did he find this hypothesis reasonable? He had the advantage of several years more data (until 1895) and his graphs used different scales, which helped him to see things differently (as he pointed out in a footnote on p. 105). He apparently saw two periods of stability in marriage rates, with a sudden drop between the two, and so favored a different correction than that made by Ogle.[5]

The problem of presenting the fall in trade as the cause of the fall in marriage rates was that when trade recovered, marriage rates continued at the lower rate. The real cause was still missing. Hooker hit on an alternative explanation: marriages were being delayed to a later age. There was some evidence in the official data on marriage of minors which suggested that his hypothesis might be correct. Though the data were far from reliable, he calculated the implied constant increase in the age of marriage necessary to maintain the lower marriage rate observed in his graph. As he said of his arguments:

> the foregoing is not an absolutely strict mathematical proof: it belongs to that class of demonstrations which depend for their verification upon the coincidence of hypothesis and observed facts, to the exclusion of all other hypotheses. We are here given certain facts—represented by the actual curve of the marriage-rate. We know that this particular form of the curve will follow from a certain change in the average marriage-age; and we find further from another set of data that in all probability this particular change in the average marriage-age has actually occurred (the Registrar-General's average marriage-age being shown to be inaccurate). It is, of course, not impossible that this may not be the true cause; it is conceivable that some other phenomenon may produce an identical result upon the marriage-rate. . . . Failing such other explanation, therefore, we are entitled to regard the increased average marriage-age as the sole cause of the decreased marriage-rate. It follows, at all events, quite certainly that if there are any other causes, be they large or small, they must exactly neutralise each other. (Hooker 1898, 113–14)

(Notice here the final resort to the old statistical argument that causes may cancel out in the aggregate.)

Hooker was then faced with the problem of accounting for the sudden increase in marriage age from the 1870s. He rejected Ogle's thesis of agricultural depression as wrong on grounds of timing, and his standard of living argument as being relevant throughout the whole period. Hooker argued that better wages dominated the decision to marry for the period up to the 1870s, when increased earnings justified the high "normal" rate of marriage, but that increased education levels, which delayed marriage through prudence, explained the lower "normal" rate of marriage post-1870s.

Although most of Hooker's causal arguments related to the question of explaining the changes between two periods of stable marriage rates, one of the novelties of his approach was a discussion of short-run fluctuations, particularly the lag between the two series. By carefully matching the kinks in the two curves, he found that the trade kink sometimes preceded and sometimes followed the marriage-rate kink. This did not faze Hooker, who suggested that people's expectations about the future increase or decline in trade could explain why marriages should sometimes advance ahead of trade.

> It offers an explanation, at all events, of phenomena which in many cases precede events of which they are usually the effect. (Hooker 1898, 107)

The relationship between the marriage rate and trade was also treated as one of several examples by Bowley in his masterly 1901 textbook treatment of the graphical method discussed earlier. Recall his general advice

> In order to show direct connection, we shall try to make one line lie as nearly as possible over the other. (Bowley 1901, 172)

To do this required a series of operations:

> Draw a preliminary diagram in which both lines are entered on any scales; this will suggest the resemblances to be tested. Notice in what period the fluctuations are greatest; this in general should be the period to be taken, for it is here that the causal relations have had most play. (Bowley 1901, 172)

From his first graph of the marriage rate, trade, and wheat (his Figure I opposite p. 175), Bowley inferred that there was considerable agreement in movement between the marriage rate and trade, though they were not completely in line. On the other hand, the wheat price was not consistent in its relation to the marriage rate: changes in the price of wheat had first diverged from those in the marriage rate, and then accorded with them.

In an argument similar to that of Ogle (but based on data from a different time period), Bowley suggested that the double tendency in the long-term wheat price–marriage relation was to be accounted for by the change in general prosperity. But he did not resort to Ogle's explanation that appealed to freight rates, arguing instead for other (unspecified) causes operative on wheat prices:

When exports and imports are increasing in value, trade is stimulated, and in spite of rising prices, marriageable people are sanguine that the prosperity will remain and the prices fall; but when the prices fall, so do the profits and incomes, and marriageable people are more prudent. For these reasons we may expect the marriage rate and foreign trade lines to resemble each other.

Now the increase of the marriage rate corresponding to an inflation of trade, and inflation of trade to a time of rising prices in general, we shall find the price of wheat in particular, which is connected with the course of prices in general, rising when trade is inflated and falling when it is depressed, and therefore rising and falling with the marriage rate. But since the price of wheat is influenced also by special causes, it will not always correspond to the state of trade, and still less to the marriage rate, with its former tendency to opposite variations. (Bowley 1901, 176)

Bowley's next step was to take the average of both series, marriage rate and trade, and place them on a graph so that the averages coincided (his Figure II, opposite p. 175), so that "any correspondence between the two lines can be at once detected" (Bowley 1901, 172). Finally, the question of scale was resolved so that the average movement (peak to trough) for both series was equated on the graph by choice of scale. The effect of this was to replace the base line with the average line and make it possible to equate average changes (as in his Figure III, opposite p. 175, our figure 2.3). If at this stage the lines lay almost exactly over each other, we can now see that this would be the graphic equivalent of an almost perfect correlation.

Yet, despite the almost mechanical nature of Bowley's instructions for making the best graphs, inference problems remained, for there were clearly trade-offs in the methods if one wanted to make causal inferences about both the longer-term trends and the short-term fluctuations:

It is now seen that the fluctuations since 1880 lie more closely together in the two curves, but that this closeness has been obtained by the partial sacrifice of the years 1872–80, and there is now a complete disagreement before 1870. A yet shorter period, 1879–1893, would show a very close agreement; but so special a selection would vitiate any general argument.

Our conclusion is, that since 1870 the causes which affect foreign trade have also affected the marriage rate at the same dates and in the same sense, and that the more marked the effects on the one, the more marked are the effects on the other also, but that there is no law of simple proportion between them. (Bowley 1901, 177)

**FIGURE 2.3**   Marriage Rate and Imports and Exports per Head

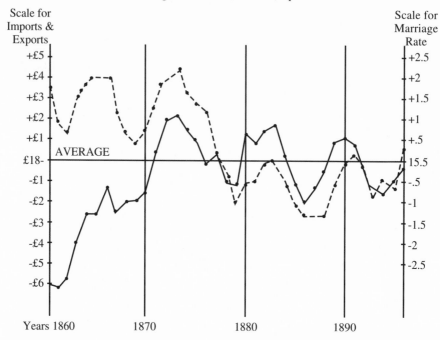

Bowley's Figure III, showing average fluctuations around the mean line for the marriage-rate per 1000 for England and Wales and the value of imports plus exports per head of the British population.
(*Source:* Reset from Bowley, 1901, opposite page 175.)

For Bowley then, the connection between the marriage rate and trade was that both were subject to a common set of causes rather than that there was a direct cause-effect relation between the two or even that trade was an indicator (a "barometer") of the marriage rate, in the sense that trade provided an accurate measuring instrument for marriage rates so that a change in one betokened a change in the other. This view, that the observed relation was the outcome of common causes, perhaps relates to Bowley's beliefs on correlation (which I discuss below).

Before I continue with the second part of this case study, let me take stock. I have described the ways in which three exemplary scholars used their graphical techniques, and, by quoting extensively from their work, have sought to show how they constructed causal arguments in conjunction with inferences drawn from relevant tables and graphs. First, as we can see, the arguments were fairly complicated: causes were taken to be unstable, dominant in some periods, unimportant in others. Causes might have large or small effects, and often effects had to be ascribed to unknown causes. Causes were seen to fit together in different ways, sometimes additively, sometimes canceling one another out, and sometimes simply disappearing in the face of more dominant ones. Second, causes were generally treated as being historically contingent: evidence for their presence was sought only in specific temporal contexts, and their inferred effects were limited to specific times and places. But within this common framework of assumptions, there was still considerable variation in how such causes were thought to work. Sometimes a graphed variable was presented as a direct cause of another variable, sometimes it was seen as operating through chains of other intervening causes. Sometimes variables were depicted as being in loose associations with each other, and sometimes one variable was seen as providing an indicator (an index or "barometer") of another variable. Third, the graphs and tables were used not merely to illustrate the arguments, but as genuine investigative tools to sort out and evaluate various causal claims. The process of working through the different tables and constructing different graphs has the quality of a sequence of experiments, though with much more limited manipulative power and without the control and measurement systems characteristic of laboratory experiments.[6] Of course, what is being manipulated is not the marriage rate or trade, but the statistical data on marriage rates and trade. In this investigative mode, the inferences themselves relied on a mix of intuitions about how to read off results from the graphs and tables, in light of common background knowledge. All this suggests that even the use of "the graphical method" implied no well-articulated, uniform conception of causal analysis or method of causal inference. Thus it is scarcely surprising that when statistical correlation techniques became available they were rather quickly exploited by social statisticians with the hope that they could be used to overcome the obvious limitations of graphical methods.

## III. Correlating the Marriage Rate and Trade

The concept and techniques of statistical correlations were first elaborated in the decade of the 1890s. They were not uncontroversial. During the following decade, the coefficient's method of calculation, its assumptions and its interpretation, were all subject to further negotiation. Stigler (1986) recounts how Pearson's methods and interpretation were grounded in biometrics and assumed a bivariate normal distribution of the variables, while Yule's interpretation and methods were developed partly on social science data and were less dependent on distributional assumptions. Pearson, of course, was a central figure in the biometrics (eugenics) group, while Mackenzie (1981) labels Hooker, Yule, and Bowley (our authors here) as "outsiders." Correlation was introduced into the search for relationships between the marriage rates and trade in 1901, both in Bowley's textbook and in another paper by Hooker.[7] (Since each referenced the other, we should not grant either of them precedence.)

Let us start with what Bowley had to say about the theory of correlation:

> It is never easy to establish the existence of a causal connection between two phenomena or series of phenomena; but a great deal of light can often be thrown by the application of algebraic probability. . . . When two quantities are so related that . . . an increase or decrease of one is found in connection with an increase or decrease (or inversely) of the other, and the greater the magnitude of the changes in the one, the greater the magnitude of the changes in the other, the quantities are said to be *correlated*. *Correlation* is a quantity which can be measured numerically. (Bowley 1901, 316)

Bowley's treatment of correlation theory assumes two variables X and Y each being "subject to variations x, y, which are due to a multitude of individually unimportant causes" (Bowley 1901, 316) so that x is the sum of all its causes and y is the sum of its causes. When the individual x, y values are grouped into pairs:

> as measurement of two quantities at the same date, or of two parts of the same organism, . . . if x and y are quite independent, none of the causes producing them are common to both (Bowley 1901, 317)

then we have a situation of no correlation. If, on the other hand, some of the causes of x coincide with the causes of y, then there will be a degree of association measured by the correlation coefficient r.

Bowley used correlation measurements to investigate further the relationships between the marriage rate, the price of wheat, and trade (imports plus exports). He employed a longer data period than in his graphic analysis and split it into two parts: 1845–1864 and 1875–1894 based on earlier discussions of the different sub-period relations evident in his graphs. (This separation into homogeneous sub-periods is the time-series equivalent of the homogeneous populations required by the authors discussed in Stephen Turner's case study.) He calculated the four correlation coefficients (r) and their probable errors (pe) (see Bowley 1901, 322). He found a "a slight negative correlation" between the marriage rate and the wheat price in the first period ($r = -.30$; pe $= .14$), but found "better evidence" for the latter period ($r = +.47$; pe $= .10$), which he judged as consistent with his earlier arguments. The marriage rate and trade were "quite uncorrelated" in the early period (r = +.007; pe = .15), and showed only "slight correlation" in the later period (r = +.25, pe = .14).

This final relationship was not so well supported as he had thought from using his graphic method of investigation, but he went on to argue in the following terms that there was a causal connection:

> the odds against the correspondence between the observed figures since 1875, arising without causal connection are only about 4 to 1, if we assume that the figures for each year are independent of the next. (Bowley 1901, 322)

We can see here the idea of causal connection being inferred from any correlation that is unlikely to have arisen by chance or coincidence. Unfortunately, none of Bowley's results would have passed his own test of causality, that the coefficient must be at least six times its probable error (a test which was used by a number of economists measuring correlations in those early years).[8] Nevertheless, regardless of their level, correlations in economic data began to take on the status of statistical law-like relationships, even though these correlation "laws" appeared to hold only in quite particular historical times and places.

Bowley justified his rather unimpressive results by drawing a comparison between biology and economics. The development of correlation theory, understood as a term referring inclusively to the set of new ideas and approaches to statistical reasoning and measurement,[9] was intimately connected with inheritance and the laws of evolution:

> The law of heredity can be only tested numerically by the theory of correlation; the effect of natural selection is easily considered with the help of the coefficient of regression. (Bowley 1901, 325)

But Bowley was writing for a different audience, one which needed to be warned that they should not expect to find in social science the same sort of probability laws which existed in biology:

> The great difficulty which the student of economics encounters when dealing with the theory of error is the apparent slightness of relation between this theory and the facts with which he deals. This slightness is only apparent; it is because the theory has not, in the form he meets it, been carried far enough to fit it to the very complex facts of human affairs that we do not get that exact correspondence we might desire. (Bowley 1901, 325–26)

Thus the equivocal results produced when correlation techniques were applied to social and economic data were ascribed not to peculiarities of the subject matter vis-à-vis biology, but to the fact that correlation theory was not yet sufficiently developed to cope with the complications of social science problems involving time-series data!

Hooker's experiments with the correlation coefficient proved more convincing when, in September 1901, he read his second paper on the "Correlation of the marriage-rate with trade" to the Economics and Statistics Section of the British Association in Glasgow. The technique was still sufficiently new that he felt obliged to introduce the correlation method to this audience, giving both a brief summary of the mathematical formulae (with reference to Bowley's text above) and a more lengthy discussion of the purpose and meaning of the coefficient. Correlation was not yet seen as a completely reliable method for time-series data which would allow investigators to jettison the graphic method. So, Hooker did not reject the use of visual inspection of graphs as a method of finding causal relations in economic data, but suggested rather that such techniques complemented correlation methods:

> The application of the theory of correlation to economic phenomena frequently presents many difficulties, more especially where the element of time is involved; and it by no means follows as a matter of course that a high correlation coefficient is a proof of causal connection between any two variables, or that a low coefficient is to be interpreted as demonstrating the absence of such connection. In many cases, indeed, the existence of a causal connection between two phenomena is more clearly deduced from mere inspection of diagrams than from mathematical calculation. But simple inspection will not enable us to measure the degree of correspondence, and many points will often remain hidden which may be revealed by a judicious use of correlation. (Hooker 1901, 485)

Note particularly here the suggestion that the correlation method will reveal hidden causes in the data. From Hooker's subsequent investigations it can be seen that two problems above all required attention: first, that the multiple causes at work have to be unraveled, and second, that causes operate in different time units (e.g., long-term trends and short-term cycles) which have to be sorted out. These two problems would spawn, in twentieth-century econometrics, two distinct approaches: the structural econometrics approach and the time-series approach. Here in Hooker however, they are still under joint investigation.[10]

He went on to note that in economic matters, explanations based on the identification of a single cause were very rare. More typically it is assumed that a relationship between two variables will involve one or more common causes operating on both variables, as well as a number of dominating non-common causes. In this situation the coefficient obtained will also be affected by the various causes, and it may be that the effect of the joint causes is masked by that of the others. (Hooker 1901, 485)

This problem certainly appeared to be true of the relationship between the marriage rate and trade, where Hooker's far more sophisticated treatment nevertheless supported Ogle's original inferences. He described the series from 1861–1895 as showing "correlated" oscillations but with a rising trade curve over the period and a falling marriage rate where "these latter movements may with safety be ascribed to outside causes" (Hooker 1901, 486). What justified this invocation of other causes? Well, the correlation of the two series showed r = +.18 (pe =.09): "indicating no connection between the two" (Hooker 1901, 486). Rather, he argued,

> the fact is that in this case the difference in the *general* movement of the two curves has completely overshadowed the minor oscillations; whereas it is only to these latter that we refer when in ordinary parlance, we speak of the marriage-rate as being dependent upon trade. (Hooker 1901, 486)

In order to support this view, he first worked out what he termed the "instantaneous average" soon to be called "the trend."[11] He then calculated correlation coefficients using deviations from the instantaneous average, to explore a number of relationships: between the marriage rate and exports, imports, total trade, the price of wheat, and bank clearings. To his evident pleasure, he found that the correlation between the marriage rate and exports in this new form was r = +.80 (pe =.04), which easily passed Bowley's causality test. (Again, the parallels between finding an appropriate time-length unit for which to specify a relationship and defining the correct aggregation level of geographical population units about which to generalize in Turner's case study are striking.) Anxious to test his earlier theses about the timing of the relationship, Hooker also worked out the table of lagged correlation coefficients and found that with a half-year lag of marriages behind exports, the correlation reached its maximum: r = +.86. Yule (1906) graphed Hooker's series and we can see how effective Hooker's work was in our figure 2.4, which compares well with Ogle's earlier efforts using only the graphic method.

Yule finally took up the challenge of explaining the variations in the marriage rate in a paper read to the Royal Statistical Society in December 1905 (reported 1906). He decided that what was needed first was to make corrections to the marriage-rate data to make it a consistent series. He believed that a number of changes had occurred over the years in the proportion of marriageable people in the population and their relative ages which had implications for the denomi-

**FIGURE 2.4** Fluctuations in (1) Marriage-rate and (2) Trade (Exports + Imports per Head) (Deviations from 9-year Means). [Data of R.H. Hooker, "Journal of the Royal Statistical Society," 1901.]

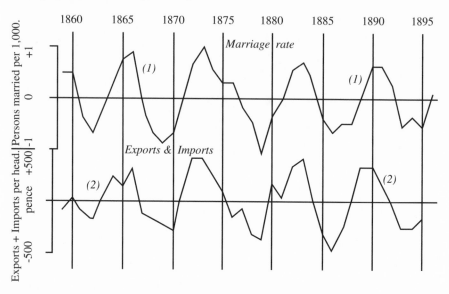

Yule's graph of Hooker's 1901 data, showing deviations from a moving average mean for the marriage-rate per 1000 for England and Wales and the value of imports plus exports per head of the British population. Hooker's reported correlation for this data set is r = +.86.
(*Source:* Reset from Yule, 1906, page 94.)

nator in the marriage-rate data. The effect of correcting for these changes so as to create a "standard population" in the denominator was to suggest a rising trend in the period up to 1873 and a much more striking fall than that commented on by Hooker over the period from 1873 to 1901.

Ever one to seek to advance an issue, Yule picked up the earlier 1898 discussion by Hooker in which he had claimed that the trend decline was due to a retardation of the age of marriage. He went on to investigate the economic factors which might influence the postponement rate. He found that the price index most closely mirrored the trend movements in the marriage rate, but that there was at least one missing "predominant" factor, which caused the decline in marriage rates since 1882–1886. His actual causal inferences pertaining to this claim were very carefully framed:

> To avoid misconception, let me say at once that I do not for a moment suppose that price-movements and this latter cause—whatever it may

be—have been the *sole* factors operative during the half-century in effecting the alterations in the marriage-rate, or anything like it. I would put the matter in a different and much more limited way: I do think that the preceding work justifies the conclusions, (1) That price-movements have been directly or indirectly a *predominant* factor during the period 1865–1900, or so closely related to a predominant factor, that they serve as an accurate index to its changes; (2) That some other cause has to be taken into account as a second predominant factor during the last fifteen or twenty years (1880 or 1885 to 1900); (3) That the effects of other factors have been relatively small. (Yule 1906, 108)

He went on to point out that the decline in marriage rates was general across Europe, which gave credence to the factor of prices since that was a common factor for all the economies.

But as to the *rationale* of the connection between marriage-rate and prices I must confess myself somewhat puzzled in view of the statistical evidence. It is natural to assume that price-changes operate on the social organism chiefly *viâ* trade, or, more directly, employment. But the course of trade . . . does not . . . tally with the marriage-rate. . . . If we turn to the direct effect of price-changes on the purchasing powers of the consumer, the matter seems no clearer. . . there appears to have been a large rise in money wages between 1860–75, and the rise in the consuming power of the population has been practically continuous since 1850; it does not date only from the fall in prices. (Yule 1906, 108–9)

It is significant that the correlation measures were unable to help Yule[12] untangle the problems created by the existence of several "predominant" causes and the fact that they appeared to combine in complex ways. These issues remained unresolved even though Yule had begun experimenting with multivariate regression (see note 15) and, in a joint paper with Hooker in the same year, had investigated the notion of "relative importance" (on which, see Kruskal 1989).

It is useful once again to take stock. Yule was one of the main developers of the correlation measure. Yet like Hooker and Bowley, who were also early users of the new statistical techniques, he relied at least as much on the graphic method for teasing out causal factors as had the earlier work without correlations by Hooker and Ogle. Computing correlations allowed the measurement of relationships between variables, but measuring correlations did not resolve disagreement, for researchers could still tell different causal stories on the basis of the same evidence, by using different methods of preparing the data, by choosing different data periods or by focusing on different time components (short-run, cycles, long-run, etc). The graphic method had encouraged the search for causes to go hand-in-hand with the search for regularities in time-series elements, since each component regularity in time-series data appeared to require its own causal explanation. This tradition continued with the introduction of correlation theory, since correlations in long-run trends were typically different from those evident in short-run fluctuations, thus giving rise to different causal inferences. Appeal to correlations made some difference, for measuring their strength enabled investigators to reject some hypothesized relationships. But it did not succeed in sorting out causes as well as had been hoped. The new arguments and inferences still have much of the *ad hoc* quality of those they replaced. As before, investigators continued to work with a variety of different conceptions of what constituted a proper causal relationship.

## IV. Correlation Moves into Economics

At this point, having demonstrated that correlation did not immediately solve the problems of causal inferences for the statisticians discussing the complicated question of the relationship between the marriage rate and trade or significantly alter the forms of the arguments, I will move on to examine how certain economists enthusiastically adopted the correlation method. I turn to the work of two American economists, J. P. Norton and W. M. Persons, who showed a firm commitment to correlation as a measure of relationship which does away with ambiguity.[13] Their rather different usage of the measure perhaps lies in the fact that they were both confident young American economists picking up the method as a new tool which they freely adapted into economics compared to the rather cautious British statisticians, associated with its development period. In this American work, both the measure and its interpretation have clearly "moved" into economics, but what correlation meant in this new environment was still open to considerable interpretation, as these two different examples show.

J. P. Norton's 1902 book is not concerned with the marriage rate and trade, but rather with the New York money market, where his expressed aim was to apply both the correlation and graphical methods of interpolation to statistical data, and where he presented both techniques as great steps forward. The usual (old-fashioned) method of dealing with statistical data was to calculate averages, averages for every conceivable time period, but Norton was not impressed with the average:

> An average is easy, often useful, but many times a senseless thing. In many cases, it must be confessed, the average tells very little. (Norton 1902, 15)

After extensive use of diagrams to sort out trends and cycles, Norton finally got to correlation coefficients as a measure of relationship and used them to determine the closest lag length between two variables as well as to argue for causal associations between variables.

He finished his main chapter on correlation with a table (Table 30, p. 96) comparing the values of the correlation coefficients from his financial series with those taken from Pearson's biometric correlations among human characteristics. The economic work showed much higher correlations (and with lower probable errors) than those of Pearson. Norton concluded:

> It is apparent by the above table that the highest degree of correlation in the biological table, the correlation between the right and left femur bones, is equalled by the correlation which we have found to exist between the reserve and loan periods.

> If the biologist can point to these correlations as satisfactory scientific laws, it is hardly more than fair to grant the same privilege to the economist. In short, it seems reasonable to conclude that economics may become almost as *exact* as biology. (Norton 1902, 96–97)

What correlation did for this American economist (and the British statisticians discussed in the previous section) was not only to provide a measure of a relationship in the correlation coefficient, but also, in the "probable error," a measure of certainty or uncertainty about that relationship. The "probable error" was an important part of this new correlation technology. Though the economic commentators did not devote much explicit attention to what this probable error meant, my reading of their practical applications suggests that by giving the uncertainty a numerical measure, it was "pinned down" and made more exact (seen most clearly in Norton's work). The uncertainty in these cases arose both from the variety of causes, known and unknown, and from the variability with which they combined. Correlation and regression did not "tame chance" in economic statistics (as Hacking claims for other domains) because chance wasn't firmly rooted there in the first place. What has been tamed here is not chance, but ignorance, for something which can be codified and measured becomes partly known and no longer so worrisome.

A rather different interpretation of correlation was used by Persons, who in 1910, appeared as firmly committed to the correlation measure as Norton. In an article for the American Statistical Association, Persons (then an assistant professor of economics at Dartmouth College) made a strong plea for the particular relevance that correlation could have for economics. Correlation was the new way of measuring the "tendencies" that economic theorists were so fond of:

> The cause and effect relation existing between economic events is especially difficult to ascertain because of the presence of innumerable variable elements. In solving his problems the economist can not, like the physicist or chemist, eliminate all causes except one and then by experiment determine the effect of that one. Causes must be dealt with *en masse*. Since any effect is the result of many combined causes the economist is never sure that a given effect will follow a given cause. In stating an economic law he always has to postulate "other things remaining the same," with, perhaps, little appreciation of what the other things may be. It is rarely, if ever, possible for the economist to state more than "such and such a cause *tends* to produce such and such an effect." Events can only be stated to be more or less probable. He is dealing mainly, therefore, with correlation and not with simple causation. (Persons 1910, 287)

Persons' case for the similarity between biology and economics was made in terms of the idea that both sciences had to deal with multiple causes which could and did combine to produce multiple effects that could not be fully specified or distinguished. He quoted Karl Pearson regarding our ability only to find out the most "probable" effects. And like the biometricians, he argued that economists too should treat the correlations with which they were confronted as having a causal significance, but not one allowing exact predictions.

> Just as the biologists cannot predict a man's height or color of eyes or temper or combativeness by knowing those qualities in his ancestors, so economists cannot predict that a definite call rate in Wall Street will go with a given percentage of reserves to deposits in New York banks or that a given supply of wheat will result in a definite price per bushel. (Persons 1910, 288)

Nevertheless, such relationships have been observed and needed to be measured: the graphic method had to give way to correlation.

> The commonly used method of measuring the amount of correlation between any two series of economic statistics is to represent the two series graphically upon the same sheet of cross-section paper and then compare the fluctuations of one series with those of the other. . . . All the questions to be tested by the statistics collected are questions of correlation. . . . The graphic method of comparing fluctuations is well enough as a preliminary, *but does it enable anyone to tell anything of the extent of the correlation between the series of figures being considered? . . . The charts do not answer the questions proposed.* (Persons 1910, 289, 290 and 294)

The correlation coefficient was going to answer his questions. But what did Persons think the correlation coefficient measured? Here he relied on Bowley's textbook treatment (discussed above), explaining that the coefficient was derived by assuming that "a large number of independent causes operate upon each of the two series X and Y, producing normal distributions in both cases. Upon the assumption that the set of causes operating upon the series X is *not independent* of the set of causes operating upon the series Y" (Persons 1910, 298–99) then a correlation r is obtained, which takes the value zero only "when the operating causes are absolutely independent" (p. 299).

He then stated that if there is a linear relationship between X and Y, the correlation will be either +1 or –1 thereby making a crucial move away from the idea of independence or non-independence of the causes of both X and Y, to the correlation coefficient as a measure of the relationship between X and Y. Thus the correlation coefficient became a measure of the closeness of the data points to a causal or functional linear relationship (the line of regression) between X and Y, rather than a measure of the relative independence of sets of multiple causes shared by the two variables, the viewpoint explicitly assumed at the outset of his discussion.[14] The parallel movement in the biometric case was from using correlation as a way of measuring the association between the size of two different bones in the body (say) because of shared common causes, to using it as a measure of the closeness of observations to a line of regression (e.g., the law of inheritance, relating fathers' heights directly to sons' heights).

This move is much more significant in the context of economic time-series data, where graphically it represents a shift from time-series graphs to two-dimensional scatter diagrams in which time is omitted. In using correlation as a measure of common causes between two sets of time-series observations, the time-specific (or historical) elements remain important. The common causes are taken to operate on each pair of events observed throughout the relevant period, and the explanation or interpretation offered with the correlation coefficient clearly refers to a particular temporal period. It is in time-series correlations thus construed that we would find the economic equivalent of the autonomous statistical laws of biometrics, where the correlation does not necessarily require any deeper causal analysis. From this perspective, the fact that the marriage rate was highly correlated with trade in the short term provided a firm associative relationship, which though it might well depend on the existence of some unspecified mechanisms could be accepted on the evidence available as a well-supported causal law, just as the biometric law of regression to the mean stood well supported, and was accepted as a causal law, in the absence of any specific knowledge of the "gemmules" that might constitute the inheritance mechanism linking the heights of fathers with those of their sons.

By contrast, correlation as a measure of closeness to a direct relation between two variables represented on a scatter diagram abstracts from the temporality or history of the observations. Correlation is here interpreted in terms of the strength of an atemporal cause-effect or functional relationship. These atemporal correlations came to be more closely tied to theoretic models of economic phenomena and their associated causal mechanisms, as we can see in work on the econometrics of demand in the early twentieth century, where correlation measured the degree to which a price-quantity cause-effect relationship was supported by the data (see Morgan 1990, Part II).

This critical shift in the meaning assigned to the correlation coefficient has been discussed in some detail not because I want to claim that Persons was historically responsible for the way in which many economists came to interpret the correlation coefficient. Rather, I discuss this move because I want to show how the correlation coefficient, as a loose associative notion depending on common cause arguments in the temporal domain, can easily be reinterpreted to serve as a measure of a direct cause-effect relationship which dispenses with both the notion of common causes and historical contingency.

We can see both these types of usage of correlation in Persons' own work. The latter, atemporal, form of correlation was adopted in his 1910 paper, and the practical and persuasive way in which Persons used the correlation coefficient here was almost sufficient to overcome the qualms which his typical economist readers were likely to have had about this way of employing the correlation concept. He discussed a number of recent economic works in which inferences about causal relations were based on graphic methods. He then systematically destroyed a number of these inferences by calculating the correlation coefficients between the relevant variables. For example, a causal relationship between money and prices apparently supported by a time-dated graph (due to Kemmerer) was not supported by Persons' calculations of the correlation coefficient ($r = +.23$; $pe = .13$). Although many of his experiments with the coefficient were destructive, he wrote approvingly of the work on the question of marriage rates versus trade. He demonstrated how the short and long run were separated out, gave correlation coefficients (and probable errors) for some of the earlier results, but did not attempt to extend this previous work.

Persons' later 1919 work, for which he became famous, involved the creation of the Harvard A-B-C "business barometers." These barometers were based on correlations of time-series data, and involved a loose associative connection between the economic variables. This work built on the earlier graphic methods of decomposition of time series into trends and cycles. Correlation became the acceptable way of establishing relationships which were law-like in the sense of implying regular co-variations in the time-series data. These might imply causal mechanisms, or they might simply be phenomenological regularities.

Graphs remained important in this temporal approach, as evident from both Persons' and Norton's work. Graphing data was a necessary preliminary to taking correlation measures; it helped to decide what correlations to calculate. It was an early form of specification searching, appropriate to the pre-computer age, when calculation was very time-consuming. This combination of graphic and correlation methods continued to play an important role in the economic analysis of time-series data and in attempts to specify and explain lagged relationships in economics. For Persons, as for Norton, the practical results of correlation were to help establish regularities and to add certainty to judgments about hypothesized associations.

## V. Historical Explanation *versus* Causal Structures

There was no simple change effected by the introduction of correlation—the changes were, in fact, both quite subtle and complex. Let me try to put them in perspective by again making use of our case study and by bringing into the account some reference to later developments in econometrics that were then only on the horizon.

Before correlation, the arguments based on graphs were employed to justify explanations of specific historical time series. Causal stories were endemic, but general causal claims purporting to encompass all the phenomena represented in a particular set of time-series data were rare. More often causal claims were related to specific periods or specific patterns in the data, such as long-term trends or cyclical patterns. Searching for economic relations in a welter of statistical data meant exactly that: looking at, and then judging visually, the significance of patterns that emerged in graphical representations of the data.

The introduction of correlation provided a new way of searching for causal relations in time-series data—giving rise to what later came to be called "correlation hunting," calculating the statistics called correlation coefficients until "satisfactory" ones were obtained. But though this method of supporting inferences was more formal, and despite the fact that investigators could now obtain a measure that pertained to a whole data set and for both short- and long-run relationships, the new technique had little effect on the focus of the arguments. Problems posed by the different values of a correlation coefficient when calculated for different relationships in trends and in cycles were still there, and different causal stories could still be advanced and supported by appeal to the same set of data. Moreover, failures of supposed causal mechanisms could easily be gotten around by such stratagems as, e.g., partitioning the data set in a particular way. Correlation coefficients, by providing a common way to measure the strength of alternative causal hypotheses, could serve effectively to reject some explanations, while providing a measure of objective support for others, and thus could be used quite effectively in support of specific causal arguments. Nevertheless, as we have seen, use of the new technique was unable to bring to closure earlier disagreements and debates about the significance of the relationship between the marriage rate and trade.

What is evident from *both* parts of the case study is that these early workers had almost no way of coping with those problems that we now typically take to provide the main task of causal modeling in the social sciences: modeling and measuring the interaction among multiple causes in a complex structure and the pathways linking successive causes. We find the notion of multiple causes being used both before and after the introduction of correlation theory, but the ways in which they were conceptualized in each period differed. The graphical method allowed investigators to search for, and occasionally to identify, a sequence of causes running through several connected variables, but they had no graphical or visual technique capable of representing their combination. Correlation was interpreted initially as yielding information only about whether two variables of interest were associated with a common or non-common set of multiple causes—yielding at best a simple division into two classes of causes. No other way of either linking up multiple causes or of distinguishing their effects was readily available (see Stephen Turner's essay in this volume for a discussion of the latter difficulty). The problem is not just that these statisticians and economists lacked the resources provided by multivariate analysis to cope with multiple causes; the deeper problem is that they also lacked what I suppose we would now call systematic or structural models capable of linking up causes in more complex ways, and thus facilitating analysis of their combination, interaction, and time sequencing.[15] Simple correlation, by itself, was insufficient to solve the problems with which they were confronted.

A second issue remains to be explored. I suggested earlier that while the introduction of the correlation coefficient could be regarded as in some sense a significant technical advance, most economic statisticians continued to use graphic methods extensively. Why was this the case? The answer is fairly simple, I think. We might easily echo Norton's comment about the average:

> A correlation is easy, often useful, but many times a senseless thing. In many cases, it must be confessed, the correlation tells very little.

Consider how much knowledge actually can be gained from a correlation coefficient of the marriage rate and trade on its own, without the accompanying graph of the two series. The correlation coefficient is informative only about the overall relationship; it does not allow you to retrieve any knowledge about what has happened to the two individual series over extended periods of time of the sort that can easily be gained from visual inspection of the time-series graph. Correlation does not tell you much. It does not tell you the history of the variable or help you to explain what has happened, nor even help you predict a time series; it is a complementary, not a replacement, technology.

In order to tie up these two points: the problem of how to take into account the combination of multiple causes with the continued use of the graphic method, we perhaps need to look ahead a little to see where all this was going. As the practice of correlation moved further out of statistical hands into the hands of the economists, the derivation of correlation as a measure of common causes lost its power, and economists came to think of the correlation statistic as a measure of the strength of an atemporal causal relation between variables. Thus the important tradition in econometrics stemming from the work of Moore and others in the early twentieth century largely dispensed with "historical time–based" explanations and moved to the "logical time" explanations discussed by Judy Klein (1994). In this context, the correlation coefficient was interpreted as a measure of strength of causal relationship (see Tinbergen 1939). The loss of temporal explanation was matched by a gain in pinning down atemporal relations. Once correlation was introduced, economists no longer relied solely on evidence based on the visual inspection of graphs to make inferences about causal relations in historical data, though the graphic method remained valuable for attempting temporal or historical explanation and for some time-series decompositions. But sometimes measured correlations could not easily be related to the historical phenomena at hand, or, otherwise highly plausible historical explanations could be shown to be at odds with the correlations. We saw hints of these problems in Hooker's comments about the difficulty of making reliable causal inferences from either graphs or correlations (see quote, page 64 above).

We can see Hooker's claims on behalf of correlation as a tool suitable for "digging below the surface" of the data to uncover hidden causal relations as one early response to this problem. Yet time-series graphs were routinely used and reported in statistical studies along with correlation and regression in economic work up into the 1930s, only fading out of use when the notion of a structure, representing the underlying causal mechanisms, began to take a firm hold in econometrics. Tinbergen was one of the originators of this notion in econometrics, but also one of the most inventive exponents of the graphic methods of statistics (see Morgan 1990, chap. 4). For Tinbergen, working on the business cycle, the causal mechanism which created the cycles in the economy was best represented by a mathematical model, a set of interrelated equations joining up the variables at different points in time and incorporating disturbing shocks from outside the system into the causal process (a representation of such a causal model is given in Wold 1949, p. 14). The general causal mechanism was now understood to be embedded in the relationships between the individual equations of the model and, as such, could not be seen in the graphs of the individual variables. Such structural models provided the basis for a causal story in terms of causal processes which defined the time order and interaction of the multiple causes at work.

But understanding causes in terms of a structural model and the measurements that accompanied its construction did not necessarily give effective explanations for any particular events in the history of the economy. So at the same time, Tinbergen made a point of setting out the time-series graphs of every regression equation, showing both the dependent variable and the roles of the individual causal variables at each point in time (an example is shown in figure 2.5). These were an essential part of his historical explanation of what had happened in particular periods, such as the Great Crash of 1929 and the 1930s' depression. These time-series graphs showed the specific multiple causes at work at any given point in time for the whole period under study, thereby providing critical information relevant to testing the ability of his model to explain the history of the economy. Thus, the trade cycle was formulated in a mathematical model intended to represent its underlying causal structure, but this still coexisted with historical explanations in graphic form of what had actually happened (an example of Tinbergen's historical explanation is shown in figure 2.5[16])

The notion of hidden or underlying structure reached its zenith in econometrics in the 1940s, and dominated work in the field through the 1970s. During this period, it was held that fundamental causal mechanisms (the combinations of causes) *could not be seen in* graphs charting the behavior of individual variables over time, as these were now regarded as displaying only surface phenomena; *nor could they be revealed* with simple correlation methods. Since the whole structure of interaction among variables provided the new focus for the study of causal mechanisms and for determining relevant measurements, graphs could be considered at best unnecessary, and at worst seriously misleading. Thus graphs, both as tools with which to investigate causes and as resources capable of providing illustrative explanations, disappeared from econometrics as the conviction took hold that genuine explanations required an appeal to systems of multiple causes combining in ways which were hidden deep inside the time-series data. Those multiple causes could only be accessed, it was held, by formulating models of hidden structure consisting of systems of simultaneous equations, and then applying equally complex multivariate statistical techniques.

These methods in their turn have proven to be more problematic than anticipated. And this may explain, at least in part, the recent resurgence of interest in searching for structure in the time domain (see note 12). Just as statistical correlation techniques did not quite solve the problem of making causal inferences

**FIGURE 2.5** "Explanation" of Pig-Iron Production. United States, 1919–1937.

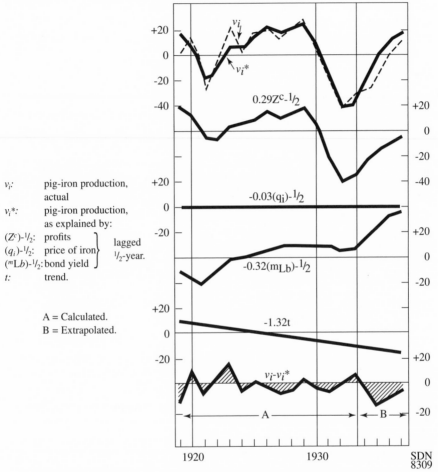

Tinbergen's Graph III.8 is an example of a time-series graph showing the regression equation through time for "pig-iron" production. The top line shows the dependent variable, actual and fitted; the bottom line shows the residuals. The intervening lines show the regressors and the coefficients in the equation.
(*Source:* Reset from Tinbergen, 1939, page 73.)

from time-series data, thus making it still necessary to rely on graphs to achieve certain sorts of understanding, so too structural methods have not yet succeeded in letting us unproblematically access causal structures in temporal data. Simple old-fashioned correlation and regression remain in the tool box.

## NOTES

I thank Stephen Turner and participants at the "Causality in Crisis" conference, Reilly Center for Science, Technology, and Values, University of Notre Dame, October 15–17th, 1993, for their comments when this essay was first presented. Seminar participants at the Tinbergen Institute, Amsterdam in January 1994 and at the University of Bristol in December 1993 asked questions which further improved my understanding of the case study implications. There are three readers to whom I owe special thanks: Raine Daston who read the paper before it was first presented, Judy Klein who provided her usual perceptive criticisms and further references when the paper was presented at the History of Economics Meeting, Babson College, June 1994, and finally Ted Porter who instilled sense into the final version. Thanks go also to Francesca Carnevalli for research help. I am responsible for any remaining confusions and errors.

1. The use of the graphic method for economic statistics was itself of comparatively recent vintage. Apart from the original work by Playfair (1796) there had been little use of the method until an explosion of interest occurred in the 1880s. This was marked by a special collection of papers on the method in the Jubilee issue of the *Journal of the Royal Statistical Society* in 1885; see particularly contributions by Levasseur and Marshall.

2. Most of the work using graphs was descriptive in intent, the graphs and tables providing illustrations to arguments in the texts (for examples on the marriage rate and trade, see Longstaff [1891] and Beveridge [1912]). Klein (forthcoming) notes a change from graphs of data to graphs of results in the scientific work, though my own account suggests a less clear-cut distinction (see Morgan 1990, 70–72). In any event, in populist economic commentary, time-series graphs function as illustrations of causal reasoning well into our own time.

3. Ogle dismissed other aspects of the Malthusian theory, namely the importance of war and pestilence, as having little effect on marriage rates in England and Wales in the period 1820–1888.

4. Ogle took the crude marriage rate per thousand and looked for exogenous causes for its decline. In the discussion following his paper, he rejected the suggestion that internal changes such as in the age distribution of the population were responsible for the fall in marriage rates. Yule (1906) took up this point explicitly in making corrections to standardize the population.

5. He made step-by-step adjustments for the three years 1877–1879 and then recalculated the series at the older (higher) "stable" value for the later years.

6. There is, of course, a long tradition which relates causal inference to our ability to manipulate, control, and intervene (see, for example, discussions by Woodward in this volume and Cartwright [1989]).

7. Both Hooker and Bowley felt they had to argue for the appropriateness of probability-based methods to socioeconomic time-series data. This was not something which could be taken for granted, and in the interwar period, economists rejected probability-based methods (see Morgan 1990, chapter 8). Bowley went through a lengthy

justification (see his 1901, 297–98) for using probability reasoning in dealing with socioeconomic, particularly time-series, data and concluded that the question of whether probability could be applied was one to be answered empirically.

8. Bowley (1901), 320: "When r is not greater than its probable error $\{.67 (1 - r_2)/\sqrt{n} \}$ we have no evidence that there is any correlation, for the observed phenomena might easily arise from totally unconnected causes, but when r is greater than, say, 6 times its probable error, we may be practically certain that the phenomena are not independent of each other, for the chance that the observed results would be obtained from unconnected causes is practically zero." See for example Persons (1919) or Thomas (1925) for justifications of the view that socioeconomic data had to pass a stricter test to justify inferences of causality because such data did not meet the conditions of independent drawings from normal populations.

9. As Mackenzie (1981) makes clear, "correlation" meant the general field of correlation and regression as well as the measure itself.

10. By the 1920s, the time-series approach was either to decompose the time series into interpretable elements (such as trends, cycles, etc.) or to undertake a spectral analysis of the data to uncover hidden periodicities. Meanwhile the structural econometrics approach had begun to use multivariate analysis. These various approaches, in their application to business cycle data, are discussed in Morgan (1990), Part I.

11. This instantaneous average was in fact a moving average based on the trade cycle, taken (on the basis of the data) to be nine years. See Klein (forthcoming, chapter 9), who claims that Hooker was responsible for coining the term "trend."

12. Following Yule's paper, there was the usual discussion. Most discussants did not argue about the correlation methods. The exception was Edgeworth, who commented in typically dry fashion, but entirely to the point when he said that he "was not sanguine about discovering the cause why rates did not comply with the rule appropriate to games of chance. Doubtless there was some correlation between adjacent events, so that they did not behave like separate balls drawn from a bag" (Edgeworth, discussing Yule, 1906, 141).

As it turned out, some of Yule's greatest contributions to statistics lay in his 1920s development of analytical methods to deal with this time-series dependence (see Klein, forthcoming, chapter 10). These developments are best described as searching for regularities in the time-series structure, an approach evident also in Herman Wold's work in the 1930s (see the discussion in the introduction to Hendry and Morgan 1995), and one which resurfaced in the 1960s.

13. Another good American example is the case of Moore (1911).

14. John Aldrich (1994) points out that this "causes in common" interpretation forms the basis of factor analysis in psychometrics.

15. This problem was evident in Yule's parallel work on socioeconomic data. His treatment of the pauperism question in 1895 contained the first use of correlation on socioeconomic data; and he went on in 1899 to measure a multiple regression of three causes affecting changes in pauperism, with a graph of the four variables to show the time-series pattern involving the four series. Nevertheless, his multiple regression was done as a cross-sectional study (taking all data as representing decadal differences, for two

separate periods), which obviated the immediate need to provide a way of linking the causes in a time-dependent path or process.

16. I thank the League of Nations Archive, UN Library, Geneva, for permission to reprint Tinbergen's Graph III.8 from his *Statistical Testing of Business-Cycle Theories, Vol. I: A Method and Its Application to Investment Activity* (Series of League of Nations Publications, II. Economic and Financial, 1938. II.A.23, Geneva, 1939).

# II

## CAUSAL MODELING AND CAUSAL INFERENCE: CAN THE FORMER JUSTIFY THE LATTER?

# The Regression Method of Causal Inference and a Dilemma Confronting This Method

## Clifford C. Clogg and Adamantios Haritou

### 1. The Problem

Much empirical research in the social sciences is concerned with the general problem of drawing causal inferences or estimating the size of causal effects. In the majority of cases, causal inferences are drawn from regression or regression-type models applied to nonexperimental data. In nonexperimental data levels of causal variables are not randomly assigned or manipulated in any fashion. Some examples of nonexperimental data are (a) large-scale social or economic surveys, such as labor-force surveys or attitude surveys; (b) time-series measurements on the population units for which causal inferences are sought, such as annual observations on the inputs and outputs of states or schools; (c) cross-sectional observations on the population units, such as import/export flows for nation states; (d) panel or other longitudinal data where cases initially sampled (individuals, families, firms) are followed through time, giving either the time sequence of variable levels or the timing of transitions among those levels ("event-history data"). In short, the data used most often derive from uncontrolled observation of the variables supposed to be both causes and outcomes.

We shall call the use of regression models applied to nonexperimental data for the purpose of drawing causal inferences the *regression method of causal inference,* RMCI for short. The most basic tenet of RMCI is that experimental manipulation or control through randomization can be replaced by *statistical control* or *partialing* with a regression model, along with a few assumptions that seem benign to most researchers. This essay is largely about those assumptions.

We take these assumptions as found in the social science literature on causal analysis. Whether these assumptions provide a proper basis for causal inference is a question raised implicitly throughout. In spite of their almost axiomatic status in the literature, we question their use as a logical foundation for causal inference. These assumptions have a great influence on inferences and they cannot be checked or validated in straightforward ways (also see Freedman, this volume). For alternative formulations of assumptions for causal inference, see Holland (1986; 1988), Rosenbaum and Rubin (1983), and Freedman (this volume). No attempt is made here to adjudicate between RMCI logic and the logic associated with the Rubin model given in these other sources.

Statistical control in most uses of RMCI simply means that other predictors are included in the model besides the predictor variable whose causal effect is to be estimated. These other variables are often called "control variables"; the terminology conveys the logic involved. The main questions that need to be considered with the RMCI method are these:(a) How closely does "statistical control" or partialing as carried out with regression analysis approximate experimental control accomplished through randomization? (b) Can we ever know whether statistical control via regression modeling has been successful? We try to answer these questions in the context of the ordinary multiple regression model, but we hasten to add that the same questions arise with multiple-equation models or other modern variants of the linear model such as logistic regression.

## 2. The Ubiquity of RMCI Logic in Social Research

RMCI logic is firmly entrenched in the methodology of social research. It is one of the main things taught in statistics and methods courses offered in the social sciences, economics included. RMCI logic is always at least implicit in research output found in journals reporting empirical social research. For example, RMCI logic is embedded in statements, found throughout the literature, that a partial regression coefficient estimates "the effect" of a given variable "when other variables are held constant" and hence "controlled statistically." Partial regression coefficients or analogous quantities are assumed to be the same as causal effects when the right controls (additional predictors) are included in the model. No referee report is more damaging than a statement that such and such a factor was not "controlled" (i.e., not included as a predictor). And one of the best arguments for publication is often that this or that factor has been "controlled" (i.e., included as a predictor or covariate).

RMCI logic is just as important for theoretical social science as for empirical social science because of the interplay between theory and research. Many research areas in sociology and related fields are defined in terms of the presence, absence, or size of the effects of one or more causal variables on some outcome of interest. Different theories posit different causal variables and are often distinguished from each other solely in terms of what causal variables are included or excluded from models that ostensibly represent the theories. Evidence about particular theories is most often summarized statistically in terms of regression models applied to several different sets of data covering different points in time, different groups, and different units of analysis. Research literature in an area accumulates in part because different samples, models, and measures of the variables are tried out in order to replicate findings. Of course, the empirical literature grows also because there are many possible "control" variables, many ways to specify how they should be included, many possible regression models that can be considered, and many possible ways to reconcile cross-sectional and longitudinal analyses using those causal variables.

It is important to recognize also that RMCI logic has major implications for data collection. The Inter-University Consortium for Political and Social Research (ICPSR) at the University of Michigan, the largest data archive for social research in the world, distributed over 45,000 data sets in 1993, about 855 gigabytes of stuff. Almost all of these data are nonexperimental in nature, and much of it is survey data. These data are analyzed routinely with regression methods to draw causal inferences. Many of the variables in data sets distributed by the ICPSR and others were measured because some researcher believed these variables were important "controls" in regression models to be used to draw causal inferences. Of course, many variables in standard surveys are regarded as causal variables, not mere control variables. It would be difficult to overstate the importance of RMCI logic for data collection, which is naturally tied to the kinds of data analysis (regression methods) that are used.

RMCI logic is implicit in many other methods that appear to differ from regression. For example, regression logic is embedded in the so-called causal models or causal-search algorithms put forward in this book and elsewhere. The distinguishing feature of many of these other methods is that sets of regression equations define a multiple-equation "causal model." When multiple-equation models are used, RMCI logic applies to each of the regression equations in the overall model, although this fact is often overlooked. It also applies to many uses of econometric and psychometric models, including conventional path analysis and structural equation modeling (Bollen 1989). Researchers using specialized regression models for the analysis of event-history data or log-linear models for contingency tables (Agresti 1990) are employing this logic also whenever they draw causal inferences from such models. We cannot understand the difference between what is assumed and what is inferred from multiple-equation super-models without first understanding the single-equation regression model. That is why we concentrate on this case here.

Readers interested in examples using both ordinary linear models and gen-eralizations can turn to the August 1993 issue of the *American Sociological Review*. Every article in this particular issue of the leading journal in sociology depends on some form of regression analysis to develop causal inferences, in all cases but one using nonexperimental data. Often the language of cause-and-effect is not used explicitly in the literature. Instead we often find terms such as "structural model," "structural coefficient," "determinants and consequences," "elaboration of correlations," "fixed-effect models for pooled cross-sections," "impacts" of this or that variable, "independent effects," "intervening variables," "spurious (i.e., noncausal) correlation," "direct and indirect influences," or "net effects" and "total effects." Such terms surely represent causal language and causal thinking. It is interesting to speculate why these ostensible synonyms for causal effect have arisen. The list of terms in the causal vocabulary could be ex-panded. Berk (1988), Lieberson (1985), and Smith (1990), among other social researchers, have also commented on the vague vocabulary of causal inference. Language and terminology are important, and several recommendations for re-moving slang appear below.

## 3. Regression as Prediction: The Conditional Regression Model

The regression model as it is usually viewed in statistics is considered first. The goal is to examine the inferences that are sought, diagnosed, and quantified in the statistical analysis of regression models. Convenient general references include Neter, Wasserman, and Kutner (1989), Montgomery and Peck (1982), and Mos-teller and Tukey (1977). Of course, there are scores of popular texts or mono-graphs on regression analysis and thousands of articles on the topic in statistics journals. The account given below is fairly consistent with regression logic in the statistics literature, not just in popular textbook accounts. It should be noted that this account carries over with little modification if generalized linear models (McCullagh and Nelder 1989), nonparametric regression models (Hardle 1990), or other modern relatives of the classical regression model are considered.

### 3.1 The Conditional Regression Model

A regression model links a variable Y to a set of predictors. The primary goal is to determine how well Y can be predicted from the predictors; hence the term "predictors." We denote the predictors as X ( for just one predictor), as X and $Z_1$ (when two predictors are considered), or as X and $Z_1, \ldots, Z_K$ when there are many predictors. Let $x$ and $z_1, \ldots, z_K$ denote the values that these predictor variables take on in the sample. Nothing essential is lost if we assume that these variables are centered around their means, and we use this convention throughout. The terminology is deliberate. In our judgment it is preferable to other terminology popular especially in the social sciences.[1] The terminology does not preclude considering how well X can be predicted from Y and $Z_1$ or how well $Z_1$ can be predicted from X and Y. Sometimes these so-called reverse regressions are of interest even when Y is obviously the thing most likely to be the "dependent" variable.

We suppose that $n$ observations on these variables are available for a random (or "representative") sample from some well-specified population, with the case or unit indexed by $i = 1, \ldots, n$. The "representativeness" of the sample is obviously important, but to illustrate the logic we simply assume that the available sample is appropriate for making inferences about some population, perhaps with appropriate weighting for disproportionate sampling of some units. Because these variables pertain to values observed in the sample, not values assigned or manipulated, they will be correlated with each other. This means that X and the Z variables will be intercorrelated, not just that these are correlated with Y. These correlations are not controlled.

The simple regression model using X as a predictor of Y can be written as

$$M_0 : Y_i = \beta_{yx} X_i + \varepsilon_i, i = 1, \ldots, n, \tag{1}$$

where we use the popular notation $\beta_{yx}$ for the *simple* regression coefficient (Blalock 1979), and $\varepsilon_i$ is the error term.[2] We call this model $M_0$ to denote that only X is included as a predictor (no other predictors are used). For a given sample, this model could just as well be written as

$$E(Y|X = x) = \beta_{yx}x, \tag{2}$$

where we have suppressed the case subscript for simplicity. This can be expressed equivalently as the assumption that $E(\varepsilon|X = x) = 0$. By conditioning on the value that X takes on (x), the predictor is regarded as a constant. Because of this, $E(\varepsilon|X = x) = 0$ is usually translated as $E(\varepsilon) = 0$, but when this is done it is nevertheless important to realize that the logic is that of conditioning on X values taken on in the sample for which predictions are sought.[3]

The ordinary linear regression thus estimates the conditional expectation of Y given X. Viewing the regression in this way shows that the model describes how one summary measure of the conditional distribution of Y given X (the mean or expectation) is produced by the specified linear function of the predictor. There are other types of regression besides the mean regression, e.g., regression models that predict other moments or other summaries of the cumulative distribution of Y given X (Manski 1991). But most accounts of regression in statistics as well as in the social sciences refer to prediction of the mean (Arminger 1994). In this setup, the variance of the error term should be understood as the conditional variance of Y given X (i.e., $V(Y|X = x) = V(\varepsilon)$). When regression-type models such as logit models are considered, the model is written formally in terms of some function of the conditional expectation of Y and error terms are not included. For generalized linear models, it is more convenient to always think of conditional means, conditional variances, or analogous quantities.

When predictor $Z_1$ is added to the model, we obtain

$$M_1 : Y = \beta_{yx.1} X + \gamma_1 Z_1 + v_1, \tag{3}$$

and this also can be written in conditional-expectation form as $E(Y|X = x, Z_1 = z_1) = \beta_{yx.1} x + \gamma_1 z_1$. Here $\beta_{yx.1}$ denotes the *partial* regression coefficient "controlling for $Z_1$." The term derives its name from either of the following mathematical facts. The partial derivative of $E(Y|X, Z_1)$ with respect to X is $\beta_{yx.1}$, so the term "partial" derives from the partial derivative of a function of several variables with respect to just one of those variables. Alternatively, the difference between $E(Y|X = x + 1, Z_1)$ and $E(Y|X = x, Z_1)$ is $\beta_{yx.1}$ for any setting x of the first predictor. Note that predictor $Z_1$ is "held constant" in calculating this difference or this derivative, and this is the meaning of "controlling for" $Z_1$. The coefficient $\gamma_1$ is of course also a partial coefficient, the change in the expectation of Y associated with a change in $Z_1$ "holding X constant." Finally, the equation error $v_1$ in this expression is assumed to have mean zero, and $V(v_1)$ is understood to mean the conditional variance of Y given both X and $Z_1$ (i.e., $V(Y|X = x, Z_1 = z_1)$).

### 3.2 Statistical Assumptions for the Conditional Regression Model

If linearity is taken for granted, the usual assumptions applied for the statistical consideration of conditional regression models are as follows. These are most often represented as assumptions about error terms and we follow this convention here. It is customary also to describe these assumptions as if the values that the predictor variables take on are constants (i.e., not random variables). Only slight changes would be required if the randomness in these were taken into account, assuming of course that the goal of the analysis is prediction.[4] The statistical assumptions are as follows:

(1) The error term has average value zero: $E(\varepsilon) = 0$, with an analogous statement for model $M_1$ ($E(v_1) = 0$).

(2) The variance of the error (or the conditional variance of Y given X) is constant (homoscedasticity): $V(\varepsilon_i) = \sigma_\varepsilon^2$ for model $M_0$, $V(v_{i_1}) = \sigma_{v_i}^2$ for model $M_1$.

(3) The errors are independent or at least uncorrelated: $E(\varepsilon_i \varepsilon_{i'}) = 0$, or the correlation between $Y_i$ and $Y_{i'}$ given the predictor values is zero. (Note that it is not necessary to have independent observations but it is necessary that the Y values are independent or at least uncorrelated given the predictors.)

(4) There are no outlying observations or unusually influential predictor values.[5]

(5) The errors are normally distributed: $\varepsilon_i \sim N(0, \sigma_\varepsilon^2)$ or $v_{i_1} \sim N(0, \sigma_{v_1}^2)$.

All of these assumptions are made in reference to the conditional mean of Y, to the conditional variance of Y, or to other properties of the conditional distribution of Y given either X or X and $Z_1$. These assumptions define the so-called normal-theory linear regression model. With these, it follows that Y given the predictors is normally distributed, which implies that we only need the conditional mean (the linear function for $E(Y|X)$ say) and the conditional variance to completely describe the prediction problem.[6]

The way that the two models $M_0$ and $M_1$ or other models are compared to each other is very important. It is useful to note that the error term in the model $M_0$ is algebraically related to the additional terms in the model $M_1$ as $\varepsilon = \gamma_1 Z_1 + v_1$ if the coefficient of X is the same in the two models. This fact will be exploited extensively below to show why the coefficient of X can change from model to model and to examine the implications of this fact.

The proper statistical analysis of the above models when assumptions (1)–(5) are valid amounts to estimation of both models with ordinary least squares. Call the least squares estimators $\hat{\beta}_{yx}$, $\hat{\beta}_{yx.1}$. The former is usually referred to as the (sample) simple regression coefficient, and the latter is usually referred to as the (sample) partial regression coefficient "controlling for $Z_1$." The two estimators can differ if the correlation between X and Z is nonzero (i.e., $r_{xz} \neq 0$). If this correlation is nonzero, then they will differ if the estimated coefficient for Z ($\hat{\gamma}$) is nonzero. For nonexperimental data, nonzero correlations can be assumed, at least in relatively large samples. This means that the problem of model selection is essentially the same thing as determining whether $\gamma$ is nonzero or whether $\hat{\gamma}$ is significantly different from zero. An equivalent procedure is checking whether the estimate of error variance from model $M_1$ is significantly less than the error variance from model $M_0$. Note that this is exactly the same as a significant difference between the estimates of $V(Y|X)$ and $V(Y|X, Z_1)$ when the assumptions above hold. Of course, it might be the case that once $Z_1$ is included, X is no longer an important predictor. In such a case, model $M_0$ is replaced by the model that includes $Z_1$ only; and we would normally then say that $Z_1$ and not X is a predictor of Y.

### 3.3 Model Selection as Selection of Predictors

Model selection refers to many things: finding the "best" transformation of Y or the predictors, detecting and correcting departures from the statistical assumptions laid out above, residual analysis and diagnostic checking, validating or cross-validating predictions with parts of the sample or new samples, checking for extra variation or dispersion not accounted for by a normal-error assumption, assessing goodness-of-fit, and so on. These are all important aspects of model selection. But model selection in the social sciences is often taken primarily as a problem of selecting predictors, and we follow that approach here.

If model $M_1$ provides an improvement in prediction relative to $M_0$, then this model is picked and model $M_0$ is discarded. If model $M_1$ is picked as a good prediction model, then statistical inferences (and predictions) are usually made conditional on this model. This means that the analyst has decided that there is information in both predictors X and $Z_1$, that inferences or predictions of Y are improved when both predictors are used, and that therefore we ought to "assume that $M_1$ is true" for the purpose of prediction, unless other information becomes available that casts doubt on this model. The "truth" of either of these models is relative and depends solely on how well the model predicts Y.

There are many ways to assess "improvement in prediction," not all of which reduce to a test of the significance of $\hat{\gamma}_1$ or to a test of the significance of the difference in the two estimated error variances. For example, we might be concerned with whether the level of correlation between X and $Z_1$ weakens the inference (decreases precision) as a direct result of the colinearity. If predictions are better "statistically" with $M_1$ than with the simpler model $M_0$ taking account of complications like these, however, then $M_1$ is picked; and predictions are assessed most of the time assuming that this is the model that generated the data. The fact that $\hat{\beta}_{yx}$ and $\hat{\beta}_{yx.1}$ differ (and differ significantly if $Z_1$ is a significant predictor—see Clogg, Petkova, and Shihadeh [1992]) is of little interest in itself. The success of the regression analysis depends on whether the estimator of $E(Y|X, Z_1)$ is better (smaller variance or mean squared error, say) than the estimator of $E(Y|X)$. It is the comparison of the estimates or the estimators of these two conditional expectations, which are linear combinations of the predictor variables included in those models, that frames the analysis of conditional regression models throughout statistics.

Ordinary statistical analysis of prediction with regression models does not preclude making inferences or statements like the following: (a) "The regression analysis shows that X and not Z is an important predictor (or Z and not X is an important predictor)." Or, (b) "The analysis shows that X and Z together provide better predictions of Y than when either is used singly." Questions concerning the relative importance of the two predictors are difficult if not impossible to answer if X and $Z_1$ are correlated. If both X and $Z_1$ are measured in the same units or have been rescaled to have the same units (as with standardized versions of both variables giving standardized regression coefficients), then we can compare the sizes of the partial regression coefficients (i.e., compare $\hat{\beta}_{yx.1}$ with $\hat{\gamma}_1$). Such comparisons are relatively unambiguous as comparisons of the size of the two partial coefficients. Variability "due to X alone" (variability inferred from $M_0$) compared to variability "due to X given Z" (incremental variability) can also be assessed. This reduces to the comparison of so-called sequential and partial sums of squares.

### 3.4 Multiple Regression: Many Predictors

Now suppose that additional predictors are available. That is, suppose that we have predictors $Z_1, \ldots, Z_K$ in addition to X. These predictors represent additional information that might be useful for the purpose of predicting Y.

We consider two specific versions as well as the general case. The first is

$$M_2: Y = \beta_{yx.12} X + \gamma_1 Z_1 + \gamma_2 Z_2 + v_2, \tag{4}$$

which should be compared directly to model $M_1$ in (3). A regression model incorporating all predictors can be represented as

$$M_K: Y = \beta_{yx.1\ldots K} X + \sum_{k=1}^{K} \gamma_k Z_k + v_K. \tag{5}$$

There are many possible models that can be selected. A model with $0 \leq k \leq K$ additional predictors (components of Z) can be formed in $\binom{K}{k}$ ways. The number of regression models including X is thus

$$\sum_{k=0}^{K} \binom{K}{k} = 2^K,$$

which is a large number. (We count the model with X only as one of these.) Call one possible selection of Z variables $z(s)$, and call the model with this combination of Z variables $M_{z(s)}$. With $K = 10$ additional predictors, the number of possible regressions that always include X is $2^{10} = 1024$. In contemporary social research, regression models that are judged to be the most credible usually include between 10 and 20 predictors; $2^{20} = 1,048,576$, which is very large. Regression routines for "all possible subsets" can be used to pick the models that are best using some criterion of prediction, at least with $K$ of moderate size.

There are thus $2^K$ possible regressions involving X and the other predictors. It is useful also to regard this as the number of possible data sets that could be formed from a set of $K + 2$ possible variables with Y and X always included. It is conceivable, indeed probable with large samples, that each of the many data sets that can be formed for analysis would lead to different "final" regression models for many or most of the possible sets that could be imagined. For the statistical analysis of conditional regressions, this is not a logical difficulty at all. The "best" regression is defined in terms of the prediction of Y for the set of predictors available. It is possible that equally good predictions (conditional regressions) could have completely different predictor variables.

Corresponding to each of the $2^K$ possible regressions (or $2^K$ data sets), there is a potentially different partial regression coefficient for X, say $\beta_{yx.z(s)}$, where the "control" denoting what is partialed refers to one possible set $z(s)$ of the Z variables. Regression modeling in statistics, or as taught in standard statistics references, is simply not concerned with the variability among these partial coefficients. That is, the fact that the partial coefficient $\beta_{yx.z(s)}$ or the estimate $\hat{\beta}_{yx.z(s)}$ varies from model to model depending on which set of Z variables has been included (in the model or the data set) is not a logical difficulty. All of these regression models are more or less satisfactory depending solely on how well Y is predicted using the *linear combination* of X and the Z variables. It is the linear combination and not the specific "weights" (or regression coefficients) in that linear combination that is the most important output from the conditional regression model. Of course, various statistical definitions of the quality or the precision of the prediction are used to assess this, and there are various "optimal" definitions even for the linear regression model.

## 4. Regression Models and Causal Inference

We now consider regression modeling as it is applied to obtain causal inferences from nonexperimental data. We distinguish sharply between statistical analysis of the conditional regression, where the linear combination of selected predictors is studied, and the causal analysis of a related "unconditional" regression where the sign and size of partial regression coefficients is the primary concern. The unconditional regression to which we refer pertains to the (uncontrolled) joint distribution of all variables included in a model, but where one is specified as the "dependent" variable and the others are specified as "independent" variables or causal variables. To draw causal inferences from such data, assumptions about the unobserved error terms in these models are required. Usually these assumptions are expressed in terms of unconditional expectations involving the predictors and the error terms, as described below.

To illustrate RMCI logic, suppose that the only goal is to estimate the causal effect of X on Y. Linearity is taken for granted as before, and of course this is a strong condition. The models presented in the previous section can be used to illustrate the RMCI logic under these stipulations.

RMCI logic is presented in one form or another in standard statistics texts for social scientists (e.g., Blalock 1979; Blalock 1982; Hanushek and Jackson 1977), in standard econometrics texts (e.g., Gujarati 1988), and in almost all standard references on covariance-structure analysis (e.g., Goldberger and Duncan 1973; Bollen 1989). Duncan (1975) is a standard reference on the use of regression models for causal inference. Classical statements of RMCI logic can be found in Simon (1954), Blalock (1964), and Duncan (1966). In many settings, the logic of analyzing surveys using regression methods, developed by Lazarsfeld and his co-workers during the 1940–1970 period (see Rosenberg 1968; cf. Goodman 1973), is virtually synonymous with RMCI logic.

### 4.1 The Two-Variable Case

Consider now a seemingly benign modification of model $M_0$ presented in (1). This model includes predictor X only, now regarded as a causal variable, and X is considered to be random, not fixed. The "causal effect" of X is now called $\beta$, which is not necessarily the same as the regression coefficient $\beta_{yx}$ used above. The causal model that is related to $M_0$ is

$$Y = \beta X + \varepsilon. \tag{6}$$

Now the *unconditional* expectation is taken, producing $E(Y) = \beta E(X) + E(\varepsilon)$, or $E(Y) = \beta E(X)$ assuming that the equation error has mean zero. The unconditional variance is also important; this is $V(Y) = \beta^2 V(X) + V(\varepsilon) + 2\beta E(X\varepsilon)$. An important change has taken place: the expectation and variance calculations have changed from conditional to unconditional. All inferences change as a result given the way that the latter is used with RMCI logic. The distinction between the two kinds of expectations, or the two kinds of regression inferences that they imply, is fundamental to RMCI logic throughout the social sciences, in spite of the fact that this difference is seldom noted except in passing.[7] Assuming that X precedes Y and that X can indeed be a causal variable, the regression coefficient

$\beta_{yx}$ is said to denote the causal effect of X on Y *if* X is uncorrelated with the error term. This is equivalent to the condition,

$$E(X\varepsilon) = 0. \tag{7}$$

Note that $E(X\varepsilon)$ is the covariance between X and the equation error $\varepsilon$, because we have assumed that X is centered, and this term is included in the (unconditional) variance calculation given above. The condition in (7) is thus an assumption that X and the equation error in the causal model of (6) are uncorrelated. We are not endorsing this as a sufficient condition for causal inference, and Freedman (this volume) gives many reasons why this condition is not sufficient.

The algebra of causal analysis based on the unconditional version of model $M_0$ given in (6) is essentially as follows. Multiply both sides by X (recall that variables are centered). Taking the expectation produces

$$\sigma_{xy} = \beta\sigma_x^2 + \sigma_{x\varepsilon}, \tag{8}$$

where $\sigma_{xy}$ is the covariance between X and Y and $\sigma_{x\varepsilon}$ is the covariance between X and $\varepsilon$. *If* the assumption in (7) is true ($E(X\varepsilon) = \sigma_{x\varepsilon} = 0$), then $\beta = \beta_{yx} = \sigma_{xy}/\sigma_x^2$. The sample analogue is the least-squares estimator of the simple regression coefficient. It is said that the sample regression coefficient (least squares estimator) is an unbiased estimator of the causal effect when this assumption holds true. See Goldberger (1973) and Duncan (1975).

Note that the "causal effect" $\beta$ is not identifiable (or cannot be equated to the regression coefficient $\beta_{yx}$) if the assumption of zero correlation or zero covariance between X and $\varepsilon$ is not true. That is, $\sigma_{x\varepsilon} = \rho_{x\varepsilon} = 0$ has been assumed in order to equate the regression coefficient and the causal effect. We do not know that this covariance is zero, and we cannot determine the true value of this covariance from the data.

We shall call the assumption in (7) a *causal assumption.* It is not a statistical assumption in the same sense as the assumptions for conditional regression were statistical assumptions. The statistical assumptions given earlier can be checked, at least in principle. The causal assumption above *cannot* be checked with the data used for the regression analysis. Suppose that just X and Y are available in the sample; nothing else is known except perhaps that X precedes Y temporally or logically in order to rule out silly causal inferences. We cannot check whether $\sigma_{x\varepsilon} = 0$. If least squares methods are used, for example, then the correlation between X and $e = Y - \hat{Y}$ is always zero, so residuals are not informative with respect to this assumption. Or, assume that X and Y follow a bivariate normal distribution, an assumption that can be checked. If this is the case, there is no information in the residuals or any other quantity estimated from the data that could be brought to bear on the validity of this key assumption. Indeed, we could just as well assume that the "causal effect" $\beta = 0$ and then $\sigma_{x\varepsilon} = \sigma_{yx}$. Virtually any value for the "causal effect" can be assumed and the equation can be solved for the unknown covariance, or vice versa.

## 4.2 Can the Causal Effect be Bounded in the Two-Variable Model?

Suppose in this subsection only that the variables are standardized; call them $X^*$ and $Y^*$. Let $\beta^*_{yx}$ denote the standardized regression coefficient, which needs to be distinguished from the "true" causal effect. In the two-variable model $M_0$, $\beta^*_{yx} = E(Y^* X^*) = \rho_{yx}$, the correlation between Y and X. Now let $\beta^*$ denote the true causal effect defined in terms of the standardized variables. (To obtain the true causal effect in terms of the original units, $\beta^*$ needs to be multiplied by $\sigma_y/\sigma_x$.) Now the causal model analogous to $M_0$ can be written as

$$Y^* = \beta^* X^* + \varepsilon^*. \tag{9}$$

Multiplying both sides by $X^*$ and taking the expectation (noting that $Y^*$ and $X^*$ are standardized) gives $\rho_{yx} = \beta^* + \sigma_{x\varepsilon^*}$, where $\sigma_{x\varepsilon^*}$ is the covariance between $X^*$ and the equation error $\varepsilon^*$. Now

$$V(\varepsilon^*) = V(Y^* - \beta^* X^*) = 1 + \beta^{*2} - 2\beta^* \rho_{yx}$$

using the fact that the variables are standardized. Because $\sigma_{x\varepsilon^*} = \rho_{x\varepsilon}\,\sigma_{\varepsilon^*}\,\sigma_{x^*}$ and $\sigma_{x^*} = 1$, we obtain $\sigma_{x\varepsilon^*} = \rho_{x\varepsilon}(1 + \beta^{*2} - 2\beta^* \rho_{yx})^{1/2}$. (Note that $\rho_{x\varepsilon} = \rho_{x\varepsilon^*}$, where $\varepsilon$ is the equation error for the unstandardized version of the model in (9).) $\rho_{x\varepsilon}$ can range over the $(-1, +1)$ interval. Direct algebra gives

$$\beta^* = \rho_{yx} - \rho_{x\varepsilon}\,[1 + \beta^{*2} - 2\beta^* \rho_{yx}]^{1/2}. \tag{10}$$

If $\rho_{x\varepsilon}$ were known, then $\beta^*$ can be determined from this expression. It is not known, however. Suppose that $\rho_{yx} = 0$, or that there is no covariation between Y and X. In this case (10) reduces to $\beta^* = -\rho_{x\varepsilon}(1 + \beta^{*2})^{1/2}$, and solving for $\beta^*$ gives

$$\beta^* = \pm \rho_{x\varepsilon}/(1 - \rho_{x\varepsilon}^2)^{1/2}.$$

As $\rho_{x\varepsilon}$ approaches $\pm$ 1, the "causal effect" $\beta^*$ approaches $\pm \infty$, so the causal effect is unbounded. In short, the causal effect could be anything even when the correlation between the observed variables is zero.

The general expression for $\beta^*$ as a function of the two correlations, obtained by solving the quadratic, is

$$\beta^* = \rho_{yx} \pm (1 - \rho_{yx}^2)^{1/2} \rho_{x\varepsilon} / (1 - \rho_{x\varepsilon}^2)^{1/2}. \tag{11}$$

Note that the right-hand side ranges between $\pm \infty$, as before. We conclude that nothing can be learned about the causal effect of X on Y from the correlation between X and Y unless something is really known about the correlation between X and $\varepsilon$. It would be helpful in our view if researchers would exploit the above by thinking about *plausible* values for the unknown correlation instead of blindly assuming that it is zero; however, this is difficult to do because the equation error here ($\varepsilon$) is a composite of left-out causes (see Section 4.4 below).

To summarize, very little or nothing can be known about the magnitude or even the sign of the causal effect from the correlation between the two variables. Any assertions to the contrary are based on assumptions. We refer the reader to Manski (1993) who gives bounds on certain quantities, some of them causal effects in the above sense but in the context of models that place other restrictions on the regression function. For other regression models with multiple predictors covered below, the bounds are essentially infinite also.

### 4.3 The Three-Variable Case

Now consider model $M_1$, i.e., the model including both X (the causal variable) and $Z_1$ (another causal variable or a "control" variable). This model along with $M_0$ and a supplemental regression involving either X regressed on $Z_1$ or $Z_1$ regressed on X is the basic setup used in Simon (1954). This "recursive" model forms the basis for the causal analysis of three-variable systems that has played a prominent role in virtually all accounts of RMCI logic in the social sciences (cf. Duncan 1975, ch. 2; Heise 1975; Bollen 1989). We restrict attention to the logic of inferring the causal effect of X on Y, and hence disregard the possible causal effects of $Z_1$ on Y (or of $Z_1$ on X or of X on $Z_1$).

Once model $M_1$ is considered along with $M_0$, RMCI logic amounts to this: If $\beta_{yx.1}$ is zero but $\beta_{yx}$ is not zero, then it is said that the simple regression coefficient (or simple correlation between X and Y) is spurious (i.e., noncausal). If $\beta_{yx}$ is zero but $\beta_{yx.1}$ is nonzero, then the simple regression coefficient (or simple correlation) is said to be a spurious noncorrelation. Path analysis is mostly about relating the sizes of these two coefficients, in either standardized or unstandardized form. Finally, in the context of the three-variable system represented by $M_1$, the coefficient $\beta_{yx.1}$ is said to be the causal effect of X on Y (or the "direct effect"), not simply the partial regression coefficient.

## 4.4 Assumptions for Causal Inference

The "causal assumptions" for the RMCI method are as follows. Although we are primarily interested in the logic for inferring the causal effect of X on Y, we shall see that assumptions necessary for this multiply with the number of predictors included, which is an important fact. That is, the same causal assumptions have to be made for the "control" variables as for the causal variable X, except in some special cases. These assumptions are very different from the statistical assumptions given previously for the conditional regression model. We hasten to add that these causal assumptions are not sufficient, but they are assumed to be necessary for making causal inferences in the literature on RMCI.

As stated previously, the assumption—call it $CA_0$—for the simple regression model $M_0$ is

$$CA_0: \ E(X\varepsilon) = 0.$$

This is usually stated in causal language as "X is not correlated with left-out causes." But $\varepsilon$ is a composite of all "left-out" variables, not a single "variable," and it can be difficult to think about the size or the sign of this covariance when $\varepsilon$ is understood as a composite of other variables. No serious researcher would believe that causal inferences can be made with this simple model. On the other hand, this model uses only one causal assumption, so this model as a causal model should not be excluded from consideration just because it includes only one predictor or causal variable. For other models the number of untestable assumptions will multiply even if the goal is to estimate just the causal effect of X on Y.

For models $M_1, M_2, M_K$, the causal assumptions are

$$CA_1: \ E(Xv_1) = 0, E(Z_1v_1) = 0;$$
$$CA_2: \ E(Xv_2) = 0, E(Z_1v_2) = E(Z_2v_2) = 0;$$

and

$$CA_K: \ E(Xv_K) = 0, E(Z_kv_K) = 0, k = 1, \ldots, K.$$

Finally, for a given model $M_{z(s)}$ having one set of the $2^K$ possible selections from the complete set of predictors, $Z_1, \ldots, Z_K$, the assumptions are

$$CA_{z(s)}: \ E(Xv_{z(s)}) = 0, E(Z_kv_{z(s)}) = 0, k \ \varepsilon \ z(s).$$

Note that for $M_1$ there are two causal assumptions, both untestable; there are three causal assumptions for $M_2$, each untestable; there are $K + 1$ causal assumptions for $M_K$, all untestable; and so on. Each selection of $Z$ variables (or each possible data set) produces a different set of causal assumptions, and it is helpful in our judgment to examine the given problem as a sampling of possible predictors (or alternatively as a sampling of possible data sets). The important thing is to recognize that the number of causal assumptions multiplies as the number of predictors multiplies.

In other words, these assumptions amount to a "specification" that the included variables are each uncorrelated with the error term for the given model. Sometimes these or similar assumptions will be referred to as the "maintained assumptions" (see Manski 1993). These assumptions are synonymous with the econometric *definition* of the term *exogenous,* although more complex definitions are often used. See, e.g., Goldberger and Duncan (1973), Gujarati (1988), and Bollen (1989).[8] Hausman (1978) and others (e.g., Godfrey 1988; White 1980) propose various "tests of exogeneity" that ostensibly check whether these correlations are zero. These tests are based on other assumptions that cannot be tested, however. For example, some tests assume that some variables are "known" to be exogenous or assume that some special nonlinear function of excluded variables is known.

It is sometimes mistakenly assumed that in order to estimate the causal effect of X on Y it is only necessary to "specify" that the analogue of $CA_0$ holds. This would mean, for example, that it must be assumed that X is uncorrelated with the equation error in the given model but that it need not be assumed that the other predictors are each uncorrelated with the equation error. *This is incorrect as a general rule.* To see this, consider model $M_1$ with variables standardized, but where the assumptions in $CA_1$ are not applied. The model can be written as

$$Y^* = \beta^* X^* + \gamma^* Z_1^* + \nu^*$$

Because we have not assumed that predictors are uncorrelated with the equation error $\nu^*$, the quantities $\beta^*$ and $\gamma^*$ are not the same as (population) partial regression coefficients. They are, however, related to them as a function of both (nonzero) covariances. Multipying both sides by $X^*$ and taking expectations gives

$$\rho_{yx} = \beta^* + \gamma^* \rho_{xz} + \sigma_{x\nu^*}.$$

Repeating the operation with Z gives

$$\rho_{yz} = \beta^* \rho_{xz} + \gamma^* + \sigma_{z\nu^*}.$$

Solving for the "causal effect" $\beta^*$ gives

$$\beta^* = \beta^*_{yx.z} + (\sigma_{zv*}\,\rho_{xz} - \sigma_{xv*})/(1 - \rho^2_{xz}), \tag{12}$$

where $\beta^*_{yx.z} = (\rho_{yx} - \rho_{yz}\,\rho_{xz})/(1 - \rho^2_{xz})$ is the ordinary standardized partial regression coefficient associated with variable X. (A similar expression can be obtained for the coefficient $\gamma^*$.) Now the "causal effect" of X, namely $\beta^*$, depends on two unknowns if X and Z are correlated, and these are the *two* covariances of the predictors with the equation error, $\sigma_{xv*}$, $\sigma_{zv*}$. If X and Z are correlated, as will be the case with nonexperimental data, then the relationship between the causal effect $\beta^*$ and the standardized partial regression coefficient $\beta^*_{yx.z}$ cannot be determined unless something is assumed about the two unknown covariances in (10). Thus, we can never hope to estimate the causal effect of X on Y by just "controlling" for the right variables. The variables "controlled for" must be "causal variables" in the same sense that X is a causal variable, and they must be uncorrelated with the equation error also. Assumptions about the correlation between any other predictor and the equation error are no less important than assumptions about the relationship between X and the equation error, even when the only goal is to estimate the causal effect of X.

The "identification problem" can be restated compactly as follows. For any regression model let $\Sigma_{yX}$ denote the vector of covariances between Y and each predictor. Let $\Sigma_{XX}$ denote the variance-covariance matrix of the predictors and let $\Sigma_{X\varepsilon}$ denote the vector of covariances between each predictor and the equation error. (X here stands for all predictors.) Finally, let $\beta$ denote the vector of "causal effects" for all predictors (not the same as the vector of partial regression coefficients). The moment equation is

$$\Sigma_{yX} = \Sigma_{XX}\,\beta + \Sigma_{X\varepsilon}. \tag{13}$$

Setting $\Sigma_{X\varepsilon} = 0$ gives $\beta = (\Sigma_{XX})^{-1}\,\Sigma_{yX}$; the sample analogue is the least squares estimator. We could just as well assume that $\beta = 0$ (or $\beta$ equal to any other value) and solve for $\Sigma_{X\varepsilon}$, subject to the constraint that this covariance matrix is positive definite. If $\beta = 0$ is assumed, then $\Sigma_{X\varepsilon} = \Sigma_{yX}$, and we can even do away with the matrix inversion.

### 4.5 The Logical Fallacy of RMCI in the Three-Variable Case

Let us now contrast model $M_0$ and model $M_1$. If $Z_1$ is an important predictor, then it will almost always be assumed that this model is true "causally," i.e., that $CA_1$ is true. This means that the partial regression coefficient (i.e., $\beta_{yx.1}$) associated with X in model $M_1$ including both X and $Z_1$ would be assumed to be the causal effect of X. If $CA_1$ is true, the estimator from model $M_0$, $\hat{\beta}_{yx}$, is said to be biased because variable $Z_1$ has been omitted. This means simply that

$E(\hat{\beta}_{yx}) \neq \beta_{yx.1}$, or that the simple regression coefficient does not estimate the partial regression coefficient.[9] Indeed, we find that

$$E(X\varepsilon) = \gamma_1 E(XZ_1) + E(Xv_1),$$

and because $CA_1$ says that $E(Xv_1) = 0$, $CA_0$ cannot be true if $Z_1$ is an important predictor ($\gamma_1 \neq 0$) and X and $Z_1$ are correlated ($E(XZ_1) \neq 0$). The "bias" is a function of the nonzero term above. Omitting $Z_1$ will indeed lead to larger *variance* in the estimation of the conditional mean of Y, but the inference about omitted-variable *bias* depends literally and completely on what has been assumed, namely, $CA_1$.

Suppose instead we assume that $CA_0$ is true, i.e., that $E(X\varepsilon) = 0$. Because $\varepsilon = \gamma_1 Z_1 + v_1$, $CA_0$ means simply that

$$E(X\varepsilon) = \gamma_1 E(XZ_1) + E(Xv_1) = 0,$$

or $\gamma_1 E(XZ_1) = - E(Xv_1)$. *This assumption is just as logical as $CA_1$, perhaps more logical because less has been assumed.* Although model $M_1$ would lead to better prediction of Y than model $M_0$, *it does not follow that $CA_1$ is true. $CA_0$ might be true instead.* That is, without additional knowledge, we cannot tell whether the first model or the second model is "right causally." We cannot tell which of the two models gives an unbiased estimator of the causal effect, or even if either model gives an unbiased estimator.

If on the other hand $CA_0$ is true, then estimating model $M_1$ leads to an "included variable bias" in spite of the fact that this model may very well lead to reduction in variance of the prediction. This term is conspicuous by its absence in the literature. But it is just as logical to talk of included variable bias as omitted variable bias once the term "bias" (not variance) is recognized for what it is, and once we recognize that it is just as logical to assume $CA_0$ instead of $CA_1$. Neither assumption can be checked with the data, and it is simply not clear why one would be more valid than the other.

To summarize, although model $M_1$ will have smaller variance (greater precision) in predicting Y when compared to model $M_0$, we simply do not know and cannot find out with the data whether $CA_0$ is true ($\hat{\beta}_{yx}$ is unbiased for estimating the causal effect) or $CA_1$ is true ($\hat{\beta}_{yx.1}$ is unbiased for estimating the causal effect). Indeed, we cannot tell whether either assumption is true. Researchers using RMCI logic have unwittingly assumed that whatever improves the conditional regression automatically validates the causal assumptions, and hence automatically validates drawing causal inferences from models that predict well. RMCI logic leads to unbiased estimation of the causal effect of X only as a consequence of the *assumption* that model $M_1$ is true causally (i.e., $CA_1$ is true). In short, causal inferences are driven more by assumptions than by the data.[10] In our judgment, social researchers using the RMCI logic have deduced little more than they have assumed. They have mistakenly assumed that because a predictor correlated with X is important for the purpose of predicting Y, it follows that bias is removed or reduced if we "control" for this variable by adding it to the regression.

### 4.6 The Apparent Justification for the Causal Assumptions

The reasons why the causal assumptions above have always been associated with RMCI logic are difficult to find in the literature. Apparently the rationale has been as follows. In a classical statistical experiment designed to estimate causal effects, levels of the causal or treatment variable X are randomly assigned to randomly chosen subjects. When this is done, X is uncorrelated with the error terms in *any* of the regression models above. Of course, in a given sample or experiment randomization may not be fully effective, and in a given sample X might very well be correlated with one or more of the Z predictor variables. If randomization was effective, however, these correlations will be zero "in the population," approximately zero "in the sample." To explore this further, consider the error term in model $M_0$ expressed in terms of all of the information available in model $M_K$. This error term can be represented as

$$\varepsilon = \sum_{k=1}^{K} \gamma_k Z_k + v_K \tag{14}$$

The key result for linear models can be expressed as a logical extension of the above, namely,

$$E(X\varepsilon) = \sum_{k=1}^{K} \gamma_k E(XZ_k) + E(X v_K). \tag{15}$$

Call the terms in the summation the "covariance-weighted linear combination of partial regression coefficients." (Recall that all variables are measured as deviations from their means.)

Randomization automatically guarantees the validity of all of the causal assumptions given earlier, if linearity is taken for granted. That is, randomization implies $E(XZ_k) = 0$ for each $Z_k$ and $E(X v_K) = 0$ also. This is the meaning of "control through randomization." It follows that $E(X\varepsilon) = 0$ also where $\varepsilon$ is the equation error for model $M_0$. Note that experimental (or randomization-based) control produces zero correlation with any excluded variable, or any variable to be included as a predictor, which can be called "component-wise" control.

The algebraic identity in (15) has been translated into RMCI logic with the goal of adjusting for nonzero covariances. This has led to prescriptions for causal analysis that amount to the following: (1) find important left-out predictors (the

$Z_k$ variables with nonzero coefficients $\gamma_k$) that are correlated with the variable X whose causal effect is to be estimated (i.e., a predictor $Z_{k'}$ with $E(XZ_{k'}) \neq 0$), (2) add these to the model, (3) assume that the model with the added predictors satisfies the causal assumptions (as if an experiment had been simulated), and (4) call the estimated partial regression coefficients causal effects. Social research is replete with statements that such and such an important predictor (with non-zero $\gamma_k$) is correlated with X and thus has to be "controlled" in order to estimate the causal effect of X.

### 4.7 The Logical Fallacy of RMCI in the General Case

We now assume that all variables are important predictors, i.e., that $\beta_{yx.1\ldots K} \neq 0$ and $\gamma_k \neq 0$ for $k = 1, \ldots, K$. That is, model $M_K$ is the best model for prediction or the best conditional regression. If model $M_1$ is true causally, in the sense that assumption $CA_1$ is true, then assumption $CA_0$ for model $M_0$ cannot be true, as before.

Now suppose that model $M_2$ is true causally, in the sense that the assumptions in $CA_2$ are true. In this case, it is erroneous to base a causal inference on model $M_1$, *but it could be the case that an inference based on model $M_0$ is appropriate*. To see this, note that assumption $CA_2$ means that $E(Xv_2) = 0$, and since $\varepsilon = \gamma_1 Z_1 + \gamma_2 Z_2 + v_2$ in this case, the truth of $CA_2$ implies

$$E(X\varepsilon) = \gamma_1 E(XZ_1) + \gamma_2 E(XZ_2), \qquad (16)$$

because $E(Xv_2) = 0$ by $CA_2$. That is, assumption $CA_0$ is true if

$$\gamma_1 E(XZ_1) = -\gamma_2 E(XZ_2).$$

In words, if the two variables left out of model $M_0$ have compensating covariance-weighted partials, then either model, $M_0$ or $M_2$, can be used to estimate the causal effect of interest. For example, suppose that variables are measured in standardized form, so that the regression coefficients are standardized regression coefficients and the expectations above are correlations. Then $\gamma_1 = .3$, $\gamma_2 = -.45$, $\rho_{x,z_1} = .6$, $\rho_{x,z_2} = .4$ satisfies the compensating partials condition. (That is, $.3 \times .6 = -(-.45) \times .4$.)[11]

The algebraic exercise can be continued. Refer now to models $M_0$, $M_1$, $M_2$. For model $M_0$, the causal assumption $CA_0$ can be true even if model $M_2$ is the best conditional regression. That is, because

$$E(X\varepsilon) = \gamma_1 E(XZ_1) + \gamma_2 E(XZ_2) + E(Xv_2),$$

$CA_0$ can be true when the weighted partials compensate for each other as $\gamma_1 E(XZ_1) + \gamma_2 E(XZ_2) = -E(Xv_2)$. For model $M_1$, the causal assumption $CA_1$ can be true even if model $M_2$ is the best conditional regression also. That is,

$E(Xv_1) = \gamma_2 E(XZ_2) + E(Xv_2) = 0,$

and this means that $\gamma_2 E(XZ_2) = -E(Xv_2)$. Algebraic relations like this can be used to derive the conditions where any one of the three models is true causally and any other model is (incorrectly) used to estimate the causal effect. We maintain that it is impossible to know from the data which causal assumption is true, which is of course very different from knowing which of the three models produces the best prediction of Y (smallest variance or mean-squared error). That is, if model $M_0$ is true causally, then including either or both other predictors can lead to included-variable bias in spite of the fact that the variance of the prediction would be reduced. If model $M_1$ is true causally, then including $Z_2$ could lead to an included variable bias and deleting $Z_1$ can lead to an omitted variable bias. We cannot know which of the causal assumptions are true from the data, or whether the estimators from any model are biased due to omitted variables or biased due to included variables. Residuals are uncorrelated with the predictors and so are uninformative. Under the ideal situation of multivariate normality, nothing in the data could inform us about the validity of the causal assumptions for any of the three models.

To see the full implications of the above algebraic exercise, consider next model $M_K$. Now take one selection of the Z variables $z(s)$ and hence a model $M_{z(s)}$. The causal assumptions for this model are given by $CA_{z(s)}$, and the part of this that pertains to X translates as

$$\sum_{k \notin z(s)} \gamma_k E(XZ_k) + E(Xv_K) = 0. \tag{17}$$

This follows because $v_{z(s)} = \Sigma_{k \notin z(s)} \gamma_k Z_k + v_K$, where $v_K$ is the equation error for model $M_K$. Now note that if the excluded Z variables are important predictors, this linear combination could be identically zero for the same reason as in the special cases given immediately above. That is, the part of $CA_{z(s)}$ that pertains to X will be satisfied when

$$E(Xv_K) = -\sum_{k \notin z(s)} \gamma_k E(XZ_k). \tag{18}$$

This is a generalization of the "compensating weighted partials" result used earlier for special cases. Note that model $M_{z(s)}$ will carry along $z(s)$ other compensating-weighted–partials conditions, one for each predictor included besides X.

We can never know whether including one or more *important predictors* adds to or detracts from the validity of the causal assumption that X is uncorrelated with the equation error. And in the process of adding predictors, we must assume also that these added variables are uncorrelated with the equation error. Yet adding variables can make matters worse by introducing correlations with the equation error. "Controlling" for additional factors increases the number of untestable assumptions. Without assumptions or knowledge that cannot be determined from or validated with the data in hand, we cannot tell whether a given model gives a biased estimator of the causal effect of X on Y due to omitted variables or due to included variables, or whether any partial regression coefficient is unbiased, or indeed whether any of the possible models permit unbiased estimation of any of the causal effects. To us, this is the fundamental dilemma of RMCI.

## 4.8 The Dilemma Restated

The setup above implies that there will be a total of $2^K$ possible (partial) regression coefficients for X, including the simple regression coefficient associated with model $M_0$. While we can pick models that give the smallest variance or mean-squared error of the linear combination, causal inference depends on picking the model that eliminates the *bias* in the partial regression coefficient to be used as the estimator of the causal effect. We claim that there is no rational basis for choosing any single one of these $2^K$ partial regression coefficients as the causal effect of X on Y. Covariance-weighted partials of left-out variables can compensate for each other in ways that force us to regard the possiblity of included-variables bias in the same way that we normally regard the possibility of excluded-variables bias. Because we cannot know which model or models of the many possible models satisfy the causal assumptions, finding the "best" statistical (conditional) regression has no bearing whatsoever on the problem of making a causal inference about the effect of X on Y.

This dilemma has other implications. Suppose that only $K'$ of the Z variables have been measured or are available. For most outcome variables of interest in social research, K is probably very large, and for most data sets for which regression models would be used to estimate "causal effects" $K'$ will be much smaller than K.[12] We can easily apply standard regression methods to pick the best predictor of Y for the $K'$ available predictors, but this has nothing at all to do with picking the model that gives the right (unbiased) estimator of the causal effect.

It is commonly assumed that regression diagnostics or so-called specification tests can be used to validate assumptions necessary for causal inference (see, e.g., Hausman 1978; Heckman and Hotz, 1989; Godfrey 1988). These might very well be useful to diagnose a conditional regression ("best predictor of Y"), but they cannot tell us whether the key assumptions ($E(Xv_{z(s)}) = 0$ and the corresponding assumptions for each predictor included) are satisfied.[13] Suppose all variables, observed and unobserved, follow a multivariate normal distribution. All diagnostics are uninformative about the causal assumptions that would be "specified" or "maintained" with any model that might be estimated.

## 5. Faith in Assumptions

Causal modeling procedures, even highly automated ones, amount to finding models that predict well or fit the data well as judged by various criteria. The tacit assumption in these procedures is that the causal assumptions hold for whatever model or models provide acceptable levels of fit or prediction. With just a small measure of faith, the size of a mustard seed perhaps, mountain-sized causal models giving causal inferences for all variables in the data set can be obtained, or so we are led to believe. Faith is a good thing in both science and Sunday School, but normally we try to reduce reliance on faith in the former as much as possible. But faith in the causal assumptions that validate drawing causal inferences from whatever model is fitted to the given set of data must be very strong indeed.

Return once again to the setup of the previous section. If only X and Y are available, the faith required can be expressed as certain belief in assumption $CA_0$. If X, Y, and $Z_1$ are available, then the faith required amounts to certain belief in $CA_1$, which includes the additional condition that $E(Z_1 v_1) = 0$. The latter condition is applied in order to claim that the partial regression coefficient for $Z_1$ is the causal effect of $Z_1$, but as we showed earlier it was necessary to estimate the causal effect of X also. Now suppose that in addition to X and Y, some set $z(s)$ of the predictors in Z is available. That is, we imagine having any one of the $2^K$ possible data sets that could be formed. The faith required here means that condition $CA_{z(s)}$ is "specified" or "maintained" or "assumed." This condition is a set of assumptions about correlations between predictors and the equation error of the given model. If the single-equation model is replaced by a multiple-equation model, then the number of untestable causal assumptions essentially multiplies as a function of the total number of variables taken as predictors in all of the equations.

Imagine that we are free to select any one $z(s)$ of the $2^K$ subsets of variables that always contain X. In addition, suppose that the sample size is very large so that all possible regression coefficients are nonzero. We, of course, do not know whether a selection of this kind leads to a model where each predictor is uncorrelated with the equation error. If all coefficients in the model are nonzero and the sample is large enough to detect this, all we would ever produce is an estimate of the partial regression coefficient $\beta_{yx.z(s)}$, which would be called the causal effect of X on Y. We cannot know whether this estimator is more or less biased as the causal effect of X when it is compared to any other partial regression coefficient that could be obtained with other models, even models that do not predict well or fit well. Indeed, we cannot know whether this estimator is less biased than the simple regression coefficient $\beta_{yx}$. We cannot know whether the causal effect is large or small, positive or negative, present or absent without *additional knowledge* that cannot be obtained from the data.

Surely arguments like these have occurred to others, although we are not aware of any explicit statement like that above. Sobel (1994) is certainly aware of this dilemma, and Freedman (this volume) gives a compact summary of the omitted-variables problem that indicates this as well. Although others have not stated the fundamental dilemma in exactly the way that we have, careful methodologists appear to be aware of it also (Smith 1990; Berk 1988). But how could a method that amounts to nothing more than "add important predictors" have been construed as a recipe for causal inference? In our judgment, there are three chief reasons.

The first factor is undoubtedly the logic of omitted-variables bias. This topic is covered in virtually all standard sources on RMCI logic. The assumptions built into the usual treatment of this concept beg the question. If a model without a particular predictor $Z_{k*}$ is not true causally in the sense that X and/or other variables are correlated with the equation error in this model, and if it is assumed that a model adding variable $Z_{k*}$ is true causally (i.e., X and all other predictors are uncorrelated with equation error in the new model), then *of course* the causal effect can be identified and estimated without bias with the new model. And *of course* the model without this predictor is likely to lead to biased estimators. The problem reduces to mere algebra along with standard statistical estimation of the regression, plus the assumption. But there is no good reason to assume that a model without predictor $Z_{k*}$ is "causally" wrong and a model with predictor $Z_{k*}$ is "causally" right. An assumption of included-variable bias is just as logical as an assumption of omitted-variable bias.

Second, researchers have confused prediction with causation (cf. Holland 1986). Adding important predictors reduces variance (see Section 3), but it does not follow that reducing variance reduces or eliminates bias (in the sense described in Section 4). It may well be the case that $V(\hat{\beta}_{yx.z(s)}) < V(\hat{\beta}_{yx})$, and if the predictors are important then $\hat{\sigma}^2_{v_{z(s)}} < \hat{\sigma}^2_{\varepsilon}$ also, at least in standard situations. Statistical criteria for prediction or selection of predictors are based on these facts or on criteria related to them. But an adequate model for prediction could lead to a small variance of a biased estimator of the causal effect; the magnitude of the bias cannot be detected from regression analysis.

Third, social researchers have been captivated by complex models and algorithms that estimate these models. It is difficult to distinguish between what is assumed and what is inferred from complex "supermodels" (Clogg and Dajani 1991; Clogg 1992; Clogg and Arminger 1993). The degree to which assumptions drive inferences about causal effects can be seen easily for the single-equation regression model. It is not so easy to determine how assumptions affect causal inferences in complex supermodels.

## 6. An Alternative Methodology for Quantifying Uncertainty in Causal Inference

In many cases it is impossible to carry out an experiment to determine causal effects. Convincing quasi-experimental data where causal variables are actually manipulated may not exist either, at least for many of the causal questions posed in social research. For perhaps the majority of cases of interest it may still be necessary to draw causal inferences from regression models. It should be clear that we believe researchers should be much more cautious in doing this, and serious sensitivity analysis would be helpful. Here we propose a method that ought to be helpful in quantifying the uncertainty in causal inferences from such models; this method also sheds additional light on RMCI logic.

Suppose that (a) only the causal effect of X on Y is sought and (b) almost all of the other predictors that could be *relevant* have been included in the data available for regression analysis. By "relevant" we mean that the researcher suspects that for at least some subsets of the other predictors, the causal assumptions $(E(Xv) = E(Zv) = 0)$ are satisfied. If in a given area there is substantial disagreement about what these predictors are, then a large set that contains as many of them as possible should be used. Suppose further that there are K of these Z variables. We acknowledge that the faith involved in framing the problem in this way far exceeds the mustard seed in size. On the other hand, we are not assuming that the causal assumptions are satisfied automatically for every predictor included in some final model, just because the predictors in the final model happen to be "significant" or because the model "fits the data." For cases where the number K of components of Z is large, their dimension might be reduced by taking principal components.

Corresponding to this data set with uncontrolled correlations among all variables, we have $M = 2^K$ possible partial regression coefficients, say $\beta_{yx.k'}$, $k' = 1, \ldots, M$, with sample estimators $\hat{\beta}_{yx.k'}$, say least squares estimators. The previous arguments imply that we are totally ignorant about which of these should be used to estimate the causal effect, or which of these is less biased than the others. The task is to combine information from this large group of estimators in a way that reflects the uncertainty in the causal inference due to the sampling of predictors (or possible data sets), not just the sampling variance in estimators from conditional regressions.

An *unweighted* mean "effect" can be defined as

$$\hat{\beta}^{uw} = \sum_{k'=1}^{M} \hat{\beta}_{yx.k'}/M, \tag{19}$$

where $M = 2^K$. Corresponding to this is a measure of variability or uncertainty in the quantity,

$$U(\hat{\beta}^{uw}) = \sum_{k'=1}^{M} (\hat{\beta}_{yx.k'} - \hat{\beta}^{uw})^2/(M-1). \tag{20}$$

Although this quantity is formally the same as the variance of the partial regression coefficients, we use different notation to reflect the fact that variability *over models* is summarized by the quantity.

Various weighted-mean "effect" estimators can be proposed. Suppose that $w_{k'}$ is the weight associated with the $k'$-th partial regression coefficient with $\Sigma_{k'} w_{k'} = 1$. An estimator taking account of these is

$$\hat{\beta}^w = \sum_{k'=1}^{M} w_k \hat{\beta}_{yx.k'}. \tag{21}$$

Researchers using RMCI are in fact applying a method of weighting evidence that can be summarized compactly in terms of this formulation. First, pick a model for the data on the basis of criteria of prediction or goodness-of-fit. Second, ignore all other models that might have been estimated. Call the regression coefficient associated with X in the selected model $\hat{\beta}_{k*}$. Then the causal effect is determined from (21) by setting $w_{k*} = 1$, $w_k = 0$ for $k \neq k*$. That is, the fitted model is assumed to be correct (or the causal assumptions are assumed valid for the fitted model) and all other models are assumed to be wrong (or the causal assumptions are assumed to be wrong for all models discarded). The arbitrariness of existing methodology is now easy to see.

There are variations of the weighted estimator that may be of special interest. A variance–weighted mean effect would use the variance of each coefficient under a proper full model. Suppose that this model is model $M_K$, the model with all predictors included. Call the error variance estimator for this model $\sigma_{vK}^2$, the mean squared error for the "full" model. As in Clogg, Petkova, and Shihadeh (1992) or Clogg, Petkova, and Haritou (1995), the variance of any regression coefficient under this full model (and conditioning on the predictor values observed) can be estimated as

$$\hat{\sigma}_{k'}^2 = \hat{V}(\hat{\beta}_{yx.k'}|M_K) = s^2(\hat{\beta}_{yx.k'})\,(\hat{\sigma}_{vk}^2/\hat{\sigma}_{v_k}^2), \tag{22}$$

where $s^2(\hat{\beta}_{yx.k'})$ is the squared standard error of the given regression coefficient, as estimated in the given model. A variance-weighted estimator would use $w_k = \hat{\sigma}_{\cdot}^2/\hat{\sigma}_{k'}^2$, where $\hat{\sigma}_{\cdot}^2$ is the average variance.

Finally, an estimator of the causal effect that reflects a prior distribution of the models can be defined. The $w_k$ could denote this prior distribution. If we are completely ignorant concerning which model gives the best estimator of the causal effect of X, it would be natural to set $w_k = 1/M$. An average taking account of this prior leads to the unweighted effect given first. See Raftery (1995) for a formal account, where a prior like this is used for picking the "best" predictive model.[14] This prior might be best for most social-science efforts with which we are familiar. A variance-weighted version can easily be defined. Corresponding to each of these other estimators, we can summarize the uncertainty in the inference with a measure analogous to (20).[15]

The logic associated with the above estimators is Bayesian, but it is not related to the ordinary Bayesian estimation of the regression model (cf. Raftery 1995). In fact, we have assumed that each model gives a "random" draw from the distribution of estimators of causal effects, or that each model gives one of M plausible values or imputations from this distribution. The uncertainty measure is merely the between-imputation variability of the draws. See Rubin (1987). These procedures formalize what many good researchers actually do: assess the stability in regression coefficients under alternative specifications. Something of this kind ought to be seriously considered in order to quantify the uncertainty in causal inference.

As an example, we simulated $n = 1000$ observations from a multivariate normal having seven components corresponding to Y, X, and $Z_1, \ldots, Z_5$. We used a large sample size so that sampling error in the usual sense would be small. The variables were standardized to have mean zero and variance one. The correlations ranged from a low of $-.2$ to a high of $.6$, and most were of "moderate" size according to standards in social research. In this case, there are 32 partial regression coefficients for the Y-X relationship. This example is used to illustrate the relative magnitudes of the two sources of uncertainty in the causal inference that could be made from a regression. These two sources are sampling variance and the uncertainty associated with not knowing which of the many possible models gives an unbiased estimator of the causal effect.

The average value of the regression coefficients for Y-X was $\hat{\beta}^{uw} = .489$; the smallest coefficient was .284 and the largest was .686. The ordinary standard errors for the coefficients range from about .024 to .031 and the average squared standard error is about .0008. Now the uncertainty measure of (20) is .020761 (square root or standard deviation is .144). In other words, the uncertainty associated with the regression coefficients as estimators of causal effects is about 26 times the sampling variance! If we called the average coefficient the causal effect of X on Y, then the Bayesian estimate of the standard deviation would be the square root of the sum of the average sampling variance of the coefficients and the between-coefficients variance, i.e., $s(\hat{\beta}^{uw}) = (.0008 + .020761)^{1/2} = .147$. Using two standard deviations as bounds we thus obtain $.489 \pm .294$, an interval estimate that reflects both sampling variance and uncertainty in model selection. If we were making a causal inference with such data, we could say that the "effect" of X on Y is at least positive taking account of the uncertainty in model selection. It is likely that the uncertainty due to model selection will be much greater than the sampling error in many standard uses of the RMCI method.

## 7. Conclusion

There are many variations of RMCI logic in social research. Indeed, this logic is applied whenever any regression-type model is applied to nonexperimental data to draw causal inferences. Similar logic is applied routinely when, say, logistic regression models instead of ordinary linear models are used. RMCI logic is implicit in so-called causal models for covariance matrices as well. Whether model search is automated or not, virtually all methods for the causal analysis of covariance matrices (Glymour, this volume; Bollen and Long 1993) amount to finding a model that predicts well or reproduces the observed covariance matrix acceptably well. The fitted covariance-structure model would most often be described as a set of regression equations, not just one regression equation. The causal assumptions (or faithfulness conditions) implicit when parameter values in such models are given causal interpretations can be restated in terms analogous to those given above, but it is a complex affair to state these so that most users can understand them. The assumptions are not only complex, there are many of them.

To reiterate, finding models that predict well or fit the data well has little or nothing to do with estimating the presence, absence, or size of causal effects. The uncertainty in making causal inferences from regression-type models needs to be appreciated more fully, and we hope that the methodology presented immediately above will be useful to do this. Developing more complex models, more elaborate definitions of exogeneity (or causal assumptions), and automated model-search procedures are worthwhile objectives, but these activities do not validate the logic of causal inference by themselves.

The causal questions that social researchers ask are important ones that we ought to try to answer. If they can only be answered in the context of nonexperimental data, then a method that conveys the uncertainty inherent in the enterprise ought to be sought. We believe that the uncertainty in causal assumptions, not the uncertainty in statistical assumptions and certainly not sampling error, is the most important fact of life in this enterprise.

## NOTES

An earlier version of this essay was presented in the conference, "Causality in Crisis?" held at the University of Notre Dame, October 15–17, 1993. Parts of this paper were presented as the Keynote Address at the annual meeting of the South African Statistical

Association, held in Kruger Park, South Africa, November 1–5, 1994. This research was supported in part by Grant No. SBR–9310101 from the National Science Foundation. The authors are indebted to Glenn Firebaugh, David Freedman, Jacques Hagenaars, Bruce Lindsay, Vaughn McKim, and Judea Pearl for helpful comments.

1. The more usual terminology—"dependent" variable ( for Y) and "independent" variables ( for X and Z)—is misleading because it is loaded with causal imagery. Of course, this terminology is well entrenched in social research as well as in other areas. We have similar objections to the econometric terminology of "exogenous" and "endogenous" variables, as will be clear in the next two sections. The other most common designation of the two types of variables, in statistics at least, is "response" ( for Y) and "factor" (for X or Z), but this implies experimentation which should not be associated with regression analysis of nonexperimental data.

2. Because we have used centered variables, the intercept can be excluded.

3. There is some subtlety here. We can regard the conditional expectation in (2) as a function of the random variable X also, and when this is done we assume that the distribution of X, say $g(X)$, does not depend on the parameters in the regression model. It is useful also to note that the unconditional variance of Y can always be factored as $V(Y) = V_X(E(Y|X)) + E_X(V(Y|X))$, which should be read as "unconditional variance of Y" = "variance (with respect to X) of the conditional mean of Y" + "average (with respect to X) conditional variance of Y". For the simple linear regression model with *random* X, this gives $\sigma_y^2 = V_X(\beta_{yx}X) + E_X(\sigma_\varepsilon^2) = \beta_{yx}^2\sigma_x^2 + \sigma_\varepsilon^2$. The sample analogue of this is the usual "analysis-of-variance" table.

4. For example, to make inferences about the conditional regression relationship over all random samples of (Y, X) or (Y, X, $Z_1$) or (Y, X, $Z_1, \ldots, Z_K$), we assume that the parameters describing the joint distribution of the predictors do not depend on the parameters in the regression models.

5. Of course, if the assumptions in (1), (2), (3), and (5) are true, assumption (4) is automatically satisfied. Assumption (5) is less important in large samples.

6. It is often taken as an assumption also that there is no measurement error in the predictor X. If the predictions of interest pertain to X values observed or to be observed in the future, however, then it is proper to base all inferences on $E(Y|X)$. Measurement error in predictor variables can be ignored when this is the case (Berkson 1950). For many uses of regression for prediction, measurement error in the predictors can be ignored, and we do not regard this assumption in the same way that we regard the others above. An entirely different setup is required if the task is to estimate the *functional* relationship between Y and X, in which case the conditional regression requires some modification. In statistics, this problem is usually regarded in the following way. If predictions are to be made on the basis of the observed values of X or future observed values of X, say temperature as recorded by a ( fallible) thermostat, then measurement error can be ignored and the setup is as given above. If the goal is to estimate the *functional* relationship between true temperature and some outcome, then measurement error in X needs to be taken into account. See Fuller (1987) for details.

7. One of the most influential books on the RMCI method is Duncan (1975). With the exception of an exercise on p. 8, the distinction between conditional and unconditional regression (or expectation) is not made in the entire book.

8. Duncan (1975) employs these assumptions throughout his important work on structural equation models. On p. 52, for example, he provides the following parenthetical remark: "We note, for future reference, that in both recursive and nonrecursive models, the *usual specification* is zero covariance of predetermined variables in an equation with the disturbance of that equation" (emphasis added). There is no further reference to this assumption other than as a "usual specification." In other references, these assumptions will be referred to as "specifications" or as "maintained assumptions" or as "conditions."

9. Asymptotic expectation or consistency can be substituted for expectation here and elsewhere, and "bias" is to be understood in these cases as asymptotic bias or inconsistency.

10. If model $M_0$ were true causally, then we could still use information from model $M_1$ to increase the precision of estimation, but using conditional regression logic. The standard error of the estimator $\hat{\beta}_{yx}$ under model $M_1$ can be estimated as $s(\hat{\beta}_{yx})$ $(MSE_1/MSE_0)^{1/2}$, where "MSE" refers to the mean-squared error under the given model and $s(\hat{\beta}_{yx})$ is the standard error of the simple regression coefficient as estimated under model $M_0$. See Clogg, Petkova, and Haritou (1995).

11. So-called faithfulness conditions elaborated below rule out this sort of exact compensation by assumption. We comment further on these conditions later.

12. We are assuming that most outcome variables of interest in social research have many causes and that not all or even most of the causes can be measured in a given set of data.

13. The literature on specification tests has to be considered carefully. For example, Hausman (1978) begins his influential paper on the subject by giving two important assumptions that need to be checked. These translate as (i) equation errors are independent of predictors and (ii) errors are homoscedastic and normally distributed. Failure of the first assumption creates bias in the coefficient estimators while the failure of the second creates bias in the variance estimators (not bias in the coefficient estimators). For causal inference bias in the coefficient estimators is much more important.

14. Raftery (1995) applies a vague prior for each of many possible predictive models that might be formulated. He then excludes models that obviously do not predict well and then selects models from the reduced set that maximize the posterior odds that one model instead of another is correct (as a predictive model). In contrast to Raftery's methods for model selection, we believe that all possible models, not just all models that predict at least moderately well, need to be used to define the uncertainty in causal inferences.

15. A possibly better strategy would be to base the weighting and the assessment of uncertainty on the variance-covariance matrix of the $\hat{\beta}_{k'}$. Arguments in Clogg, Petkova, and Shihadeh (1992) and Clogg, Petkova, and Haritou (1995) can be used to determine this matrix, conditional on some full models. The weighted averages above use only the diagonal elements of this matrix.

# From Association to Causation via Regression

## David A. Freedman

For nearly a century, investigators in the social sciences have used regression models to deduce cause-and-effect relationships from patterns of association. Path models and automated search procedures are more recent developments. In my view, this enterprise has not been successful. The models tend to neglect the difficulties in establishing causal relations, and the mathematical complexities tend to obscure rather than clarify the assumptions on which the analysis is based.

Formal statistical inference is, by its nature, conditional. If maintained hypotheses A, B, C, . . . hold, then H can be tested against the data. However, if A, B, C, . . . remain in doubt, so must inferences about H. Careful scrutiny of maintained hypotheses should therefore be a critical part of empirical work—a principle honored more often in the breach than the observance.

This essay focuses on modeling techniques that seem to convert association into causation. The object is to clarify the differences among the various uses of regression, as well as the source of the difficulty in making causal inferences by modeling. The discussion will proceed mainly by examples, ranging from Yule (1899) to Spirtes, Glymour, and Scheines (1993).

## 1. Outline

Many treatments of regression seem to take for granted that the investigator knows the relevant variables, their causal order, and the functional form of the relationships among them; measurements of the independent variables are assumed to be without error. Indeed, Gauss developed and used regression in physical science contexts where these conditions hold, at least to a very good approximation.[1] Today, the textbook theorems that justify regression are proved on the basis of such assumptions.

In the social sciences, the situation seems quite different. Regression is used to discover relationships or to disentangle cause and effect. However, investigators have only vague ideas as to the relevant variables and their causal order;

functional forms are chosen on the basis of convenience or familiarity; serious problems of measurement are often encountered.

Regression may offer useful ways of summarizing the data and making predictions. Investigators may be able to use summaries and predictions to draw substantive conclusions. However, I see no cases in which regression equations, let alone the more complex methods, have succeeded as engines for discovering causal relationships. Of course, there may be success stories that I have not found; nor does a track record of failure necessarily project into the future.

One of the first applications of regression techniques to social science is Yule (1899). Recent examples will be found in Spirtes, Glymour, and Scheines (1993), to be cited here as SGS. (The SGS theory is summarized in Glymour's essay in this volume, cited as CG.) SGS have attracted considerable attention in the philosophy of science because they have developed computerized algorithms that search for path models. With their algorithms, SGS claim to make rigorous inferences of causation from association. This is a bold claim, which does not survive examination.

The balance of this essay is organized as follows. Section 2 discusses Yule's work. Sections 3 and 4 explain the critical idea of "exogeneity." Section 5 describes a contemporary regression model. Sections 6–10 review SGS and reanalyze some of their examples. Sections 11–12 canvass some mathematical issues. Possible responses to my critique will be found in section 13. There is a brief review of the literature in section 14, and conclusions are presented in section 15. For ease of reference, standard formulas for regression are given in an appendix. I have tried to make most of the essay accessible to non-statistical readers, particularly if they will permit the occasional undefined technical term; sections 11 and 12 are more specialized.

## 2. Yule's Regression Model for Pauperism

One of the first regression models in social science was developed by Yule— "An Investigation into the Causes of Changes in Pauperism in England, Chiefly During the Last Two Intercensal Decades."[2] In late-nineteenth-century England, poor people could be supported either inside the poor house or outside. Did provision of support outside the poor house increase the number of poor people?

To address this issue, Yule used data from the censuses of 1871, 1881, and 1891. (In England, the census is taken in years that end with 1.) He considered the periods 1871–81 and 1881–91, relating changes in the number of paupers to changes in the "outrelief ratio," that is, the ratio between the number of paupers supported outside the poor house and inside. He used regression to control for two confounders—changes in the population and its age structure.

His equation can be written as follows:

(1)   $\Delta$Paup $= a + b \times \Delta$Out $+ c \times \Delta$Pop $+ d \times \Delta$Old $+$ error.

Here, $\Delta$ stands for percentage difference, *Paup* for the number of paupers, *Out* for the outrelief ratio, *Pop* for population size, and *Old* for the proportion of people aged 65 and over.

Yule's unit of analysis was the "union," which seems to have been a small geographical area like a county.[3] He had four kinds of areas: rural, mixed, urban, metropolitan. He used "Ordinary Least Squares" (OLS) to estimate the coefficients from the data, with a "50 cm. Gravet" slide rule to do the arithmetic.

To be more specific, Yule estimated a separate equation for each time period (1871–81 and 1881–91) and each kind of area. There were 2 time periods and 4 kinds of areas, thus, $2 \times 4 = 8$ equations. Within a time period, all areas of the same kind—for instance, all rural unions—are governed by one equation. (By coincidence, there are 4 coefficients in each equation, and 4 kinds of areas.)

Yule was looking for the "Hooke's Law of Poverty." Nature ran an experiment, with lots of variation over time and geography, and Yule analyzed the results. Regression was needed to control for the confounding effects of change in population and age structure. The equations were held to show that, other things being equal, changes in the outrelief ratio create corresponding changes in the number of paupers. Indeed, if you increase the outrelief ratio by one percentage point but hold the other factors constant, you will increase the number of paupers by *b* percent, *b* being the coefficient of outrelief in equation (1). More qualitatively, if *b* is positive, welfare creates paupers.

For a moment, I turn from Yule to methodology. A regression equation like Yule's is usually written as

(2)   $Y = X\beta + \varepsilon.$

In this equation, the vector *Y* represents the dependent variable, like pauperism; the matrix *X* represents the explanatory (or "independent") variables, like the outrelief ratio, population, and age structure. These are observable. The vector $\beta$ represents parameters, which are not observable but may be estimated from the data: parameters are "social constants," which characterize the process that generated the data. In Yule's equation, $\beta$ has four components—the parameters *a, b, c, d* in equation (1). The error or "disturbance" term $\varepsilon$ is also unobservable, and represents the impact of chance factors unrelated to *X*. Statistical inferences are often based on "stochastic assumptions" about $\varepsilon$, e.g., $\varepsilon$ is independent of *X*, its components are independent and identically distributed with mean 0. For details, see the appendix below.

Three possible uses for regression equations are as follows:

(i) to summarize data, or
(ii) to predict values of the dependent variable, or
(iii) to predict the results of interventions.

Yule could certainly have summarized his data by saying that for a given time period and unions of a specific type, with certain values of the explanatory variables, the change in pauperism was about so much and so much. In other words, he could have used his equations to estimate the average value of $Y$, given the values of $X$. This use of regression may run into technical problems if there are outliers, or nonlinearities in the regression surface. However, at least in principle, there do seem to be technical fixes for such problems. Furthermore, stochastic assumptions about the disturbance term play almost no role. Therefore, like most statisticians, I believe that regression can be quite helpful in summarizing large data sets.

For prediction, there is a *ceteris paribus* assumption: the system will remain stable. Prediction is already more complicated than description. On the other hand, if you make a series of predictions and test them against data, it may be possible to show that the system is stable, or sufficiently stable for regression to be quite helpful.[4] Again, any particular use of regression to make predictions may go off the rails, but there do not seem to be essential difficulties of principle involved.

Causal inference is different, because a change in the system is contemplated: for example, there will be an intervention. Descriptive statistics tell you about the correlations that happen to hold in the data; causal models claim to tell you what will happen to $Y$ if you change $X$. Indeed, regression is often used to make counterfactual inferences about the past: what would $Y$ have been if $X$ had been different? This use of regression to make causal inferences is the most intriguing—and the most problematic. Difficulties are created by omitted variables, incorrect functional form, etc. Of course, if the results of causal modeling were with any frequency checked against the results of interventions, the balance of argument might be very different.[5]

For description and prediction, the numerical values of the individual coefficients fade into the background: it is the whole linear combination on the right-hand side of the equation that matters. For causal inference, it is the individual coefficients that do the trick. In equation (1), for example, it is $b$ that should tell you what happens to pauperism when the outrelief ratio is manipulated.

At this remove, the flaws in Yule's argument may be apparent. For example, there seem to be some important variables missing from the equation, including variables that measure economic activity. Here is Yule's comment on the last-named factor:

> A good deal of time and labour was spent in making trial of this idea, but the results proved unsatisfactory, and finally the measure was abandoned altogether. [p. 253]

Yule seems to have used the rate of population growth—$\Delta Pop$ in equation (1)—as a proxy for economic activity, although that creates ambiguity. Other things being equal, population growth will by itself add to the number of paupers; in its role as proxy, however, population growth should reduce pauperism.

The equations for metropolitan unions are shown below, for 1871–81 and 1881–91:[6]

(1871–81)    $\Delta$Paup = 13.19 + 0.755 × $\Delta$Out − 0.322 × $\Delta$Pop
$$- 0.022 \times \Delta\text{Old} + \text{residual.}$$

(1881–91)    $\Delta$Paup = 1.36 + 0.324 × $\Delta$Out − 0.369 × $\Delta$Pop
$$+ 1.37 \times \Delta\text{Old} + \text{residual.}$$

For example, one metropolitan union is Westminster. Over the period 1871–81, the percentage changes in *Out, Pop* and *Old* are −73, −9, and 5, respectively. The percentage change in *Paup* predicted from the regression equation is

$$13.19 + 0.755 \times (-73) - 0.322 \times (-9) - 0.022 \times 5 = -39.$$

The actual percentage change in *Paup* is −48. The "residual" is

$$\text{residual} = \text{actual} - \text{predicted} = -48 - (-39) = -9.$$

The coefficients in the regression equation are estimated so as to minimize the size of the residuals. (Technically, it is the sum of the squares that is minimized—hence the term "least squares.") The linear combination of explanatory variables on the right side of the equation has therefore been optimized; but there is no guarantee that individual coefficients will make much sense.

There are some noticeable inconsistencies in Yule's coefficients, over time and across the various kinds of geography. Nor are the signs of the coefficients entirely reasonable. These inconsistencies may not by themselves be fatal, but certainly raise the question of whether the equations hold true for any well-defined population of times and places. If the coefficients do not have a life of their own—outside Yule's particular data set—they cannot be used to answer questions of the form, "What would happen if you change the outrelief ratio?"

The coefficients may be useful for descriptive purposes, but not for causal inference or even prediction.

Moreover, there are familiar difficulties of interpretation. At best, Yule showed that changes in pauperism and the outrelief ratio were associated, even after adjusting for changes in the population and its age structure. The direction of the causal arrow, however, is by no means clear. Yule's theory is that outrelief is the cause, and pauperism the effect. That is a reasonable view. However, the opposite idea seems equally tenable—a union that is flooded with paupers may not be able to build poor houses fast enough, and resorts to outrelief. If so, pauperism causes outrelief. Also, Governor Pete Wilson's theory may have some plausibility for nineteenth-century England if not twentieth-century California: unions that provide generous outrelief attract paupers from elsewhere.[7]

Yule must have been aware of these problems. After allocating the changes in pauperism to their various causes (including the residual), he withdraws all causal claims with one deft sentence:

Strictly, for "due to" read "associated with." ( footnote 25, p. 270)

Yule's paper is quite modern in spirit, with two exceptions: he did not rely on statistical significance, and he did not use a graph. Figure 4.1 brings him up to date.

## 3. Regression Estimates and Conditional Expectations

In the regression model (2), $Y$ is the dependent variable, like pauperism; $X$ represents the explanatory variables, like the outrelief ratio, population, and age structure. If all goes well, the regression equation will estimate the "conditional expectation" of $Y$ given $X = x$, that is, the average value of $Y$ corresponding to given values for the explanatory variables.

To clarify the definitions, consider two procedures:

*Procedure #1*. Select subjects with $X = x$; look at the average of their $Y$'s.

*Procedure #2*. Intervene and set $X = x$ for some subjects; look at the average of their $Y$'s.

These procedures are quite different. The first involves the data set as you find it. The second involves an intervention.

Regression does seem to let you move from selection to intervention; that is why the technique is so popular. However, regression approximates the selection procedure, rather than intervention. Nor does the statistical analysis prove that the two procedures give the same results: how could it? Instead, causal inferences are made by *assuming* that selection tells you what would happen if you were to intervene.

**FIGURE 4.1**   Yule's model for pauperism. The figure represents equation (1) in graphical form. The asterisks denote a high degree of statistical significance.

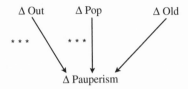

*Notes:* To determine the asterisks, I recomputed Yule's regression for the metropolitan unions over the period 1871–81, using data in his Table XIX. I replicated his coefficients, as shown in the display, although roundoff error is quite large.

$$\Delta Paup = 12.884 + 0.752 \times \Delta Out - 0.311 \times \Delta Pop + 0.056 \times \Delta Old + \text{residual}.$$

|  | 10.367 | 0.135 | 0.067 | 0.223 |
|---|---|---|---|---|
|  | 1.24 | 5.57 | −4.65 | 0.25 |

Under the coefficients are standard errors (SEs) and *t*-statistics. The SE indicates the likely size of the difference between an estimated coefficient and its true value. The *t*-statistic is the ratio of an estimate to its SE. Generally, a *t*-statistic above 2 or 3 in absolute value indicates that the corresponding parameter is unlikely to be truly 0. The parameters are features of the model, and the SEs are computed on the basis of the stochastic assumptions in the model; for details, see the appendix. (Of course, Yule's model is open to serious question.) In figure 4.1, the explanatory variables are correlated; such correlations are often signaled by curved, double-headed arrows; error terms are not shown either.

The phrase "*X* is exogenous" is often taken to mean that selecting on *X* will produce the same results as intervening to set the value of *X*—the basic assumption in many analyses. Exogeneity also has weaker meanings, to be taken up later. The ambiguity is unfortunate, because analysts may assume exogeneity in a weak sense, and proceed as if they had established something more. It is only exogeneity in the strong sense defined above that enables you to predict the results of interventions from nonexperimental data. The distinction between selection and intervention is acknowledged even by the modelers:

> Formally speaking, probabilistic analysis is indeed sensitive only to covariations, so it can never distinguish genuine causal dependencies from spurious correlations. . . . (Pearl 1988, 396)

Such admissions—like Yule's footnote 25—are fatal to the enterprise. Of course, Pearl does not give up. For instance, he goes on to say that experiments just provide the opportunity to observe yet more correlations, a move he attributes to Simon (1980).

Figure 4.2 is Pearl's. On the left, it seems that *X* and *Z* cause *Y*; manipulating *X* or *Z* will change *Y*. However, if only we had measured the variables *U* and *V*, we might have seen that they were the joint causes of *X*, *Y*, and *Z*, as in the right-hand panel. If so, manipulating *X* and *Z* will not change *Y* at all. No amount of statistical analysis on the observables—on *X*, *Y*, and *Z*—can tell us which panel

**FIGURE 4.2**   After Judea Pearl (1988, 397). Causation cannot be inferred from association by using causal models.

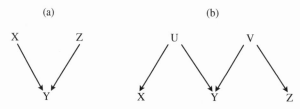

*Notes:* In panel (a), $X$ and $Z$ are assumed to be independent. In panel (b), $U$ and $V$ are assumed to be independent; it may be shown in consequence that $X$ and $Z$ are independent. Also see Duncan (1975, 113–27).

expresses the right theory. Indeed, matters can be arranged so that both theories lead to the same joint distribution for the observables.

## 4. Two Ideas of Conditional Probabilities

The distinction between the two ideas of conditioning—selecting subjects with $X = x$, or intervening to set $X = x$—seems fundamental. A concrete example may help, and conditional probabilities are easier to deal with than conditional expectations.

Many studies have demonstrated an association between cervical cancer and exposure to two sexually transmitted diseases—herpes and chlamydia. Suppose we had data as shown in table 4.1. The incidence rate of cervical cancer is 200 per 100,000 for women exposed to herpes and chlamydia (top left); 116 per 100,000 for women exposed to herpes but not chlamydia; and 130 per 100,000 for those exposed to herpes, the two exposure categories for chlamydia being combined. Other cells may be read in a similar way.

With sample data, there is a role for technical statistics in estimation and testing—for instance to see if the rates within a row are constant across columns.

**TABLE 4.1**   Rate of cervical cancer cases per 100,000 women, by exposure to chlamydia and herpes. Data are hypothetical.

|  |  | *Chlamydia* | | |
|---|---|---|---|---|
|  |  | *Yes* | *No* | *Combined* |
| Herpes | Yes | 200 | 116 | 130 |
|  | No | 180 | 80 | 87 |
|  | Combined | 190 | 90 | 100 |

However, the real question is not association but causation. Does herpes cause cervical cancer? What about chlamydia? Biotechnology might find a way to eliminate *Herpes simplex* as well as *Chlamydia trachomatis*. That would be a great relief, but would it reduce the incidence rate of cervical cancer?

To consider the issue of causality more directly, suppose that we actually know the rates for the population of interest, as shown in table 4.1. Statistical testing must now fade into the background. The overall incidence rate is 100 cervical cancers per 100,000 women (table 4.1, bottom right); among women exposed neither to herpes nor to chlamydia, the rate is lower—80 per 100,000. If cervical cancer is caused by herpes and chlamydia, eliminating the microorganisms responsible for those diseases should reduce the incidence rate of cervical cancer from 100 to 80 per 100,000. On the other hand, if the relationship is not causal, eliminating those microorganisms will have little effect on the incidence rate of the cancer.

To be more explicit, 80/100,000 has been found by selecting women who are exposed to neither herpes nor chlamydia and computing the incidence rate of cervical cancer for that group: one interpretation of conditional probability. If we intervene and eliminate the two diseases, we want to know the rate after the intervention: that is another interpretation. The two interpretations are different, because the underlying procedures are different. Statistical analysis of the numbers in the table, however refined or complex, cannot prove that a hypothetical intervention will give the same results as selection. This may seem obvious, even banal; but if you grant the point, the causal modeling game is largely over.

What is the situation for table 4.1? The story is far from certain. Current epidemiological opinion favors the idea that cervical cancer is caused by certain strains of human papilloma virus (HPV); herpes and chlamydia have no etiologic role, but serve only as markers for exposure to HPV. If that opinion is correct, wiping out herpes and chlamydia will have no impact on rates of cervical cancer.

Due in part to the rarity of cervical cancer, cohort studies do not seem to be available. (The numbers in table 4.1, although hypothetical, are not unreasonable.) My point is even stronger for the real studies of the association between cervical cancer and herpes or chlamydia. Problems created by incomplete data cannot simplify the task of inferring causation from association.[8]

## 5. Another Regression Example

Rindfuss et al. (1980) propose a model to explain the process by which a woman decides how much education to get and when to have her first child. The model illustrates many features of contemporary technique.[9] Before we take up the model, let the authors say what they were trying to do:

The interplay between education and fertility has a significant influence on the roles women occupy, when in their life cycle they occupy these roles, and the length of time spent in these roles. . . . This paper explores the theoretical linkages between education and fertility. . . . It is found that the reciprocal relationship between education and age at first birth is dominated by the effect from education to age at first birth with only a trivial effect in the other direction. [Abstract]

No factor has a greater impact on the roles women occupy than maternity. Whether a woman becomes a mother, the age at which she does so, and the timing and number of subsequent births set the conditions under which other roles are assumed. . . . Education is another prime factor conditioning female roles. [p. 431, footnote omitted]

The overall relationship between education and fertility has its roots at some unspecified point in adolescence, or perhaps even earlier. At this point aspirations for educational attainment as a goal in itself and for adult roles that have implications for educational attainment first emerge. The desire for education as a measure of status and ability in academic work may encourage women to select occupational goals that require a high level of educational attainment. Conversely, particular occupational or role aspirations may set standards of education that must be achieved. The obverse is true for those with either low educational or occupational goals. Also, occupational and educational aspirations are affected by a number of prior factors, such as mother's education, father's education, family income, intellectual ability, prior educational experience, race, and number of siblings. [p. 432, citations omitted]

The model used by Rindfuss et al. (1980) is shown in figure 4.3. The diagram corresponds to two linear equations in two unknowns, ED and AGE (variables are defined in table 4.2):

(3)  $ED = a \times AGE + A,$
(4)  $AGE = a' \times ED + A'.$

According to the model, a women chooses her educational level and age at first birth as if by solving these two equations for the two unknowns. The coefficients $a$ and $a'$ are "social constants," to be estimated from the data. The terms $A$ and $A'$ take background factors into account:

(5)  $A = A_0 + b \times DADSOCC + c_1 \times RACE + \ldots + c_7 \times YCIG$
$+ \text{random error drawn from a box,}$

(6)  $A' = A_0' + b' \times FEC + c_1' \times RACE + \ldots + c_7' \times YCIG$
$+ \text{another random error drawn from a box.}$

**FIGURE 4.3**   The model in diagram form (Rindfuss et al. 1980; SGS, 140; CG, p. 216). Variables are defined in table 4.2 below. Explanatory variables (DADSOCC, RACE, etc.) are correlated; error terms are not shown in the diagram.

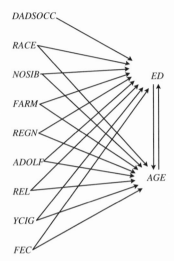

**TABLE 4.2**   Variables in the model (Rindfuss et al. 1980).

*The endogenous variables*

| | |
|---|---|
| ED | Respondent's education |
| | (Years of schooling completed at first marriage) |
| AGE | Respondent's age at first birth |

*The exogenous variables*

| | |
|---|---|
| DADSOCC | Respondent's father's occupation |
| RACE | Race of respondent (Black = 1, other = 0) |
| NOSIB | Respondent's number of siblings |
| FARM | Farm background (coded 1 if respondent grew up on a farm, else 0) |
| REGN | Region where respondent grew up (South = 1, other = 0) |
| ADOLF | Broken family (coded 0 if both parents present at age 14, else 1) |
| REL | Religion (Catholic = 1, other = 0) |
| YCIG | Smoking (coded 1 if respondent smoked before age 16, else coded 0) |
| FEC | Fecundability (coded 1 if respondent had a miscarriage before first birth; else coded 0) |

*Notes:* The data are from a probability sample of 1,766 women 35–44 years of age residing in the continental United States; the sample was restricted to ever-married women with at least one child. DADSOCC was measured on Duncan's scale, combining information on education and income; missing values were imputed at the overall mean. SGS (p. 139) give the wrong definitions for NOSIB and ADOLF.

Again, the parameters $A_0$, $b$, $c_1$, . . . are social constants to be estimated from the data. The random errors are assumed to have mean 0, to be statistically independent from woman to woman, and to be identically distributed. Correlations across equations (5) and (6) are permitted.

Equations (3–6) are not quite regression equations, due to the simultaneity of (3) and (4); fitting by OLS (ordinary least squares) would create "simultaneity bias." Thus, Rindfuss et al. use an estimation procedure called "two-stage least squares."[10] FEC does not come into equation (5), nor DADSOCC into equation (6). Graphically, there is no arrow from DADSOCC to AGE in figure 4.3; likewise, there is no arrow from FEC to ED. These behavioral assumptions are critical to the statistical enterprise. Without them, or some similar assumptions, two-stage least squares could not be used. Technically, the system would not be "identifiable" (section 11.4).

The main empirical finding: The estimated coefficient of AGE in the first equation is not "statistically significant," i.e., the coefficient $a$ in (3) could be zero. The sort of woman who drops out of school to have a child would drop out anyway. If looked at coldly, the argument may seem implausible. A critique can be given along the following lines.

(i) *Statistical assumptions.* Just why are the errors independent and identically distributed across the women? Independence may be reasonable, but heterogeneity is more plausible then homogeneity.

(ii) *The assumption of constant coefficients.* Rindfuss et al. are assuming that the same parameters apply to all women alike, from poor blacks in the cities of the Northeast to rich whites in the suburbs of the West. Why?

(iii) *Omitted variables.* Surely, important variables have been omitted from the model, including two that were identified by Rindfuss et al. themselves—aspirations and ability. Malthus thought that wealth was an important factor. Social class matters and DADSOCC measures only one of its aspects.[11]

(iv) *What about the "no arrow" assumptions, from DADSOCC to AGE and FEC to ED?*

(v) *Are FEC and DADSOCC exogenous?*

(vi) *Are the equations "structural"?*

Questions (iv–vi) will be discussed in the next section, as will the idea of "structural" equations.

## 5.1 A Thought Experiment

A simpler version of the model restricts attention to a more homogenous group of women, where the only relevant background factors are DADSOCC and FEC. To make causal inferences from the data using the model, we need to believe

**FIGURE 4.4**   A simpler version of the model.

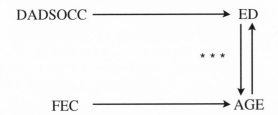

that the arrows are as shown in figure 4.4, that DADSOCC and FEC are exogenous, and that the equations are "structural." The following thought experiment may help to define the last term, and the empirical commitments behind the words.[12]

The *gedanken* experiment involves two groups of women. In both groups, fathers are randomized to jobs, and some of the daughters are chosen at random to have a miscarriage before their first child. (The statistical terminology of randomization is dry; the *gedanken* experimentalist intervenes, for instance, to make the fathers do one job rather than another: professors are caused to work as plumbers, and taxi drivers are installed as hospital anesthetists.)

> *Group I.* Daughters are randomized to the various levels of ED, and AGE is observed as the response. (The *gedanken* experimentalist strikes again, forcing some women to stay in school longer than they wish, while preventing others from continuing their education.)
>
> *Group II.* Daughters are randomized to the various levels of AGE, and ED is observed as the response. (More *gedanken* intervention is needed.)

The statistical model can now be translated. For the women in Group I, AGE should not depend on DADSOCC—the "no arrow" assumption; however, AGE should depend linearly on ED. For the women in Group II, ED should not depend on FEC—the other "no arrow" assumption; however, ED should depend linearly on DADSOCC. Rindfuss et al.'s discovery is that ED would not depend on AGE.

There is one final assumption: the equations and parameters that describe the responses of the women in the experiment must also describe the natural situation. That is what "structural" means. For instance, a woman who freely chooses her educational level and her time to bear children does so by using the same equations as a woman made to give birth at a certain age. In short, with respect to the matters at issue, life in Des Moines proceeds more or less along the same lines as life in the Gulag.

The thought experiment provides the intellectual foundation for the model, by articulating the background assumptions. These assumptions have not been subjected—cannot be subjected—to direct empirical proof. Nor can assumptions be validated by appealing to thought experiments that are almost unthinkable. Do the modelers have some other method in reserve? If the assumptions remain unvalidated, what is the logical status of their implications?

### 5.2 Exogeneity

Identifying the exogenous variables is a major problem. For example, you can obtain results quite different from those of Rindfuss et al., by using variables other than DADSOCC and FEC as "instruments."[13] Rindfuss et al. respond that estimates made by

> instrumental variables . . . require strong theoretical assumptions . . . and can give quite different results when alternative assumptions are made . . . it is usually difficult to argue that behavioral variables are truly exogenous and that they affect only one of the endogenous variables but not the other. (Rindfuss et al. 1984, 981–82).

In short, results can depend quite strongly on assumptions of exogeneity, and there is no good way to justify one set of assumptions rather than another. Also see Bartels (1991), who comments on the impact of exogeneity assumptions and the difficulty of verification.

## 6. Automated Searches for Causality

SGS (Spirtes, Glymour, and Scheines 1993) have computerized algorithms that search for path models. Using the algorithms, SGS claim to make rigorous inferences of causation from association. Their theory is summarized in Glymour (this volume), cited as CG. For present purposes, a "path model" is a recursive system of regression equations, in which the dependent variables from some equations are used as explanatory variables in later equations.[14]

The basic idea in path models is this: putative causes combine with parameters and random errors by multiplication and addition in order to produce their effects. I have discussed such models elsewhere and do not believe they offer much help in deducing causation from association, because there is little evidence to support the basic assumptions (Freedman 1987). To pursue the discussion here, a slightly more explicit definition of the models may be in order.

> *Definition:* A "path model" starts with variables at "level 0," which are exogenous in the minimal sense that they are not explained within the model. Variables "at level 1" are built up as linear combinations of level 0 variables, plus independent random errors. More generally, variables "at level $k$" are built up as linear combinations of variables at previous levels;

again, there are additive, independent random errors. Variables at level 1, level 2, . . . are "endogenous," in the sense that they are explained within the system. The path model may be presented as a "path diagram," like figure 4.1 or figure 4.5 below. Nodes represent variables in the model; if there are arrows from $X$, $Y$, . . . to $Z$, then $X$, $Y$, . . . are explanatory variables in the regression equation for $Z$. Nodes are often called "vertices," and the diagrams are referred to as "graphs" or "causal graphs."[15]

The path model may represent mere association—conditional dependence and independence relations. Or the model may represent causation. I will take that up later. For now, however, either interpretation suffices. Suppose the graph is "sparse"—each equation in the model involves relatively few variables. Suppose too there are no troublesome algebraic identities among the regression coefficients; in SGS terminology, the distribution is "faithful" to its graph (SGS, p. 35; CG p. 208; and see section 11.2 below). You have a sample—many independent realizations of variables $X$, $Y$, $Z$, . . . . You are willing to assume the distribution conforms to a path model, but do not know which model. You do not even know which variables are at level 0, which are at level 1, and so forth.

SGS claim their algorithms are likely to find the underlying path model, or a rather similar model, and in short order. Their most convincing evidence is based on simulation experiments, where the computer generates data from a path model and the SGS algorithms try to infer the model from the data (SGS, pp. 145ff, 152ff, 250ff, 320ff, 332ff); in these experiments, the algorithms do very well. Roughly speaking, the SGS algorithms are variants of "best subsets" regression, the search being over graphs rather than subsets. The data come into the SGS algorithms only through the covariance structure. The rest of the apparatus—the diagrams, the Markov property, faithfulness, etc.—consists of assumptions.

SGS seem to assert that their algorithms determine causality, as a matter of mathematics. Such assertions are not defensible. In the SGS formalism, causation is obtained not by mathematical proof but by mathematical assumption. If you assume that the arrows in the underlying path diagram represent causes, then the arrows found by the algorithms represent causes. If you assume that the underlying arrows represent mere associations, then the arrows found by the algorithms represent associations. Causation has to do with empirical reality, not with mathematical proofs based on axioms. The issue is not one of theorems, but of the connection between theorems and reality.

The SGS algorithms, like many earlier statistical procedures ( factor analysis, LISREL, etc.), proceed by analyzing the correlation matrix of a set of variables. I will call such methods "correlational." Sections 7–10 consider applications of the SGS algorithms to real examples. Sections 11–12 try to explain the key ideas in the SGS formalism, and indicate by mathematical example some of the intrinsic limitations. Before proceeding, however, I discuss the SGS statement of assumptions.

## 6.1 The SGS Statement of Assumptions

SGS discuss the role of assumptions in their theory several times (pp. 53–69, 75–81, 324–25, 351). However, the clearest statement can be found when SGS are trying to discredit the evidence that smoking causes lung cancer:

> effects **** cannot be predicted from **** sample conditional probabilities. [p. 302]

Readers may consult the original for context, to see whether the omitted material affects the meaning. The advantage of the quote is clarity. If the statement is generally applicable, then SGS—like Yule and Pearl before them—have disavowed the ability to infer causation from association.

## 7. The SGS Examples

SGS share my pessimistic views about regression. They claim, however, that their algorithms will succeed where regression has failed:

> In the absence of very strong prior causal knowledge, multiple regression should not be used to select the variables that influence an outcome or criterion variable in data from uncontrolled studies. So far as we can tell, the popular automatic regression search procedures [like stepwise regression] should not be used at all in contexts where causal inferences are at stake. Such contexts require improved versions of algorithms like those described here to select those variables whose influence on an outcome can be reliably estimated by regression. In applications, the power of the specification searches against reasonable alternative explanations of the data is easy to determine by simulation. . . . [p. 257]

At first reading, SGS seems to be filled with real examples showing the successful application of their algorithms. That is an illusion. Many of the examples are based on simulation, and I set those aside.[16] The real examples are mostly to be found on pp. 132–52 and 243–56.[17]

The main examples given in SGS are path models. But these cannot withstand scrutiny—section 5 above, sections 8–9 below. One exception is the stratification model of Blau and Duncan (1964). SGS (pp. 142–45) and CG (pp. 222–23) seem to be quite critical of this model; their current position is almost diametrically opposite to the one in Glymour et al. (1987, 33–39). Like SGS, I do not believe that the Blau-Duncan regressions are a satisfactory causal model. On the other hand, as descriptions of the data, the equations can tell us something important about our society (Freedman 1987, 122, 220). The discussion in SGS adds little to our understanding either of the model or of stratification.

SGS appear to use the health effects of smoking as a running example to illustrate their theory.[18] Again, there is an illusion. The causal diagrams are all hypotheticals, no contact is made with data, and no substantive conclusions are drawn. If the diagrams were proposed as real descriptions of causal mechanisms, they would be open to devastating criticism.

What about the substantive question: does smoking cause lung cancer, heart disease, and many other illnesses? SGS appear not to believe the epidemiological evidence. When they actually get down to arguing their case, they use a rather old-fashioned method—a literature review with arguments in ordinary English (pp. 291–302). Causal models and search algorithms have disappeared.

I approve of the method if not the implementation: the summary is wrong in some places and tendentious in others. However, the review does show the complexity of the issues. To make judgments about causation, you need to consider death certificate data, necropsy data, case control and cohort studies, twin studies, dose response curves, as well as animal experiments and human experiments. The force of the epidemiological evidence—and the SGS critique—depends on the complex interplay among these various studies and data sets.

In the end, SGS and CG do not really make bottom-line judgments on the health effects of smoking, at least so far as I can see. Their principal conclusion is methodological: nobody understood the issues.

> Neither side understood what uncontrolled studies could and could not determine about causal relations and the effects of interventions. The statisticians pretended to an understanding of causality and correlation they did not have; the epidemiologists resorted to informal and often irrelevant criteria, appeals to plausibility, and in the worst case to *ad hominem*. . . . While the statisticians didn't get the connections between causality and probability right, the . . . 'epidemiological criteria for causality' were an intellectual disgrace, and the level of argument . . . was sometimes more worthy of literary critics than scientists. [pp. 301–22]

Part of a sentence in SGS (p. 4) does seem to grant one of the major claims made by the epidemiologists, "smoking does cause lung cancer." But that only complicates the puzzle. If you don't believe the evidence, why accept the claim?

Despite SGS, the epidemiologists did have a good understanding of the issues and made a strong case against smoking. The arguments were imperfect, and some reasonable doubts may remain. But the data, taken all in all, are compelling. The epidemiological literature on smoking is far stronger than anything I have seen in the social sciences. For a survey of the evidence, see Cornfield et al. (1959); this paper is still worth reading. More recent data are reviewed in International Agency for Research on Cancer (1986).

SGS elected not to use their analytical machinery on the smoking data—a remarkable omission. When applied to the examples that SGS actually chose, the algorithms produce one small disaster after another, as will now be seen. In sum, SGS claim to have developed techniques for generating causal models; but they do not have any success stories.

## 8. Using the SGS Search Procedure

The SGS search procedures are embodied in a computer program called TETRAD. Version 2.1 of this program was kindly provided by Richard Scheines and Peter Spirtes. The BUILD module is the part of TETRAD used to discover path models with no latent variables. I ran BUILD on two examples—Rindfuss et al. and AFQT (to be discussed in section 9).

### 8.1 Rindfuss et al.

To explain AGE (age at first birth) in the Rindfuss et al. example, the SGS algorithms select the variables shown in table 4.3. Regression estimates for the coefficients, based on summary data in SGS, are reported in the first three columns of the table. The coefficients for ADOLF (the indicator for women from broken homes) and YCIG (an indicator for smoking by age 16) have positive signs. That is paradoxical: women from broken homes and women who smoke should be having children earlier, not later.[19] The signs should be negative, not positive. SGS do not comment on this issue.

Rindfuss et al. (1980) give standard deviations and correlations for their data; SGS (p. 139) used these statistics to compute a covariance matrix, but reversed some of the signs. The last three columns of table 4.3 report regression estimates computed from the correct covariances: the problem with YCIG disappears, but the sign for ADOLF stays positive. Anyone can make a mistake entering data; ignoring paradoxical signs in a causal model is quite another matter.

SGS report only a graphical version of their model. They say,

> Given the prior information that ED and AGE are not causes of the other variables, the PC algorithm (using the .05 significance level for tests) directly finds the model [in figure 4.5(a)] where connections among the regressors are not pictured. (SGS, p. 139; CG, pp. 215–16)

However, connections among regressors can be of interest. Although TETRAD is supposed to discover the causal ordering of explanatory variables, it produces the very strange model shown in figure 4.5(b). For example, the model says that race and religion cause region of residence. Comments on the sociology may be out of place, but consider the statistics. The equation is

(7)  REGN = $a + b \times$ RACE $+ c \times$ REL $+ \varepsilon$.

REGN is a dummy variable, coded 1 for respondents who grew up in the South, 0 for others; RACE is 1 for black respondents and 0 for others; REL is 1 for Catholics, 0 for others; $\varepsilon$ is normally distributed. In consequence, this equation forces impossible values on REGN: the left-hand side is 0 or 1, the right-hand

**TABLE 4.3**   The SGS model for age at first birth, computed using the SGS covariance matrix or the Rindfuss et al. covariance matrix. (Intercepts are not reported; OLS estimates.)

|  | *SGS covariance* | | | *Rindfuss et al. covariance* | | |
|---|---|---|---|---|---|---|
|  | $R^2 = 0.27$ | | | $R^2 = 0.24$ | | |
|  | *Estimate* | *SE* | *t* | *Estimate* | *SE* | *t* |
| RACE | −1.66 | .30 | −5.50 | −1.66 | .30 | −5.46 |
| REGN | −0.56 | .19 | −3.01 | −0.63 | .19 | −3.35 |
| ADOLF | 1.89 | .22 | 8.60 | 2.01 | .22 | 8.98 |
| YCIG | 2.14 | .25 | 8.63 | −0.89 | .25 | −3.53 |
| FEC | 2.72 | .28 | 9.70 | 2.77 | .28 | 9.27 |
| ED | 0.67 | .04 | 18.00 | 0.60 | .04 | 15.72 |

*Notes:* (i) The first column in table 4.3 shows parameter estimates. The second shows standard errors, or SEs, which indicate the likely size of the differences between the estimates and the true parameter values. The *t*-statistics in the third column are the ratios of estimates to SEs. Generally, a *t*-statistic above 2 or 3 in absolute value indicates that the corresponding parameter is unlikely to be truly 0. For details, see the appendix.

(ii) The parameters are features of the model, and the SEs are computed using the model. If you do not believe in the existence of the parameters apart from the data, or do not accept the statistical assumptions in the model, the SEs and *t*-statistics are likely to be meaningless. In any case, performing multiple tests—as in a search algorithm—complicates the interpretation of the *t*-statistics (Freedman 1983; CG, pp. 245–46).

(iii) $R^2$ is generally interpretable as a descriptive statistic, whether or not the assumptions of the model hold true. An $R^2$ of 0.27 indicates that about 27% of the variance in AGE has been explained; that isn't much, and models in the social science literature often have even less explanatory power. For a critical discussion of $R^2$, see Freedman and Lane (1981, 78–81).

(iv) According to current epidemiological opinion, smoking does have some biological effect, delaying conception by several weeks. However, the women who choose to smoke are different from the non-smokers, and have their first child almost a year earlier. This effect remains even after controlling for the measured background factors in the regression: the coefficient of YCIG is –0.89 years.

**FIGURE 4.5**   The left-hand panel shows the model reported by SGS. The right-hand panel also shows connections among the regressors, as determined by the SGS search program TETRAD.

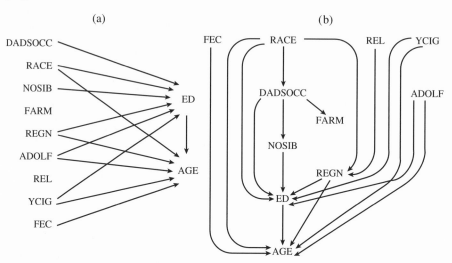

*Notes:* BUILD indicates that latent variables are present, i.e., errors are correlated across equations. BUILD asks whether it should assume "causal sufficiency"; without this assumption (SGS, p. 45), the program output is uninformative. Therefore, I told BUILD to make the assumption; I believe that is what SGS did for the Rindfuss example. Also see Spirtes, Scheines, Glymour, and Meek (1993, 13–15). I told BUILD that ED and AGE could not cause the remaining variables, following (SGS, p. 139). However, SGS actually made the stronger assumption that (i) FEC, ED and AGE could not cause YCIG, and (ii) FEC, ED, AGE and YCIG could not cause the remaining variables. With the assumption of causal sufficiency, BUILD seems to use the PC algorithm; without the assumption, the FCI algorithm comes into play. Much of this information comes from Richard Scheines (personal communication). Data are from Rindfuss et al. (1980), not SGS; with the SGS covariance matrix, FARM causes REGN and YCIG causes ADOLF.

side varies from $-\infty$ to $+\infty$. Now $R^2$ is only 0.16, so $\varepsilon$ contributes most of the variance: equation (7) can hardly be defended as an approximation. Having dummy variables in the middle of path diagrams is a blunder. (FARM creates a similar problem; so does NOSIB, although less extreme.) In short, the SGS algorithms have produced a model that fails the most basic test—internal consistency.

## 9. The Armed Forces Qualification Test

SGS discuss an example based on the Armed Forces Qualification Test (AFQT).[20] The AFQT is a linear combination with fixed weights of scores on certain subtests. Some of these subtests, as well as subtests that are not part of the AFQT, are listed in table 4.4. The problem is to decide which subtests go into the AFQT and which do not.

The problem may be stated more algebraically, as follows:

(8)   AFQT score $= a_1 \times$ NO $+ a_2 \times$ WK $+ \ldots + a_7 \times$ GS
$$+ b_1 \times \text{UN}_1 + \ldots + b_n \times \text{UN}_n$$

where $\text{UN}_1, \ldots, \text{UN}_n$ are unobservable. Some of the $a$'s are zero, and the challenge is to figure out which ones.

We have data on 6,224 subjects, summarized as a covariance matrix. SGS say:

> a linear multiple regression of AFQT on the other seven variables gives significant regression coefficients to all seven and thus fails to distinguish the tests that are in fact linear components of AFQT. . . . Given the prior information that AFQT is not a cause of any of the other variables, the PC algorithm in TETRAD II correctly picks out {AR, NO, WK} as the only . . . variables that can be components of AFQT. . . . (SGS, pp. 243–44, also see CG, p. 217)

To test the claims about regression, I ran AFQT on all the observable subtests. As table 4.5 shows, EI and MC are related to AFQT only at the chance level. Moreover, MK and GS have negative coefficients, but psychometric practice frowns on subtests that are negatively related to overall test scores. It is a natural conjecture that NO, WK, and AR go into AFQT while the other four subtests do not. Contrary to the claims of SGS, the AFQT can be handled by ordinary statistical methods.

The AFQT problem is in some ways quite easy. By definition, the "causes" or subtests combine linearly with the parameters to produce the AFQT as an "effect." Joint normality of test scores seems to follow from the procedures used to construct the tests: consequently, scores on any one subtest can be presented

**TABLE 4.4**   Subtests analyzed by SGS. Some go into the AFQT and some do not.

| | |
|---|---|
| 1. Numerical Operations | NO |
| 2. Word Knowledge | WK |
| 3. Arithmetical Reasoning | AR |
| 4. Mathematical Knowledge | MK |
| 5. Electronics Information | EI |
| 6. Mechanical Comprehension | MC |
| 7. General Science | GS |

as a linear combination of other subtest scores, with additive random errors. Thus, critical issues in most empirical studies have disappeared.[21]

## 9.1 TETRAD

According to SGS, given the prior information that AFQT does not cause the other variables, TETRAD correctly picks out AR, NO, and WK as the components of the AFQT.[22] Without that prior information, however, TETRAD declares AFQT to be the *cause* of these subtests, rather than the *effect*. With the prior information, TETRAD produces the strange results shown in figure 4.6.[23] Now, for instance, the subtest NO may "cause" the overall test score AFQT, but it can hardly cause the other subtests AR or MK. Furthermore, there is a cycle in the figure:

$$MC \to AR \to WK \to GS \to MC.$$

In principle, such cycles were excluded by prior assumption, as well they might be. Subtests should not cause themselves, even indirectly. To sum up:

(i)  ordinary least squares techniques pick out NO, AR, and WK for the probable components of the AFQT, just as TETRAD does;

(ii)  TETRAD produces the curious model in figure 4.6.

## 10. Foreign Investment and Political Oppression

As noted in section 7, SGS are quite pessimistic about typical social science applications of regression. While I agree with the bottom line, their specific objections seem misplaced. One example is enough to make the point. Timberlake

**TABLE 4.5**  Regression of AFQT on all the observable subtests.

|     | Estimate | SE | t |
| --- | --- | --- | --- |
| NO | 0.24 | .022 | 10.8 |
| WK | 1.17 | .029 | 40.5 |
| AR | 1.03 | .028 | 36.4 |
| MK | −0.24 | .028 | −8.7 |
| EI | −0.03 | .024 | −1.3 |
| MC | 0.03 | .024 | 1.3 |
| GS | −0.13 | .029 | −4.6 |

*Note:* Variables were centered at their means.

**FIGURE 4.6** AFQT and its subtests arranged in causal order by the SGS search program TETRAD.

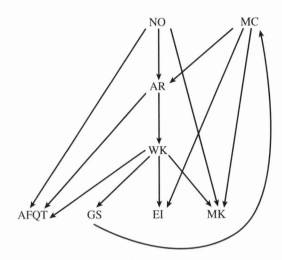

*Notes:* I believe SGS used BUILD, with the assumption of causal sufficiency, on pp. 243–44 for the AFQT example. Also see Spirtes, Scheines, Glymour, and Meek (1993, 8–11). The program indicates there are latent variables, i.e., correlations in the errors.

and Williams (1984) offer a regression model to explain political exclusion (PO) in terms of foreign investment (FI), energy development (EN), and civil liberties (CV). High values of PO correspond to authoritarian regimes that exclude most citizens from political participation; high values of CV indicate few civil liberties. Data come from seventy-two countries. Correlations among the Timberlake-Williams variables are shown in table 4.6.

**TABLE 4.6** The Timberlake and Williams data. Correlation matrix for political oppression (PO), foreign investment (FI), energy development (EN), and civil liberties (CV).

|      | *PO*   | *FI*   | *EN*   | *CV*   |
|------|--------|--------|--------|--------|
| PO   | 1.000  | −.175  | −.480  | .868   |
| FI   | −.175  | 1.000  | .330   | −.391  |
| EN   | −.480  | .330   | 1.000  | −.430  |
| CV   | .868   | −.391  | −.430  | 1.000  |

*Source:* SGS, p. 249.

The equation proposed by Timberlake and Williams is

(9)   $PO = a + b \times FI + c \times EN + d \times CV + error.$

Empirical results are shown in the first three columns of table 4.7. The estimated coefficient of FI is significantly positive and is interpreted as measuring the effect of foreign investment on political exclusion (Timberlake and Williams 1984, 143).

SGS discuss this example (pp. 248–50), suggesting that Timberlake and Williams have confused cause and effect. The alternative causal sequence is not spelled out. Presumably, the idea is that dictators 'cause' foreign investment in the sense that investors think dictatorial regimes offer greater stability, etc. The main step in the SGS statistical argument comes down to this: the correlation of −.175 between political exclusion and foreign investment is at the chance level. The calculation rides on two assumptions: (i) the seventy-two countries in the data set are a random sample from some much larger set of countries, and (ii) the variables follow a multivariate normal distribution. These time-honored but madcap assumptions are not stated explicitly by SGS, let alone justified. (Of course, the assumptions behind the statistics in Timberlake-Williams might seem equally antic.)

However, for the sake of argument, let us grant SGS their assumptions. On that basis, the standard error for the correlation in question is about $1/\sqrt{72} \approx .12$. I change the suspect correlation coefficient from its observed value of −.175 to

**TABLE 4.7**   The Timberlake and Williams model. Political exclusion (PO) is regressed on foreign investment (FI), energy development (EN), and civil liberties (CV). The first three columns show results for the observed correlation matrix (Table 4.6). The last three columns show what happens when $r(PO,FI)$ is set to 0.

|    | $R^2 = .81$ | | | $R^2 = .93$ | | |
|----|----------|------|------|----------|------|-----|
|    | Estimate | SE | t | Estimate | SE | t |
| FI | .23 | .059 | 3.9 | .44 | .036 | 12 |
| EN | −.18 | .060 | −2.9 | −.22 | .037 | −6 |
| CV | .88 | .061 | 14.4 | .95 | .038 | 25 |

*Note:* The coefficients reported by SGS on p. 249 are not standardized and therefore do not match the correlation matrix. Coefficients in table 4.7 are standardized, that is, computed from variables standardized to have mean 0 and variance 1.

the new value of 0, a difference of about 1.5 SEs. I then recompute the model (last three columns in table 4.7). The results are even better for Timberlake and Williams: the estimated coefficients are bigger and more significant; the signs stay the same; and $R^2$ moves closer to 1.[24]

I will not defend the model any further. Measurement problems are extreme, and the list of omitted variables very long. SGS may well be right, that cause and effect have been confused. But the demonstration is peculiar. The correlation matrix cannot show that FI, EN, and CV cause PO—the fatal flaw in the Timberlake-Williams model. (Of course, Timberlake and Williams are not alone in this respect.) Nor can the matrix show that FI, EN, and CV do not cause PO—the corresponding flaw in SGS. Indeed, it is trivial to construct four variables labelled FI, EN, CV, and PO, such that FI, EN, and CV do cause PO; but sample correlation matrices will look rather like the one in table 4.6. This only sharpens the basic question. What do any of these calculations tell us about the world outside the computer?

## 11. Some Mathematical Issues

Sections 11 and 12 address by mathematical example two questions:

(i) To what extent can correlational methods recover an underlying path diagram?

(ii) When can the arrows in the diagram be interpreted as indicating causation, rather than conditional independence and dependence?

The examples will indicate how SGS use the "faithfulness" assumption to help them answer such questions. Issues of identifiability and consistency will be discussed, and methodological contributions in SGS will be delineated. Sections 11 and 12 are more technical than previous material; readers can skip to section 13 without losing the thread of the argument.

The focus is on linear models. Suppose you have a covariance matrix that describes certain variables. Assume these variables are jointly normal, with mean 0; that avoids all questions of linearity, etc., and all problems created by having only finite amounts of data. However, the statistical procedures I am considering—like the SGS algorithms—will operate on that covariance matrix, and on nothing else. Such procedures may be called "correlational."

Path models were defined in section 6. Briefly, you start with variables at level 0; variables at level $k$ are linear combinations of variables at lower levels, plus independent random errors. In a path diagram, nodes represent variables. There is an arrow from $X$ to $Y$ if $X$ is used as an explanatory variable in the equation for $Y$.

Exogeneity is a critical concept. As indicated before, the term is used in at least three senses. The weakest definition is purely mechanical: exogenous variables are not explained within the model, but are supplied to the model. Variables at level 0 in a path model are exogenous in this minimal sense. A more restrictive definition: exogenous variables are statistically independent of the error terms in the equations. The third idea is the one that is relevant to causal inference: $X$ is exogenous if selecting subjects with $X = x$ gives the same results as intervening to set $X = x$.

There are tests for exogeneity in the literature, as well as model specification tests. However, these have limited relevance to causal inference. For example, Hausman (1978) assumes that certain variables are known *a priori* to be exogenous, and then tests whether other variables are exogenous; he interprets exogeneity as orthogonality to disturbance terms. He also has a test that detects correlation between errors from equations in a path model. White (1980 and 1982) focuses on similar issues—for instance, testing whether the variables have a jointly normal distribution.

Another reference in the econometric literature is Engle, Hendry, and Richard (1983). These authors distinguish several kinds of exogeneity; "strict" exogeneity means independence of variables and error terms, but only "super" exogeneity permits estimating the effects of interventions. Examples are given to illustrate the definitions (Engle, Hendry, and Richard 1983, 287–94). There is further discussion in Leamer (1985).

## 11.1  The Basic Statistical Problem

Suppose you have $n$ random variables with a jointly normal distribution; all the variables have mean 0, and you know the covariance matrix, which is positive definite. You wish to present this covariance matrix as a path model. In a sense, nothing is easier. Simply order the variables, arbitrarily, as $X_1, X_2, \ldots, X_n$. By successively applying regression, we can find coefficients $a_{ij}$ and error terms $\varepsilon_i$, such that $X_1, \varepsilon_2, \ldots, \varepsilon_n$ are all independent with mean 0, and equation (10) holds.

$$
\begin{aligned}
X_2 &= a_{21}X_1 + \varepsilon_2 \\
X_3 &= a_{31}X_1 + a_{32}X_2 + \varepsilon_3 \\
&\quad\bullet \\
(10) \qquad &\quad\bullet \\
&\quad\bullet \\
X_n &= a_{n1}X_1 + \ldots + a_{n,n-1}X_{n-1} + \varepsilon_n
\end{aligned}
$$

Then $X_1$ is presented as exogenous and the "cause" of $X_2$; next, $X_1$ and $X_2$ "cause" $X_3$; and so forth. In short, there are many ways to present a covariance matrix as a path diagram; few if any will be relevant for causal inference.[25]

**FIGURE 4.7** If two path diagrams have the same covariance matrix, correlational methods cannot tell them apart; the faithfulness assumption is made to rule out such problems. The lower case letters on the arrows denote "path coefficients," that is, standardized regression coefficients.

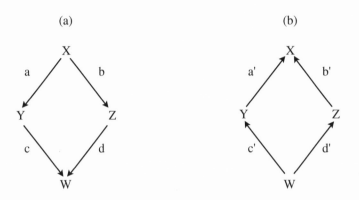

## 11.2 The Faithfulness Assumption

How can you single out one path diagram from the many that correspond to a given covariance matrix? At this point, SGS seem to use the "faithfulness" assumption; this assumption is also used to handle confounding, as discussed in section 12.1 below. Basically, a covariance matrix is faithful to a diagram provided conditional dependencies and independencies are determined by the presence or absence of arrows in the diagram, rather than specific numerical values of parameters.

By way of example, figure 4.7 shows two path diagrams. On the left, $X$ causes $W$ through the intervening variables $Y$ and $Z$; on the right, the flow of causality is reversed.[26] The lower case letters on the arrows stand for "path coefficients," that is, standardized regression coefficients. How could SGS distinguish between the two theories in the figure? Their idea seems to be as follows:

> In the left-hand diagram, $Y$ and $Z$ are conditionally independent given $X$; on the right, however, $Y$ and $Z$ are conditionally dependent given $X$.

Another contrast:

> In the left-hand diagram, $Y$ and $Z$ are conditionally dependent given $W$; on the right, however, $Y$ and $Z$ are conditionally independent given $W$.

Therefore, the pattern of conditional dependence and independence identifies the diagram. (In both diagrams, $X$ and $W$ are conditionally independent given $Y$ and $Z$.)

This idea works for many path diagrams, but fails for others. Indeed, the path coefficients can be chosen so the pattern of conditional dependence and independence is the same in the two diagrams. Even worse, both diagrams can give rise to the same covariance matrix—so correlational methods cannot tell which is right. SGS make the "faithfulness assumption" in order to rule out such indeterminacies. (The workings of the assumption will be explained below.) However, that only moves the difficulty to another place. Faithfulness is hardly an empirical fact; it is an assumption about unobservables, made to rule out situations that cannot be handled by correlational methods. The SGS analytical program can now be stated rather simply. If the arrows in a path diagram represent causation not association, and if the path diagram can be estimated from data, then SGS can indeed infer causation from association.

The balance of section 11.2 provides technical backup; readers can skip to section 11.3. The left-hand panel in figure 4.7 is described by

(11)  $Y = aX + \delta_1, \ Z = bX + \delta_2, \ W = cY + dZ + \delta_3.$

In this equation, $X, \delta_1, \delta_2, \delta_3$ are independent and normal, with mean 0; $X, Y, Z,$ $W$ all have variance 1. The covariance matrix of $X, Y, Z, W$ can be computed from the four parameters $a, b, c, d$ as shown in (12).

|      |     | $X$ | $Y$ | $Z$ | $W$ |
|------|-----|-----|-----|-----|-----|
|      | $X$ | 1   | $a$ | $b$ | $ac + bd$ |
| (12) | $Y$ | $a$ | 1   | $ab$ | $c + abd$ |
|      | $Z$ | $b$ | $ab$ | 1   | $d + abc$ |
|      | $W$ | $ac + bd$ | $c + abd$ | $d + abc$ | 1 |

It is a little theorem, which follows by a tedious calculation from (48) in the appendix below, that

(13)  $\mathrm{cov}(X, W \,|\, Y, Z) = 0.$

This is an example of a conditional independence relation forced by a graph; equation (13) holds whatever the path coefficients in figure 4.7 may be.

The diagram on the left in figure 4.7 is reversible provided

(14)   $\operatorname{cov}(Y, Z | W) = 0.$

By (48) below, equation (14) is equivalent to

(15)   $\operatorname{cov}(Y, Z) = \operatorname{cov}(Y, W) \times \operatorname{cov}(Z, W).$

By (12), this means

(16)   $ab = (c + abd)(d + abc).$

Rearranging (16) gives the quadratic equation

(17)   $cd(ab)^2 - (1 - c^2 - d^2)\, ab + cd = 0.$

One solution to (17) is

$$(18) \quad ab = \frac{1 - c^2 - d^2 - \sqrt{(1 - c^2 - d^2)^2 - 4c^2d^2}}{2cd}.$$

I chose *a, c, d* more or less at random, getting .1925, .2873 and .1245, respectively.[27] I computed *b* from (18), getting .2063. This choice forces the conditional independence relation (14), and violates the faithfulness assumption: conditional independence comes from the parameter values, not the presence or absence of arrows.

Given the values for the four parameters *a, b, c, d*, the covariance matrix (12) can be evaluated as

$$(19) \quad \begin{vmatrix} 1.0000 & 0.1925 & 0.2063 & 0.0810 \\ 0.1925 & 1.0000 & 0.0397 & 0.2922 \\ 0.2063 & 0.0397 & 1.0000 & 0.1359 \\ 0.0810 & 0.2922 & 0.1359 & 1.0000 \end{vmatrix}$$

The path coefficients in the right-hand panel of figure 4.7 are easily computed from (19):

the path coefficient from *W* to *Y* is $c' = \operatorname{cov}(Y,W) = .2922$;
the path coefficient from *W* to *Z* is $d' = \operatorname{cov}(Z,W) = .1359$;
the path coefficients from *Y* and *Z* to *X* are obtained by multiple regression, as $a' = .1846$ and $b' = .1990$.

**FIGURE 4.8**  Graphs (a) and (b) have the same covariance matrix. Both are complete: there is an arrow from every variable. The numbers on the arrows are path coefficients, that is, standardized regression coefficients.

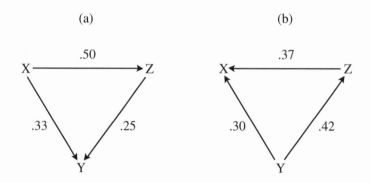

With these choices, faithfulness does not hold, and (19) can be represented by either diagram in figure 4.7. (For details on multiple regression, see the appendix.) In effect, the faithfulness assumption precludes certain algebraic identities among the parameters, like (16). Since parameters are not observable, the faithfulness assumption is not subject to direct empirical tests based on finite amounts of data.

## 11.3   Complete Graphs

Even if the covariance matrix is faithful to a graph, however, problems of indeterminacy remain—particularly if the graph is "complete" in the sense that every pair of vertices is joined by an arrow. Figure 4.8 illustrates this indeterminacy. The same covariance matrix (20) for the variables $X, Y, Z$ is represented either by the diagram in panel (a) or the one in panel (b), where the flow of "causality" is reversed.

|  |  | $X$ | $Y$ | $Z$ |
|---|---|---|---|---|
| (20) | $X$ | 1.00 | .46 | .50 |
|  | $Y$ | .46 | 1.00 | .42 |
|  | $Z$ | .50 | .42 | 1.00 |

For a second example of indeterminacy when the graph is complete, consider four variables $X, Y, Z, W$ with covariance matrix $\Sigma$ given by

$$(21) \quad \Sigma = \begin{pmatrix} 1 & 3/4 & 3/4 & 3/4 \\ 3/4 & 1 & 3/4 & 3/4 \\ 3/4 & 3/4 & 1 & 3/4 \\ 3/4 & 3/4 & 3/4 & 1 \end{pmatrix}$$

Figure 4.9 shows two complete path diagrams, both of which are compatible with the given covariance matrix. In the left-hand panel, $X$ is exogenous, and "causes" $Y$; then $X$ and $Y$ "cause" $Z$; finally $X$, $Y$, $Z$ "cause" $W$. In panel (b), the flow of "causality" is reversed. The equations corresponding to the left-hand panel are given as (22); panel (b) is described in (23).

**FIGURE 4.9**   Two complete path diagrams, and a factor analysis model, all having the same covariance matrix.

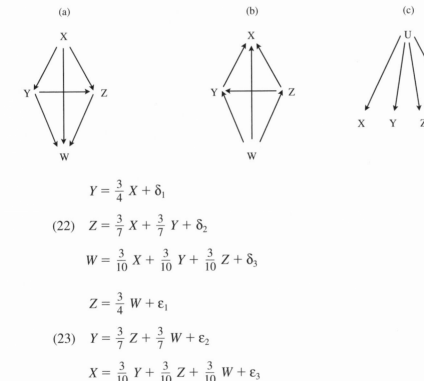

$$Y = \frac{3}{4} X + \delta_1$$

$$(22) \quad Z = \frac{3}{7} X + \frac{3}{7} Y + \delta_2$$

$$W = \frac{3}{10} X + \frac{3}{10} Y + \frac{3}{10} Z + \delta_3$$

$$Z = \frac{3}{4} W + \varepsilon_1$$

$$(23) \quad Y = \frac{3}{7} Z + \frac{3}{7} W + \varepsilon_2$$

$$X = \frac{3}{10} Y + \frac{3}{10} Z + \frac{3}{10} W + \varepsilon_3$$

The covariance matrix $\Sigma$ is also compatible with the factor analysis model (24), where the unobservable exogenous variable $U$ causes all four observables (right-hand panel of figure 4.9).

(24)　$X = U + \zeta_1,\ Y = U + \zeta_2,\ Z = U + \zeta_3,\ W = U + \zeta_4$

In each system of equations (22–23–24), the error terms are assumed to be independent and normally distributed with mean 0; error terms are independent of the exogenous variable. As a technical matter, the covariance matrix (20) is faithfully represented by both graphs in figure 4.8. Likewise, the covariance matrix (21) is faithful to figure 4.9(a) and to 4.9(b). Proofs may be based on (48) below.

To sum up, if a covariance matrix is faithful to a complete graph (with all pairs of vertices joined by arrows), it is faithful to many such graphs. Then correlational methods cannot tell the causes from the effects. SGS techniques work best when the graph is sparse, that is, relatively few pairs of vertices are joined by arrows (section 6).

### 11.4   Identifiability and Consistency

The focus continues to be on linear models. In statistical terminology, models are "identifiable" when they make different predictions about observables. For example, suppose you have two models for your data. If, for all data sets,

P(data | model 1) = P(data | model 2),

there is an obvious problem—the data cannot distinguish between the models. If a path model is complete, or the faithfulness assumption is not imposed, then the graph underlying a covariance matrix is not identifiable; that is the message of sections 11.1–3. By way of illustration, the models in figure 4.7 are identifiable only if faithfulness holds.

However, even if we assume that a covariance matrix is faithful to a graph that is not complete, there may be several such graphs (SGS, p. 89). For example, the following three graphs can generate the same covariance matrix:

$$X \to Y \to Z \qquad X \leftarrow Y \to Z \qquad X \leftarrow Y \leftarrow Z$$

Thus, SGS do not seem to have succeeded in defining a class of graphs and covariance matrices for which identifiability holds (SGS, p. 194).

In statistical terminology, estimators are "consistent" provided that, as the sample gets larger and larger, these estimators come closer and closer to the population parameters. If the parameters are not identifiable, however, consistency is problematic. SGS seem to claim that their algorithms will find all the path diagrams compatible with a given covariance matrix. However, the theorems suggest that the algorithms will at best find one such graph. SGS also seem to claim that their algorithms are consistent. However, without an identifiability theory for linear models, they cannot really be talking about consistency.

Statisticians do have the weaker notion of "Fisher consistency," named after R. A. Fisher: when applied to data for the whole population, an estimator should reproduce the population parameters exactly. Theorems like 5.1 in SGS (p. 405) seem to demonstrate the analog of Fisher consistency, rather than anything stronger. Such theorems show that, given the population covariance matrix, the algorithms will produce one graph consistent with that matrix.

## 11.5 Methodological Contributions

There is a connection between the theory of "Directed Acyclic Graphs" (DAGs) and conditional independence of random variables. (See Darroch et al. 1980; Kiiveri and Speed 1982; Speed and Kiiveri 1986; Pearl 1986; Pearl 1988; Verma and Pearl 1990a; Geiger 1990; Pearl and Verma 1991.) Much of this work is reviewed in SGS and CG. However, the mathematics of nonlinear causal diagrams seems to be irrelevant to the big question: how do we infer causation from association?

Most of the applications in SGS are linear, i.e., based on path models. The "nonlinear causal diagrams" turn out to be multinomial models for categorical data; examples are on pp. 147–51. The issues about causation are quite similar to those for linear models, although the technical details are different. The real applications in CG all seem to involve linear models.

This section will focus on path models. To describe the novelty in the SGS approach to estimation, suppose you have data from a path model and wish to estimate the model. Consider two cases:

*Case I.* You know the classification of variables as to level: that is, you know which variables are at level 0, and which are at level 1, and so forth.

*Case II.* You do not know the classification of variables as to level.

In Case I, SGS have little to tell us about estimation (as to confounding, see section 12.1). Some of their algorithms seem to be equivalent to regression, others may be less efficient. In Case II, SGS try to estimate the classification of variables as well as the path coefficients. That is the methodological contribution. To estimate the classification, SGS must impose the faithfulness assumption (section 11.2). It is disappointing that SGS do not pin down the sense in which their algorithms are successful (section 11.4).

## 12. More Examples, and Some Theory

Section 12.1 explains how the faithfulness assumption and conditional independence are supposed to eliminate confounding. Section 12.2 discusses omitted variables. Sections 12.3–5 revisit two examples from a more mathematical perspective; the idea is to show the limits of correlational methods.

### 12.1   Faithfulness, Conditional Independence, and Confounding

The problems created by unobservable variables are well known. As indicated above, SGS handle such problems by imposing the faithfulness assumption. More specifically, the assumption is used to rule out confounding. If confounding can be eliminated, the goal is in sight—association may soon be converted into causation. This section, which is based on work by James Robins (personal communication), examines the logic in more detail. Also see Pearl and Verma (1991).

With some models, exact conditional independence forces a choice:

- either there is no confounding by unmeasured common causes,
- or the faithfulness assumption is violated.

Near-independence is not good enough; associations may then be entirely spurious. Thus, causal inferences made by the SGS technique need exact conditional independence as well as the faithfulness assumption.

This use of the faithfulness assumption has some theoretical interest. However, in order to base empirical work on such mathematical ideas, it would seem necessary to resolve the following questions, which SGS have not addressed:

- Can the basic models be validated?
- Can exact conditional independence be demonstrated?
- Given exact independence, why is exact cancellation of confounded effects overwhelmingly less likely than the total absence of such effects?

As a practical matter, exact independence seems quite unusual. However, the theory is worth understanding, and an example will make the position clearer. Figure 4.10 shows a relatively simple diagram where faithfulness and conditional independence would eliminate confounding. The arrows denote causation, not mere association. Variables $X, Y, Z$ are observable; $U$ is unobservable. Such unobservables are also called "confounders" or "unmeasured common causes." The joint distribution is normal, and variables are standardized to have mean 0 and variance 1.

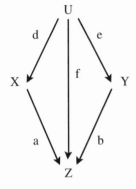

**FIGURE 4.10** The faithfulness assumption, conditional independence, and confounding. Variables $X$, $Y$, $Z$ are observable; $U$ is unobservable. Arrows represent causation, not just association. The lower-case letters on the arrows denote path coefficients. If a path coefficient vanishes, the corresponding arrow must be deleted.

|       |       | $U$            | $X$             | $Y$             | $Z$ |
|-------|-------|----------------|-----------------|-----------------|-----|
|       | $U$   | 1              |                 |                 |     |
| (25)  | $X$   | $d$            | 1               |                 |     |
|       | $Y$   | $e$            | $de$            | 1               |     |
|       | $Z$   | $f + ad + be$  | $a + bde + fd$  | $b + ade + fe$  | 1   |

The covariance matrix for all four variables is shown in (25).[28] Of course, only the covariance matrix (26) of the observables $(X, Y, Z)$ can be estimated from the data. In particular, $de$ is determined from the observables, as $\mathrm{cov}(X, Y)$.

|       |       | $X$            | $Y$             | $Z$ |
|-------|-------|----------------|-----------------|-----|
| (26)  | $X$   | 1              |                 |     |
|       | $Y$   | $de$           | 1               |     |
|       | $Z$   | $a + bde + fd$ | $b + ade + fe$  | 1   |

It may help to review the idea of faithfulness, in the context of our example. Faithfulness is an assumption about unobservables; more specifically, it is a constraint on the relationship between the full covariance matrix (25) and the graph in figure 4.10. The assumption amounts to this: independence relationships (conditional and unconditional) are determined by the presence or absence of arrows in the diagram, not specific parameter values.

In particular, if the covariance matrix (25) is faithful to the diagram in figure 4.10, you cannot set any of the path coefficients to 0, except by deleting the corresponding arrow. An arrow from $X$ to $Z$, say, entails that $X$ has some causal effect on $Z$, no matter how small that effect may turn out to be.

I return to more conventional issues. In our example, the parameter of interest is $b$, the causal effect of $Y$ on $Z$. Due to the unmeasured confounder $U$, a regression of $Z$ on $X$ and $Y$ produces a biased estimate of $b$. By a slightly tedious calculation, the coefficient of $Y$ in the regression equation is

(27)  $b + fe(1 - d^2)/(1 - d^2e^2)$.

(For details on multiple regression, see the appendix.) The bias in the regression estimate is the second term in (27). A slightly different perspective: $\text{cov}(Y, Z)$ in (26) measures the total association between $Y$ and $Z$. Part of this association is real: $b$ measures the causal effect of $Y$ on $Z$. Alas, part of the association is spurious: $ade + fe$ represents the effects of the confounder $U$.

The goal is to separate the real part of the association from the spurious part. The familiar obstacle: we have only (26), not (25). And (26) does not suffice to separate $b + ade + fe$ into its components. But, SGS might say, suppose that $X$ and $Z$ are conditionally independent given $Y$:

(28)  $\text{cov}(X, Z \mid Y) = 0$.

By (48) below, this means

(29)  $\text{cov}(X, Z) = \text{cov}(X, Y) \times \text{cov}(Y, Z)$.

A bit of algebra based on (25) shows that (29) is equivalent to

(30)  $a(1 - d^2e^2) + df = de^2f$.

Although $de$ is known and $0 < de < 1$, there are many possible ways to solve equation (30). At this point, SGS would invoke the faithfulness assumption, concluding that

(31)  $a = 0$ and $f = 0$.

The implication: we have to remove the arrow from $X$ to $Z$, as well as the arrow from $U$ to $Z$.

Confounding has now been eliminated. On this basis, $\text{cov}(Y, Z) = b$; the whole of the association is real, and regression produces an unbiased estimate for the causal effect of $Y$ on $Z$. At last, association has been converted into causation. Of course, quite a lot of causality was built into figure 4.10 from the beginning—by assumption.

Those were the implications of exact conditional independence. On the other hand, suppose we have approximate conditional independence: for instance $\text{cov}(X,Z \mid Y) = .00001$. Now the faithfulness assumption has no force. Given the covariances in (26), we can match them by suitable choice of the other parameters, even if $a = b = 0$.[29]

With approximate conditional independence, observed associations can be entirely spurious. Thus, even in the realm of mathematics, faithfulness and conditional independence preclude confounding only when the independence is exact. To make the contrast sharper, let us assume faithfulness.

- If $\text{cov}(X, Z \mid Y) = 0$, then the association between $Y$ and $Z$ is purely causal; the effects of the unmeasured common cause $U$ do not confound the relationship between $Y$ and $Z$.
- If $\text{cov}(X, Z \mid Y) = .00001$, then confounding by unmeasured common causes may account for all of the observed association between $Y$ and $Z$.

Apparently, converting association into causation is still quite a dicey affair.

A similar problem must be considered when estimating path models from data (section 11). Exact conditional independence, together with the faithfulness assumption, often permits us to identify the path diagram from the covariance matrix. However, approximate conditional independence is not enough: then, the covariance matrix will be faithful to a variety of complete graphs.

A final example is the Timberlake-Williams model (section 10 above). This model explains political exclusion (PO) in terms of foreign investment (FI), energy development (EN) and civil liberties (CV); the sample correlation matrix was shown in table 4.6. Consider three scenarios for the "true" correlation matrix $\rho$.

(i) Suppose $\rho$ happens to equal the sample correlation matrix. Then, faithfulness obtains.

(ii) Suppose the true correlation $\rho(\text{PO,FI})$ between foreign investment and political exclusion happens to vanish exactly. Then, the Timberlake-Williams model violates the faithfulness condition; presumably, that is SGS's real complaint.

(iii) If $\rho(\text{PO,FI}) = .00001$, faithfulness is restored. According to the SGS criteria, Timberlake and Williams are back in business.

Within the framework of path models, scenario (ii) cannot be rejected at conventional significance levels; neither can (iii); and (i) represents our best estimate, subject to large uncertainties. SGS seize on hypothesis (ii), the only one that legitimates their critique. They are balking at shadows.

## 12.2 Omitted Variables

The problem of omitted variables was raised by Cliff Clogg at the Notre Dame conference, and this section paraphrases one of his points. There is a response variable $Y$, with explanatory variables $X$ and $Z$; these may be construed as vectors. Suppose the data are generated according to the "true" model (32T).

$$(32T)\ Y = X\beta + Z\gamma + \varepsilon \qquad (32R)\ Y = X\beta_R + \delta$$

The parameter vectors $\beta$ and $\gamma$ are unknown, and to be estimated from data by regression; it is $\beta$ that is of primary interest. Subjects are assumed to be independent and identically distributed; $(X, Z)$ and the error term $\varepsilon$ are independent and jointly normal; all variables have expected value 0. Consider too the "restricted" model (32R), where $\beta_R$ is defined so that $E\{Y|X\} = X\beta_R$. The constituents of (32R) may be computed from the true model.[30]

In principle, the variables $X$, $Y$ and $Z$ are all observable; $X$ and $Z$ may be correlated. However, investigators who do not know that $Z$ is relevant may fit the restricted model R rather than the true model T. If so, the estimate of $\beta$ can be quite biased. In the vernacular, $\beta_R$ includes the effect of $X$ on $Y$ through $Z$. The covariance matrix of $(X, Y)$ cannot distinguish between the two models, because the matrix can be generated by either model. Therefore, no statistical procedure based on that matrix can tell you whether the restricted model is right or wrong.[31]

## 12.3 On the Direction of Causality

This section uses "cause" in its ordinary (perhaps undefinable) meaning, not as shorthand for certain kinds of covariation. I return to Judea Pearl's example, shown in figure 4.2(a). Given the covariance matrix for $X$, $Y$ and $Z$, the SGS algorithm will produce the graph shown in panel (a). If you tell the algorithm that omitted variables are a possibility, it will tell you that $Y$ cannot cause $X$ or $Z$.

In the example, $X$, $Y$ and $Z$ are the only observables, and their covariance matrix is faithful to the graph in figure 4.2(a). I claim that such information cannot by itself determine the direction of the causal flow. To substantiate this claim, I now construct two theories. In both, the observables $X$, $Y$ and $Z$ will have the same covariance matrix, faithful to the graph in figure 4.2(a). However, the direction of the causal flow will be different in the two theories.

*Theory #1:* I first generate $X$, $Z$, $U$ as independent N(0,1) variables; $U$ is an unobservable error term. (If you want to intervene and change $X$ or $Z$, now is your moment.) Then

(33)  $Y = X + Z + U.$

According to theory #1, $X$ and $Z$ cause $Y$, as suggested by figure 4.2(a).

*Theory #2:* I first generate $Y$ as N(0, 3). (If you want to intervene and change $Y$, now is your moment.) After a suitable pause, so that time's arrow will delineate the flow of causality, I generate the errors $V_1$, $V_2$ and $V_3$ as independent N(0,1/3) variables, and then produce $X$, $Z$ and $U$ according to

$$X = \frac{1}{3} Y + V_1 - V_2$$

(34)  $$Z = \frac{1}{3} Y + V_2 - V_3$$

$$U = \frac{1}{3} Y + V_3 - V_1$$

In this theory, $Y$ causes $X$ and $Z$.

As far as the observables are concerned—namely, the joint distribution of $X$, $Y$ and $Z$—theories #1 and #2 agree. Furthermore, the joint distribution is faithful to the graph in figure 4.2(a). But the direction of causality is determined neither by the data nor by the mathematics. With correlational methods, causality follows from the assumptions about the unobservables.

## 12.4   The AFQT problem

SGS seem to claim that, as a demonstrable mathematical fact, their procedures will find the right answers:

> Assuming the right variables have been measured, there is a straightforward solution to these problems: apply the PC, FCI, or other reliable algorithm, and appropriate theorems from the preceding chapters, to determine which $X$ variables influence the outcome $Y$, which do not, and for which the question cannot be answered . . . then estimate the dependencies by whatever methods seem appropriate and apply the results of the previous chapter to obtain predictions of the effect of manipulating the $X$ variables. No extra theory is required. We will give a number of illustrations. . . . (SGS, p. 242)

The first example given by SGS to illustrate this claim is AFQT (section 9 above). To demonstrate that SGS are exaggerating more than a little, I pose a sharp mathematical question with the essential features of the AFQT problem. Then, I show the question to be undecidable by correlational methods. (Of course, when applied to the real example, both SGS and ordinary least squares made the right guess.)

To set up the question, assume that $X$ and $Y$ are random variables; $X$ is a vector, $Y$ is scalar.

(35)   $Y$ is a linear combination of $X$'s, with fixed weights.

(36)   The observables are $Y$ and $V_1, \ldots, V_7$.

Some $V$'s are $X$'s, some $V$'s are ringers. (A "ringer" is a variable that does not enter into the linear combination for $Y$.) There are also unobservables, including the $X$'s that are not $V$'s. Assume too

(37)   The full joint distribution is multivariate normal, with mean 0.

You are given the covariance matrix for the observables, but not the full covariance matrix. The problem is to say which of the $V$'s are $X$'s and which are ringers. I claim this problem is not solvable, because I can produce two different theories leading to different classifications of the $V$'s, but having the same joint distribution for the observables.

*Theory #1:* I use the covariance matrix for the seven observable subtests $V_1 = $ NO, $\ldots V_7 = $ GS together with the three unobservable subtests, CS, AS and PC. (The subtests are listed in table 4.8, on page 154 below). The full distribution is defined to be jointly normal, and all variables have mean 0. Let $Y = $ $.5 \times$ NO $+$ AR $+$ WK $+$ PC, where NO, AR and WK are observable but PC is unobservable. In this theory, $V_1$, $V_2$, $V_3$ are $X$'s, the remaining $V$'s are ringers. This theory happens to have been more or less correct, prior to 1989: see equation (42) in section 12.5.

*Theory #2:* Again, I use the covariance matrix for the seven observable subtests $V_1 = $ NO, $\ldots V_7 = $ GS together with the other three unobservable subtests CS, AS, PC. I create an auxiliary variable $U$, which is independent of the ten subtests and has small variance. The distribution of these eleven variables is defined to be jointly normal, and all variables have mean 0. There are three additional unobservables, defined as follows:

(38)   $T_1 = .25(\text{AR} + \text{NO}) + .5\text{PC} + U,$
(39)   $T_2 = .25(\text{WK} + \text{NO}) + .5\text{PC} + U,$
(40)   $T_3 = .75(\text{AR} + \text{WK}) - 2U.$

Let

(41)   $Y = T_1 + T_2 + T_3.$

In theory #2, $T_1$, $T_2$, $T_3$ are the unobservables; all the $V$'s are ringers. The auxiliary variables $U$, CS, AS, PC serve only to define the joint distribution.

Theory #1 and theory #2 provide the same joint distribution for the observables. Therefore, no statistical procedure based on the joint distribution—like the SGS algorithms or any other correlational methods—can adjudicate between the two theories.

This section and the previous one demonstrate the obvious: you cannot infer cause and effect relationships by doing arithmetic on a correlation matrix, because association is not causation. The mathematical development in SGS avoids such problems only by imposing more or less arbitrary conditions (like faithfulness) on unobservable variables, as discussed in sections 11.2 and 12.1.

In the present section, neither theory #1 nor theory #2 fits into the SGS framework: $Y$ is a deterministic function of the explanatory variables, with no stochastic error term; see (35). Furthermore, if $U$ and PC are treated as variables rather than error terms in (38–39–40), the joint distribution in theory #2 is, presumably, unfaithful to its causal graph. Similar comments apply to the previous section.

### 12.5   Institutional Background on the AFQT

The "Armed Services Vocational Aptitude Battery" (ASVAB) has ten subtests, including the seven listed in table 4.4, section 9 above. All ten are shown in table 4.8 below.

Until January 1989 the AFQT was computed as follows:

(42)   $\text{AFQT} = .5 \times \text{NO} + \text{AR} + \text{WK} + \text{PC}.$

After that date, NO was replaced by MK; a "verbal" score VE was defined as $\text{VE} = \text{WK} + \text{PC}$; and terms were standardized to have mean 0 and variance 1 on some calibration data—the "NORC 1985 sample." AFQT was redefined as

(43)   $\text{AFQT} = \text{MK}_Z + \text{AR}_Z + 2 \times \text{VE}_Z,$

where the subscript Z denotes standardization. Throughout the period, raw scores were by Congressional requirement converted to percentiles based on the NORC sample. Presumably, the data used by SGS come from 1988 or before, since they pick up formula (42) rather than (43); section 9 above.[32]

## 13.   Responses

Formal statistical inference is, by its nature, conditional. If assumptions A, B, C, . . . hold, then H can be tested against the data. However, if A, B, C, . . . remain in doubt, so must inferences about H. Indeed, the statistical calculations may prove to be quite misleading.

Many assumptions are made but only a few are tested. Those made without testing are called "maintained hypotheses." They are usually statistical and often rather technical—linearity, independence, exogeneity, etc. Careful scrutiny of such assumptions would therefore seem to be a critical part of empirical work.

In the social sciences, however, statistical assumptions are rarely made explicit, let alone validated. Questions provoke reactions that cover the gamut from indignation to obscurantism. *We know all that. Nothing is perfect. Linearity has to be a good first approximation. The assumptions are reasonable. The assumptions don't matter. The assumptions are conservative. You can't prove the assumptions are wrong. The biases will cancel. We can model the biases. We're only doing what everybody else does. Now we use more sophisticated techniques. What would you do? The decision-maker has to be better off with us than without us. We all have mental models, not using a model is still a model.*

With the SGS approach, responses are more subtle but no more empirical. Proponents often seem to take a Bayesian stance: faithfulness is justified on the grounds that the exceptional cases have measure 0, and must therefore be viewed as negligible *a priori*.[33] However, the SGS approach is frequentist not Bayesian; the simulations, being done on finite-state computers, must concentrate in a set of measure 0; and the SGS class of models has measure 0 within larger classes of models. Indeed, from my perspective, the whole class of path models seems rather unlikely—given the intensity of the research effort and the paucity of convincing examples. The assumptions that diagrams are sparse and faithful stretch credibility even further.

**TABLE 4.8**    The ten subtests in ASVAB. The first seven were analyzed by SGS.

| | |
|---|---|
| 1. Numerical Operations | NO |
| 2. Word Knowledge | WK |
| 3. Arithmetical Reasoning | AR |
| 4. Mathematical Knowledge | MK |
| 5. Electronics Information | EI |
| 6. Mechanical Comprehension | MC |
| 7. General Science | GS |
| 8. Coding Speed | CS |
| 9. Auto & Shop Information | AS |
| 10. Paragraph Comprehension | PC |

*Notes:* ASVAB Form 17, July 1990.

Attempts have also been made to justify the faithfulness assumption by appeals to continuity. If a covariance matrix is unfaithful, small changes to parameter values make it faithful. However, the same argument can be turned against correlational methods. For example, if a covariance matrix is faithful to an incomplete graph, small changes to hidden parameters make the graph complete and vitiate the SGS search procedures. Section 12.1 points to another kind of instability in the SGS framework. The continuity defense (like the Bayesian argument) reflects an aesthetic judgment about modeling styles. Taste is no substitute for empirical verification.

The SGS criteria for causality may also be defended as follows—it is unlikely that anything could produce the patterns of intercorrelation identified by SGS, other than causation; thus, correlational methods shift the burden of argument. Figures 4.5 and 4.6 should dispose of this idea. In real examples, the patterns identified by the SGS search algorithms can hardly represent cause-and-effect relationships. The burden would seem to be on the modelers: how can they recommend an algorithm that gives such results?

Proponents of modeling can also be heard to argue that all of us make assumptions about unobservables. However, what is unobservable with one design may become observable with another. And some investigators still deal with unobservables the hard way—by doing the right studies. For example, take Fisher's "constitutional hypothesis": there may be a genetic factor that predisposes you to smoke and to get lung cancer, heart disease, etc.[34] This putative genetic factor is the unobservable common cause for smoking and illness.

The epidemiologists did not deal with the constitutional hypothesis by introducing special assumptions. Instead, they studied the matter empirically, using data from twin studies. For a recent report on the Swedish twin registry, see Floderus et al. (1988). On the Finnish twin registry, see Kaprio and Koskenvuo (1989). Data on the Danish twin registry are fragmentary. There are forthcoming data on the U.S. twin registry, which are quite strong (Carmelli and Page 1996). The numbers on lung cancer are suggestive, but still small—this is a rare disease, even among smokers. The data on heart disease and total mortality, however, make the constitutional hypothesis untenable.

### 13.1   A Comment from Judea Pearl

Judea Pearl writes that

Correlation-based model-searching schemes produce causal inferences with only limited guarantees. Yet such schemes have potential, if conducted under conditions that screen out accidental independencies while maintaining structural independencies—for example, longitudinal studies under slightly varying conditions. This assumes, of course, that under such varying conditions the parameters of the model will be perturbed, while its structure remains stable. Maintaining such delicate balance under changing conditions may be hard in real-life studies. However, considering the alternative of resorting to controlled, randomized experiments, such longitudinal studies are still an exciting opportunity.

Additionally, any investigator who is searching for a causal model knowing that the parameters might be tied together by some hidden equation, like (17) [section 11.2], is wasting time (and public funds). Such a model, even if correct, is bound to be useless, because without the assumption of autonomy (i.e., that each parameter can be perturbed without altering the others), the model cannot predict the effect of interventions or other changes.... [personal communication]

Also see Pearl (1993) and Pearl and Wermuth (1993).

## 14. Other Literature

There is an extensive literature on the evaluation of models, going back at least to the Keynes-Tinbergen exchange (Keynes 1939 and 1940; Tinbergen 1940). Also see Liu (1960) and Lucas (1976). For more recent discussions, with other citations to the literature, see Freedman (1987 and 1991). Many authors have tried to explain the basis for inferring causation by using regression. See, for example, Pratt and Schlaifer (1984; 1988) or Holland (1986a; 1988). Of enthusiastic views on social-science modeling, there is no shortage; see, for instance, Smelser and Gerstein (1986) or Bartels and Brady (1993). For recent discussions of causal modeling, see Cox and Wermuth (1993), Pearl (1996) or Humphreys and Freedman (1996).

## 15. Conclusions

SGS have not succeeded in clarifying the circumstances under which causal inferences can be drawn from observed associations, nor have they invented a reliable engine for performing this feat. Their algorithms have some technical interest, but will make causal inferences only when causation is assumed in the first place. To be more explicit: If we assume that the arrows in a path diagram represent causation rather than association, and we also assume that the path diagram can be estimated from data, then indeed SGS can infer causation from association. The faithfulness assumption and exact conditional independence will together eliminate certain kinds of confounding. Even so, causality is assumed into the picture at the beginning, not proved in at the end. As Nancy Cartwright says, "No causes in, no causes out."[35]

The larger problem remains. Can quantitative social scientists infer causality by applying statistical technology to correlation matrices? That is not a mathematical question, because the answer turns on the way the world is put together. As I read the record, correlational methods have not delivered the goods. We need to work on measurement, design, theory. Fancier statistics are not likely to help much.

## Appendix:   Regression and Conditioning

For ease of reference, this appendix presents the usual formulas for computing regressions, and conditional covariances. I begin with regression. Suppose $\xi$ and $\eta$ are random variables; $\xi$ may be a row vector. We seek the column vector $\beta$ of regression coefficients for $\eta$ on $\xi$. Let $C = E\{\xi'\xi\}$ and $D = E\{\xi'\eta\}$; the prime denotes matrix transposition. Assume $C$ is positive definite. Then

(44)   $\beta = C^{-1}D.$

Now $\eta = \xi\beta + u$, where $u$ is automatically orthogonal to $\xi$. The mean square of $u$ may be computed as follows:

(45)   $E(u^2) = E(\eta^2) - \beta'C\beta.$

If $\xi$ and $\eta$ have mean 0, then $C = \text{cov}(\xi)$ and $D = \text{cov}(\xi,\eta)$; also, $E(u) = 0$. Likewise, if some component of $\xi$ is a non-zero constant, $E(u) = 0$. If now the variables are jointly normal, $u$ is independent of $\xi$.

I turn to estimation. Recall equation (2), repeated here for ease of reference.

(2)   $Y = X\beta + \varepsilon.$

In this equation, $X$ is the "design matrix," representing the explanatory variables. There is one row for each unit in the study, and one column for each variable; the entry in the $i$th row and $j$th column represents the $j$th variable, as observed on the $i$th unit in the study; $X$ may include a column of 1's, if there is to be an intercept in the equation. $Y$ is a column vector representing the dependent variable, whose $i$th component represents the value of $Y$ for the $i$th unit in the study. $\varepsilon$ is also a column vector, with one component for each unit in the study, representing the impact on $Y$ of chance factors unrelated to $X$. Typically, there will be many fewer parameters than data points, so $\beta$ has relatively few components.

The ordinary least squares estimator for $\beta$ is denoted by a hat, and may be computed as

(46) $\hat{\beta} = (X'X)^{-1}X'Y$.

The covariance matrix for $\hat{\beta}$, conditional on the design matrix, is computed as

(47) $\operatorname{cov}(\hat{\beta} \mid X) = (X'X)^{-1} \operatorname{var}(\varepsilon_i \mid X)$

Of course, (46) is related to (44); this is seen by defining $(\xi, \eta)$ as a row chosen at random from $(X, Y)$.

The "predicted values" and "residuals" are defined as follows:

$\hat{Y} = X\hat{\beta}$ and $e = Y - \hat{Y}$.

The residuals are automatically orthogonal to $X$. The residual sum of squares, minimized by the choice of $\beta$, is RSS $= ||e||^2 = \Sigma_i e_i^2$. Then $\operatorname{var}(\varepsilon_i \mid X)$ in (47) may be estimated as RSS$/(n-p)$, where $n$ is the number of data points and $p$ is the number of explanatory variables. Variances will be found along the diagonal of the covariance matrix, and the standard error is computed as the square root of the variance. In deriving these formulas, it is assumed that given $X$, the components of $\varepsilon$ are conditionally independent and identically distributed, with mean 0.

Suppose the model has an intercept. Then $R^2$ may be defined as

$R^2 = \operatorname{var}\{\hat{Y}\}/\operatorname{var}\{Y\}$, where, e.g.,

$$\operatorname{var}\{Y\} = \frac{1}{n} \sum_{i=1}^{n} (Y_i - \bar{Y})^2 \quad \text{and} \quad \bar{Y} = \frac{1}{n} \sum_{i=1}^{n} Y_i.$$

If all variables have mean 0, then $R^2$ may be computed as

$\hat{\beta}' X' X \hat{\beta}/(n \times \operatorname{var}\{Y\})$.

The usual formula for computing conditional covariances may be presented as follows. Let $n > 2$. Suppose $X_1, X_2, \ldots, X_n$ are jointly normal. We seek the conditional covariance of $X_1$ and $X_2$ given $X_3, X_4, \ldots, X_n$. Let $\Sigma$ be the covariance matrix of $X_3, X_4, \ldots, X_n$. Let $\kappa_1$ be the covariance of $X_1$ with $X_3, X_4, \ldots, X_n$; let $\kappa_2$ be the covariance of $X_2$ with $X_3, X_4, \ldots, X_n$. We view $\kappa_1$ and $\kappa_2$ as $n-2 \times 1$ column vectors. The conditional covariance is given by

(48) $\operatorname{cov}(X_1, X_2 \mid X_3, \ldots, X_n) = \operatorname{cov}(X_1, X_2) - \kappa_1' \Sigma^{-1} \kappa_2$.

The prime denotes matrix transposition. Details on the material in this appendix may be found in standard texts, for instance, Rao (1973).

**NOTES**

Many useful comments were made by Dick Berk, John Cairns, Cliff Clogg, Mark Hansen, Larry Hanser, Jerome Horowitz, Paul Humphreys, Ron Lee, Tony Lin, Bill Mason, Vaughn McKim, Judea Pearl, Diana Petitti, Jamie Robins, Tom Rothenberg, Terry Speed, and Steve Turner. Amos Tversky's work on the essay amounted to collaboration.

Richard Scheines and Peter Spirtes deserve special mention. Through many exchanges on the topics of this essay, they were helpful, thoughtful, and surprisingly patient. Their public comments (pp. 163–76) have quite a different tone, but that is understandable.

Research partially supported by NSF Grant DMS 92-08677.

1. Gauss was fitting orbits to astronomical observations, with least squares to estimate the elements of the orbits (Gauss 1809). Stigler (1986, 145–46) awards priority to Legendre (1805).

2. See Yule (1899), Stigler (1986, 345–58), and Desrosières (1993).

3. There were about 600 such areas in England. A poor-law union "consisted of two more parishes combined for administrative purposes" (Stigler 1986, 346).

4. Meehl (1954) provides some well-known examples. Predictive validity is best demonstrated by making real *ex ante* forecasts in several different contexts (Ehrenberg and Bound 1993).

5. Also see Manski (1993).

6. These, and the other six equations, are reported in Yule's Table C, p. 259. His Table XIX gives data for metropolitan unions, in the form of "percentage ratios" for 1871–81 rather than differences, apparently to avoid negative numbers. The equations were fitted to data; the numerical coefficients in the displays are estimates for the corresponding parameters in (1); the residuals are observable, but are only approximations to unobservable disturbance terms.

7. According to Stigler (1986, 356–57), Pigou criticized Yule for ignoring "the non-quantitative facts of the situation. . . . It is well known that, during recent years, those unions in which out-relief has been restricted have, on the whole, enjoyed a general administration much superior to that of other unions."

Stigler responds that "Pigou's ad hoc speculation . . . could not, of course, be disproved from the data Yule used." In effect, this allows Yule to defend himself by pleading ignorance.

8. For a discussion of the epidemiology, see Cairns (1978), Peto and zur Hausen (1986), Sherman et al. (1991), Hakama et al. (1993), and Muñoz et al. (1992).

9. I use this example because it is discussed by SGS, pp. 139–40; also see CG, pp. 215–16.

10. See, e.g., Maddala (1992); for discussion, see Daggett and Freedman (1985).

11. The solution to the "omitted variable" problem may seem easy—just throw some more variables into the model. The difficulties are explored in Clogg and Haritou (this volume). Also see Freedman (1983a).

12. Also see Pearl (1994a, b).

13. See Hofferth and Moore (1979) or Moore and Hofferth (1980). An "instrument" is an exogenous variable, used as part of the two-stage least squares estimation procedure. Some investigators may draw a terminological distinction: an "instrument" is exogenous, but does not appear as an explanatory variable in the equation being estimated. For purposes of estimation, exogenous variables are assumed to be independent of error terms; this does not suffice for causal inference (section 11). Even the independence assumption is not to be made lightly (Clogg and Haritou, this volume).

14. The model used by Rindfuss et al. would not fall into this category, if ED and AGE really influenced each other. The SGS framework excludes reciprocal causation, by assumption; so do path models, as I define them. However, some authors extend the definition of path models to include simultaneous equation models for reciprocal causation.

15. SGS seem to make the strong—and quite unusual—assumption that exogenous variables are independent of each other. That may be part of the reason why their algorithms estimate such peculiar models in figures 4.5 and 4.6 below. There is another, even more esoteric, point. To estimate an equation, its error term need only be assumed independent of the explanatory variables. If so, error terms from different equations may be correlated; then standard procedures for computing the correlations among the variables will not apply (Freedman 1987, 112–14; Seneta 1987, 199). SGS seem to interpret correlated errors as indicating the presence of "latent variables." Such variables will be mentioned in notes to figures 4.5 and 4.6, below.

16. Simulations tell us how well the SGS algorithms do *if* the underlying statistical assumptions hold good; the assumptions are built into the computer code that generates the simulated data. When applying statistical algorithms to real data, a critical question is *whether* those assumptions hold. The simulations do not address such questions.

17. The parallel material in CG is on pp. 214–17 and 222–24.

18. See, e.g., SGS, p. 18 and pp. 216–37; also see CG, pp. 220–21 and 231–33.

19. Smoking, broken homes, and early childbearing seem to be correlates of social disadvantage, and indicators of personality traits. DADSOCC and RACE are quite imperfect controls for family background; therefore, YCIG and ADOLF are likely to pick up effects of background, as well as effects of omitted personality variables. See note (iv) to table 4.3. This sort of bias is discussed in section 12.2 below. Also see Clogg and Haritou (this volume).

20. SGS, p. 243, also see CG, p. 216–17. Institutional background on the AFQT will be found in section 12.5.

21. On the other hand, unobserved variables may create serious problems (section 12.4).

22. SGS, p. 243, also see CG, p. 216–17.

23. The program output is given in Spirtes et al. (1993, 10–11).

24. The new matrix is still positive definite, so it is a legitimate correlation matrix. Section 12.1 discusses the connection between the Timberlake-Williams model and the faithfulness assumption. Also see Cartwright (1989, 79–84).

25. For the construction in (10), simply choose $a_{21}$ so $E\{X_2 | X_1\} = a_{21}X_1$; choose $a_{31}$ and $a_{32}$ so $E\{X_3 | X_1, X_2\} = a_{31}X_1 + a_{32}X_2$; and so forth. For details, see the appendix below. Since the ordering of the variables in (10) is arbitrary, fitting such equations or drawing path diagrams cannot determine which variables are causes and which are effects. In particular, $X_1$ may be exogenous in the sense that it is statistically independent of disturbance terms; that by itself does not suffice to estimate the results of manipulating $X_1$, since we cannot tell whether $X_1$ is a cause or an effect.

26. In this section, I use "cause" in its ordinary (perhaps undefinable) sense. However, the technical point—about the possibility of estimating path diagrams from covariance matrices—still holds if the arrows are interpreted as merely representing association. "Causation" is then colorful shorthand (perhaps too colorful) for a certain kind of covariation.

27. There was a bit of luck here, because some values for *a, c, d* will not produce correlation matrices.

28. Covariance matrices are symmetric; only the lower triangular part is shown. Entries are assumed to be positive but less than 1. The matrix is assumed to be positive definite.

29. This matching assumes, for instance, that any two of the variables have positive covariance given the third. To avoid violating the faithfulness assumption, if you set *a* and *b* to 0, erase the corresponding arrows; if that is distasteful, set *a* and *b* to small but positive values. The SGS logic would apply to a wide variety of diagrams; however, an arrow from *Y* to *X*, no matter how small the coefficient, spoils the show.

30. Indeed, $\beta_R = \beta + \alpha$ where $\alpha$ is obtained by the regression of $Z\gamma$ on $X$. In other terms, $Z\gamma = X\alpha + \eta$, where $\eta$ is normal with mean 0, independent of $X$. Then $\delta = \epsilon + \eta$. It may be seen that $\alpha$ depends linearly on $\gamma$.

31. See Clogg and Haritou (this volume), who make the following very interesting point. Adding variables that are correlated with $\epsilon$ can also bias the estimate of $\beta$ ; this "included variable" bias can be just as troublesome as the more familiar "omitted variable" bias: the latter problem cannot be solved by throwing variables into the model. The SGS treatment of omitted variables was discussed in section 12.1 above.

32. SGS appear to be considering raw scores, and I follow suit. The material in this section was reported by Larry Hanser, personal communication; he refers to Welsh et al. (1990, esp. Table 3 on p. 5) and Eitelberg (1988, esp. p. 73).

33. The "measure" here is the uniform distribution in Euclidean space, e.g., length, area, volume. . . . Mathematicians call the uniform distribution "Lebesgue measure," in honor of Henri Lebesgue (1875–1941) who developed its mathematical foundations. The SGS argument (p. 95) seems to be a variation on Laplace's "principle of insufficient reason" (Stigler 1986, 127).

34. See SGS, pp. 298–99, CG, pp. 231–33.

35. Cartwright (1989, chapters 2 and 3). Also see Pearl and Verma (1991).

# REPLY TO FREEDMAN

## *Peter Spirtes and Richard Scheines*

In *Causation, Prediction, and Search* (Spirtes, Glymour, and Scheines 1993), we undertook a three-part project. (Henceforth we will refer to *Causation, Prediction, and Search* as *CPS*.) First, we characterized when causal models are indistinguishable by population conditional independence relations under several different assumptions relating causality to probability. Second, we proposed a number of algorithms that take sample data and optional background knowledge as input, and output a *class* of causal models compatible with the data and the background knowledge; the algorithms (with the exception of the heuristic algorithm described in chapter 11) were accompanied by proofs of their correctness given assumptions that were clearly stated in *CPS*, and that we will restate below. Finally, we offered a theory of how to predict the effects of interventions in causal structures, given only partial knowledge of causal structure. Freedman's objections are all directed against the causal inference algorithms we proposed. We do not have room here to discuss all of his criticisms, but we have answered his major points. With regard to the points we do not have room to discuss, the reader should be warned that Freedman is an unreliable interpreter of what we have written. For convenience, we have divided Freedman's objections into the following categories:

1. Freedman questions some of the assumptions on which our correctness theorems are based. Some of his criticisms are based on covariance matrices that he constructed. None of the examples he constructed in sections 11.2, 11.3, or 12.3 are counterexamples to any theorem that we stated, nor are they even germane to the question of how probable are the assumptions we make. His examples only illustrate points discussed in detail in our book (particularly in the chapter on indistinguishability), in which we give similar examples.

2. The most serious charge that Freedman makes is that the algorithms do not compute what we say they do. According to Freedman:

> SGS seem to claim that their algorithms will find all the path diagrams compatible with a given covariance matrix. However, the theorems suggest that the algorithms will at best find one such graph. SGS also seem to claim that their algorithms are consistent. However, without an identifi-

ability theory for linear models, they cannot be talking about consistency. (Freedman, p. 144–45)

Freedman's claim that our algorithms do not find all of the path diagrams "compatible" (in a sense explained in detail below) with a given covariance matrix is false, and based on a fundamental misunderstanding of the output of the algorithms and the theorems of correctness.

3. Finally, Freedman claims our empirical cases are unconvincing. He raises many substantive objections to a number of the models of real data that were presented in our book. What we said about these examples was:

> We illustrate the algorithms for simulated and real data sets. With simulated data the examples illustrate the properties of the algorithms on samples of realistic sizes. In the empirical cases we often do not know whether an algorithm produces the truth. But it is at the least very interesting that in cases in which investigators have given some care to the treatment and explanation of their data, the algorithm reproduces or nearly reproduces the published accounts of causal relations. It is also interesting that in cases without these virtues the algorithm suggests quite different explanations from those advocated in published reports. (Spirtes, Glymour, and Scheines 1993, 132–33.)

Freedman's substantive criticisms of the models produced do not in any way contradict this claim.

Freedman's criticisms are marred by some fundamental misinterpretations of what we say. On more than one occasion he criticizes what we "seem" to say, even though we have never held or stated the position he attributes to us. Of the few direct quotations from us that he gives, two are taken out of context (Freedman, pp. 128–30). In one of these two quotations, Freedman replaces key qualifiers which completely change the meaning of the quotation with "****", and remarks that "Readers may consult the original for context, to see whether the omitted material affects the meaning" (Freedman, p. 128). Later we will produce the full quote so that readers will be able to see how the omitted material and the context affects the meaning.

## 1. The PC Algorithm

A number of Freedman's objections are based on misunderstandings of the algorithms that we proposed and their accompanying correctness theorems. Hence, it will be useful to give brief descriptions of the algorithms and the correctness theorems here. Since most of Freedman's objections are directed against the PC algorithm, that is the algorithm we will discuss in the most detail.

A directed acyclic graph (DAG) can be used to represent two distinct types of objects. First, it can be used to represent a set of probability distributions, all of which share certain conditional independence relations in common. Second, it can be used to represent causal relationships between variables; given a causally sufficient set of variables **V** (i.e., every common cause of variables in **V** is also in **V**), there is an edge from *X* to *Y* in *G* if and only if *X* is a direct cause of *Y* relative to **V**. We call DAGs interpreted in this way causal DAGs. These two distinct uses of DAGs can be linked by the following assumptions:

*Causal Markov Condition*: Let *G* be a causal DAG with vertex set **V** and *P* be a probability distribution over the vertices in **V** generated by the causal structure represented by *G*. *G* and *P* satisfy the Causal Markov Condition if and only if for every *W* in **V**, *W* is independent of **V**\(**Descendants**(*W*) ∪ **Parents**(*W*)) given **Parents**(*W*).

*Faithfulness Condition*: Let *G* be a causal DAG and *P* a probability distribution generated by *G*. <*G*, *P*> satisfies the Faithfulness Condition if and only if every conditional independence relation true in *P* is entailed by the Causal Markov Condition applied to *G*.

The meaning and justification of these two conditions are discussed more fully in Glymour (this volume) and in *CPS*. However, if one makes these assumptions, then the probability distribution of data generated by a causal process represented by causal DAG *G* is a member of the set of probability distributions represented by *G*.

Sometimes different DAGs represent the same set of probability distributions. In that case we say that the two DAGs are faithfully indistinguishable. (In *CPS* we define two DAGs as faithfully indistinguishable when they entail the same set of conditional independence relations. But if they entail the same set of conditional independence relations they represent the same set of probability distributions.) For example, the two DAGs in figure 5.1 are faithfully indistinguishable.

**FIGURE 5.1**

$$(i) \qquad\qquad (ii)$$
$$X \to Y \leftarrow Z \to W \qquad X \to Y \leftarrow Z \leftarrow W$$

For each DAG *G*, there is a set of DAGs that are faithfully indistinguishable from it; we call this the faithful indistinguishability class of *G*. The faithful indistinguishability class of (i) in figure 5.1 consists of the DAGs (i) and (ii). This faithful indistinguishability class can be more compactly represented by an object that we call ( following Verma and Pearl 1990b) a *pattern*. Hence a pattern represents a set of DAGs (which in certain special cases may contain only one member). A pattern is a type of graph that can contain both undirected edges and directed edges. The pattern representing the faithful indistinguishability class of (i) in figure 5.1 is shown in figure 5.2. We will not explain here the rules for determining whether a given DAG $G$ is in the set of DAGs represented by a pattern. See *CPS* and Glymour (this volume) for more details. However, notice that when an oriented edge (e.g., $X{\rightarrow}Y$) in figure 5.2 appears in a pattern, then the edge $X{\rightarrow}Y$ occurs in *every* graph in the set represented by the pattern. In contrast, when an unoriented edge (e.g., $Z - W$) appears in a pattern, then the two variables it connects are adjacent in every DAG in the set represented by the pattern, but they may have different orientations (e.g., $Z{\rightarrow}W$ in (i) and $W{\rightarrow}Z$ in (ii)).

**FIGURE 5.2**

$X \rightarrow Y \leftarrow Z - W$

Since all of the members of a faithful indistinguishability class represent the same set of probability distributions, it is not possible to determine which member of the class generated a given distribution in the set. For that reason, the output of the PC algorithm is not a single DAG, but a faithful indistinguishability class represented by a pattern, although in some special cases, the set of DAGs represented by a pattern contains a single DAG.

The PC algorithm takes as input sample data and optional background knowledge. The algorithm relies on tests of conditional independence, so in practice a user also has to assume something about the class of distributions in which the population distribution lies. In implementations of the algorithm, we have allowed the user to assume that the population is either linear normal or discrete. Freedman dislikes the assumption of linearity in many cases, but this assumption is not essential, as long as some reasonable distributional assumptions can be made which enable the algorithm to make judgments about conditional independence. Of course, the distribution may either be completely unknown, or so complicated as to make judgments about conditional independence impossible in practice.

The correctness theorem for the PC algorithm states that assuming (i) the Causal Markov condition, (ii) the Faithfulness condition, and (iii) causal sufficiency (i.e., every common cause of a pair of measured variables is itself measured), then from conditional independence relations true in the population, the PC algorithm outputs a pattern that represents the faithful indistinguishability class of the true causal DAG. For example, given a distribution generated by the graph (i) in figure 5.1, the output of the algorithm is the pattern shown in figure 5.2.

## 2. Assumptions

According to Freedman:

> Their algorithms [those found in *CPS*] have some technical interest, but will make causal inferences only when causation is assumed in the first place. To be more explicit: If we assume that the arrows in a path diagram represent causation rather than association, and we also assume that the path diagram can be estimated from data, then indeed SGS can infer causation from association. . . . Even so, causality is assumed into the picture at the beginning, not proved in at the end. (Freedman, pp. 156–57)

It is not clear what this passage means. The output of the PC algorithm is a pattern, typically not a path diagram (or DAG). We do simply assume that the data is generated by *some* causal process. We do not simply assume that the output of the algorithm (given correct judgments about population conditional independence relations) represents a set of DAGs, one of which is a correct description of that causal process. We *derive* that conclusion from the Causal Markov and Faithfulness conditions, and causal sufficiency. (If the assumption of causal sufficiency seems unreasonable in a particular case, the FCI algorithm we proposed can be used in place of the PC algorithm.) Moreover we believe that these axioms are often widely, but not explicitly, already assumed in a variety of contexts by practicing statisticians.

Freedman complains that "causality is assumed into the picture at the beginning." It is a banality that drawing causal inferences from statistical data requires making assumptions about the relationship between statistical data and causal structure. Such assumptions are even made by experimentalists (who assume an instance of the Causal Markov condition.) We fail to see the point in replacing a discussion of two precisely stated axioms, the Causal Markov and Causal Faithfulness conditions, with complaints about something so vague as "causality is assumed into the picture at the beginning." Instances of the Causal Markov condition are assumed in a wide variety of contexts in statistics. Since Freedman does not raise objections to the Causal Markov condition, we will not discuss it further here.

The Faithfulness condition is essentially an assumption that conditional independence relations are generated because of the structure of the DAG (i.e., for all parameterizations of the DAG), rather than because of the parameters (i.e., for some, but not all, parameterizations of the DAG). In *CPS* (chapter 3) we discuss a number of ways in which the Faithfulness condition can fail. One way it can fail is if there are deterministic relationships among measured variables. The path models in figure 5.3 illustrate another way in which the Faithfulness condition can fail (where the error terms have been omitted from the DAGs).

**FIGURE 5.3**

$$\text{(i)} \qquad\qquad \text{(ii)}$$

$$X \to Y \to Z \qquad X \to Y \leftarrow Z$$

The equations for model (i) in figure 5.3 are:

$$X = \varepsilon_X$$
$$Y = a \times X + \varepsilon_Y$$
$$Z = b \times Y + c \times X + \varepsilon_Z$$

Suppose that the data were generated by a linear causal process described by DAG (i), and the error terms are independent normally distributed variables. This does not entail any conditional independence relations among the variables for all values of the linear coefficients. However, if $c = -a \times b$, then $X$ and $Y$ are independent. If the PC algorithm were applied to this data, it would incorrectly produce a pattern whose only member is model (ii). The distribution is not faithful to DAG (i) which generated it, but is faithful to DAG (ii) which did not. This is not a violation of the theorem we proved about the correctness of the PC algorithm, because the theorem assumes that the distribution is faithful to the DAG of the causal structure that generated the data (DAG (i)) which is not true in this case. The examples that Freedman constructs in his section 11.3 add nothing to our discussion in *CPS* of indistinguishability, and the examples in 11.2 and 12.3 add nothing to our discussion in *CPS* of faithfulness.

However, the mere existence of cases where the Faithfulness condition fails is of no interest in itself. To draw an analogy, estimation algorithms can fail in the large sample limit on sets of measure 0; this is not an objection to the estimation methods. Similarly, the relevant question here is not the existence of data sets on which the Faithfulness condition will fail, but how likely it is that it will fail. The examples that Freedman gives in sections 11.2 and 12.3 do not even address this issue.

In the linear case, for a given directed DAG $G$, if we parameterize the space of models by the linear coefficients and the variances of the exogenous variables, the probability of the Faithfulness condition failing in the population is of Lebesgue measure 0. This has the consequence that any Bayesian who, conditional on belief in a causal structure represented by $G$, has a distribution over the linear parameters that is absolutely continuous with Lebesgue measure, assigns measure 0 to a violation of the Faithfulness condition. Of course we cannot prescribe what priors a Bayesian ought to accept, but we think that the kind of prior that we have described is implicitly held by many people in many cases.

James Robins (personal communication) has raised questions of a more interesting type. We will consider here one simple variation of the kind of question he asks. Suppose that from a sample the population is judged to be faithful to $G_2$ but not $G_1$. One would expect that the probability of $G_2$ given the data would be boosted relative to the probability of $G_1$ given the data, because while *all* parameterizations of $G_2$ are compatible with the conditional independence relations judged to hold in the population, only a relatively small set of parameterizations of $G_1$ are compatible. But how large is this boost in probability, and might it not be overwhelmed by priors favoring $G_1$ over $G_2$? To test this, we did the following Bayesian analysis. We generated 2000 sample data points from a random parameterization of DAG (ii) in figure 5.3, where each of the variables is ternary. The PC algorithm correctly found that DAG (ii) generated the data. We assigned equal prior probabilities to DAG (i) and DAG (ii), and we assigned a Dirichlet distribution to the parameters of DAG (i) and DAG (ii). (The distribution we used assigns the same probability to any two faithfully indistinguishable DAGs and is described in Heckerman, Geiger, and Chickering 1994.) We then calculated the ratio of the posteriors of DAG (i) to DAG (ii) given the data for ten different data sets generated from random parameterizations of (ii) where each variable could take on three different values, and ten different data sets generated from random parameterizations of (ii) where each variable could take on four different values. (We did the calculations using the K2 program lent to us by Cooper.) The likelihood ratio of DAG (ii) to DAG (i) given the data for cases where each variable took on three different values averaged $10^{6.21}$, and for the cases where each variable took on four different values it averaged $10^{13.7}$. For the case where the true model is (ii), this gives some indication of how heavily the priors would have to favor $G_2$ before we would be justified in favoring the alternative to the output of the PC algorithm. More Bayesian analyses of this kind, using different graphs, latent variables, and different priors over the parameters would be useful, and we plan to perform them.

## 3. The Correctness Results

The correctness theorem for the PC algorithm states that assuming the data is generated by some causal DAG, the Causal Markov and Faithfulness conditions, and causal sufficiency (i.e., every common cause of a pair of measured variables is itself measured), then the PC algorithm outputs a pattern that represents the faithful indistinguishability class of the true causal DAG as long as its judgments about conditional independence relations in the population are correct. Although the theorem was stated under the assumption that population conditional independence facts were known, it is easy to show that the output of the PC algorithm is correct with probability 1 in the large sample limit, if instead of being given population conditional independence relations, one is given a class of distributions that the distribution over the measured variables falls in, and a test of conditional independence in that class of distributions which is correct with probability 1 in the large sample limit.

According to Freedman:

> SGS seem to claim that their algorithms will find all of the path diagrams compatible with a given covariance matrix. However, the theorems suggest that the algorithms will at best find one such graph [i.e., DAG]. (Freedman, pp. 144–45).
>
> In statistical terminology, models are "identifiable" when they make different predictions about observables. For example, suppose you have two models for your data. If, for all data sets
>
> $$P(\text{data} \mid \text{model 1}) = P(\text{data} \mid \text{model 2}),$$
>
> there is an obvious problem—the data cannot distinguish between the models. If a path model is complete, or the faithfulness assumption is not imposed, then the graph underlying a covariance matrix is not identifiable. . . . However, even if we assume that a covariance matrix is faithful to a graph that is not complete, there may be several such graphs (SGS p. 89). For example, the following three graphs can generate the same covariance matrix:
>
> $$X \rightarrow Y \rightarrow Z \qquad X \leftarrow Y \rightarrow Z \qquad X \leftarrow Y \leftarrow Z$$
>
> Thus, SGS do not seem to have succeeded in defining a class of graphs and covariance matrices for which identifiability holds. (Freedman, p. 144)

This passage reveals several confusions that Freedman has about the output of the algorithm. First, we never claimed to be able to reliably determine a unique DAG from a covariance matrix. That is why if the data is generated by any of the three causal DAGs Freedman refers to, the output of the algorithm is a pattern that represents *all three* DAGs. Moreover, each DAG in the set represented by the pattern is not a statistical model in the usual sense. That is, a DAG is not associated with one probability distribution, it is instead associated with a set of probability distributions (which under the assumption of faithfulness all share the same set of conditional independence relations.) It is only after a DAG is parameterized that it is associated with a single probability distribution. So the concept of identifiability as defined by Freedman does not even apply to the graphs that are represented by the output of the PC algorithm; P(data | graph) is not well defined. We were of course well aware of the fact (and repeatedly emphasized) that even under the assumption of faithfulness, that there exist *parameterizations* of the three graphs in Freedman's example that represent the same covariance matrix. It is a simple consequence of a number of theorems described in chapter 4 of *CPS* (based on Verma and Pearl 1990b and Frydenberg 1990.) It is for precisely this reason that the output of the PC algorithm in this case is not a single DAG, but a pattern which represents all three of the DAGs. And as previously explained, we *have* identified sets of models which cannot be distinguished from each other given just a probability distribution, assuming Causal Markov, Faithfulness, and causal sufficiency; this is just the faithful indistinguishability class of a DAG.

In section 11.3 Freedman constructs examples of models that cannot be distinguished by their correlation matrices. All of the examples of indistinguishability that he presents in this section are simple consequences of the indistinguishability theorems we state in chapter 4 of *CPS*. Moreover, if data from any of the indistinguishable models were given to the FCI algorithm, its output would be a set of graphs that contained all of the indistinguishable models.

The *relevant* questions about the output of the PC algorithm are:

1. Is the set of DAGs represented by the output of the PC algorithm too small (in the sense that it might leave out the true DAG)? The answer is no, if one assumes Faithfulness, Markov, causal sufficiency, and correct statistical decisions about conditional independence.

2. Is the set of DAGs represented by the output of the PC algorithm too large (in the sense that, under the assumptions made, some DAG represented by the output pattern does not have a parameterization that faithfully generates the true distribution, and hence could not be correct?) Again the answer is no. (In *CPS* we showed that each of the DAGs represented by the output pattern entailed the conditional independence relations that hold in the true distribution. Under the assumption of causal sufficiency, and the existence of the right conditional density functions, this entails the stronger result that each of the graphs has a parameterization that faithfully generates the true distribution.) There are parallel correctness results for the other algorithms in *CPS* (except for the algorithm described in chapter 11, which is heuristic).

In an artfully worded passage, Freedman hints that despite our claims, we ultimately have disavowed the possibility of discovering causal relations from conditional probabilities. He says:

SGS discuss the role of assumptions in their theory several times ( pp. 53–69, 75–81, 324–45, 351). However, the clearest statement can be found when SGS are trying to discredit the evidence that smoking causes lung cancer:

> effects **** cannot be predicted from **** sample conditional probabilities. [p. 302]

Readers may consult the original for context, to see whether the omitted material affects the meaning. The advantage of the quote is clarity. If the statement is generally applicable, then SGS—like Yule and Pearl before them—have disavowed the ability to infer causation from association. (Freedman, p. 128)

Here is the quote with the omitted words restored, and some more of the beginning of the sentence:

> . . . leading epidemiologists, such as Lillienfeld, seem simply not to have understood that *if the relation between smoking and cancer is confounded by one or more common causes,* the effects of abolishing smoking cannot be predicted from the "risk ratios," i.e., from sample conditional probabilities. (Spirtes, Glymour, and Scheines 1993, 302, italics added)

The conversational implication of the words removed is quite clear; if the relation between smoking and lung cancer is *not* confounded by one or more common causes, the effects of abolishing smoking *can* be predicted from sample conditional probabilities. Moreover, this latter claim is a simple application of a theory developed at length in chapter 7 of *CPS*; the goal of that theory was to specify conditions under which it is possible to calculate the effects of interventions from conditional probabilities, and the conditions under which it is not. In the context of the theory developed in chapter 7, and with the omitted words restored, it is clear that the quotation is an *application* of our theory, not a *denial* of it. (Incidentally, the Freedman quotation reveals another fundamental confusion: the quotation concerns the possibility of making quantitative predictions of the effects of causal intervention from partial knowledge about causal structure and sample probability distributions, not the possibility of inferring features of causal structure from sample probability distributions.)

## 4. Examples

At the beginning of the discussion of empirical examples in SGS we said:

> We illustrate the algorithms for simulated and real data sets. With simulated data the examples illustrate the properties of the algorithms on

samples of realistic sizes. In the empirical cases we often do not know whether an algorithm produces the truth. But it is at the least very interesting that in cases in which investigators have given some care to the treatment and explanation of their data, the algorithm reproduces or nearly reproduces the published accounts of causal relations. It is also interesting that in cases without these virtues the algorithm suggests quite different explanations from those advocated in published reports. (Spirtes, Glymour, and Scheines 1993, 132–33.)

We do not think that Freedman's substantive criticisms in any way contradict this claim. We do not have room to discuss all of his criticisms, but we will discuss some of the major ones here.

### 4.1    The Timberlake-Williams example

The point of our discussion of the Timberlake-Williams case was that the FCI algorithm produced an alternative to the Timberlake-Williams model that was consistent with background knowledge, was testable, and fit the data. We think that this sheds considerable doubt upon the Timberlake-Williams model. None of Freedman's remarks are relevant to our criticism of the Timberlake-Williams model. Freedman does not think any of the models discussed (including the one constructed from the output of the FCI algorithm) have good evidence for them, because some of the assumptions are dubious, and the sample size is small. We agree, but we never made any claim to the contrary. We even pointed out that we could not reproduce the correlation matrix reported by Timberlake and Williams from the sources they cite. Our own simulation studies indicated that the PC algorithm (which is considerably simpler than the FCI algorithm) was not very reliable at sample sizes considerably larger than that in the Timberlake-Williams studies.

### 4.2    The AFQT example

We were given a variety of test scores (which we will call the subtest scores) and an AFQT score. The AFQT score is calculated from some of the subtest scores we were given, but not others. The point of this example was to show that *given the same background knowledge* regression techniques could not find the components of the AFQT score, but the PC and FCI algorithms can. Freedman points out that given *more* background knowledge than was given to the PC and FCI algorithms, regression techniques would be able to pick out the components of the AFQT score. (In particular, his regression analysis needs two further assumptions: first, that an increase in a component score increases the AFQT score, and second, that a component score is not highly negatively correlated with another component score because of a common cause.) Both of these assumptions are reasonable in this case. But it does not take away from the fact that the PC and FCI algorithms derive the same result with less information, and in cases where these background assumptions are not so reasonable, the regression technique could not be employed. Freedman also points out that although the PC algorithm gets the right relationship among the subtest scores and the AFQT score, it gets the wrong relationship among the subtest scores. This was evident because the PC algorithm output a pattern that represented graphs that had a cycle, which does not make substantive sense in this case. This is true. However, this application of the algorithm assumed linearity. The relationship of the component scores to the AFQT score (as calculated in 1987) is *known* to be linear. There is no reason at all to believe that the relationship among the subtest scores is linear. So, where the distributional assumptions are known to be correct, the algorithm gets the right answer, and where there was no reason to believe that the assumption is correct, the algorithm gets the wrong answer.

### 4.3 Smoking and Lung Cancer

According to Freedman:

> . . . does smoking cause lung cancer, heart disease, and many other illnesses?. . . When they actually get down to arguing their case, they [SGS] use a rather old-fashioned method—a literature review with arguments in ordinary English (pp. 291–302). Causal models and search algorithms have disappeared. (Freedman, p. 129)

> SGS elected not to use their analytical machinery on the smoking data— a remarkable omission. (Freedman, p. 130)

The output of a search algorithm is one kind of evidence for a causal claim, but there are others. We discussed some of them in the case of smoking and lung cancer. The reason that we discussed the smoking and lung cancer case in ordinary English was because we felt a historical account of that case concretely illustrated how the issues concerning causal inference from observational data, and prediction of the effects of intervention, had been dealt with by the statistical community. Freedman disagrees with some of our discussions of the evidence. The relevance of any of this to the question of the reliability of the algorithms we propose escapes us.

We did not apply the algorithms to smoking and lung cancer data because we happened not to possess any such data. Without the data, we do not know which variables were measured, how they are distributed, etc. Whether the algorithms are applicable and would produce an interesting result depends upon the answers to these questions. But there is nothing in principle that prevents the application of the algorithms to such data.

### 4.4 The Rindfuss Example

Freedman points out there were some errors in the covariance matrix we used. He does not point out that those errors make no difference in the output of the algorithm.

## 5. Conclusion

Much of applied statistics is devoted to drawing causal conclusion from statistical data. But when examining causal inferences, statisticians all too often simply trot out a few examples of circumstances under which particular causal inferences cannot be made. What is needed is a rigorous examination of what causal relations can be inferred given varying kinds of background knowledge. What could we infer about causal structure if we knew the time order of the variables? What if we knew no common causes were acting? What if we have measured a large number of variables? In addition, there should be a theory about when it is possible to determine the effects of an intervention upon that causal structure, given the inferences that can be made about causal structure. *CPS* attempts to systematically examine these kinds of questions. Under a variety of assumptions there are positive results that state when two different causal structures can be distinguished given the data, and negative results that state when they cannot. There is a systematic examination of the relationship between partial knowledge about causal structures and what can be predicted about the effects of intervention. It is not complete, and there may be many other kinds of assumptions that should be investigated, but we think that it is an example of the kind of research into causal inference that should be an important part of statistics.

The space of possible causal models, even when many have been eliminated by background knowledge, is far too large to search without automated help. One part of our project was the proposal of some algorithms that use statistical data and background knowledge supplied by a user to suggest causal models compatible with the background knowledge and the data. As we have consistently emphasized (despite Freedman's claims to the contrary) it is an empirical question whether the algorithms are reliable when applied to real data. Even if the algorithms are useful in practice, informative and reliable causal inference from correlational data will not be easy. The right variables must be measured, they have to be distributed in such a way that reasonable tests of conditional independence can be made, and the causal relationships among the variables cannot be too complex. All of these caveats are clear in our book and emphasized in Glymour (this volume).

There are three kinds of evidence concerning the reliability of the algorithms. First they are asymptotically correct given the assumptions we make. Second, simulation tests indicate that when the modeling assumptions are satisfied, and the sample sizes are reasonably large, they are reliable. Third, there is some evidence from application to real data sets.

The evidence from application to real data sets is the most fragmentary. We generally did not have access to raw data, and in most cases we do not know whether the output, however reasonable it may seem, is correct. In the AFQT case, among the relationships known to be linear, it output an answer known to be correct. In the case of Spartina biomass (Spirtes, Glymour, and Scheines 1993, 244–48), the algorithm again predicted from observational data results confirmed by experiment. In a number of other examples in *CPS* not discussed by Freedman, the answers appeared reasonable, but it is not known whether they are correct. More extensive applications will provide more evidence about the reliability of the algorithms to real data sets. We are currently carrying out some applications of the algorithms to medical databases.

# REJOINDER TO SPIRTES AND SCHEINES

## *David A. Freedman*

According to Spirtes and Scheines, "The most serious charge that Freedman makes is that the algorithms do not compute what we say they do." Not at all. My main point is that the SGS algorithms are irrelevant. Of course, with enough assumptions, SGS can make causal inferences (section 12.1 of my essay); with other such assumptions, even the much-criticized regression models can succeed. The problem is that the assumptions put the algorithms and the models into the playpen. By comparison, the charge of bugs in the program is relatively minor. Here, I discuss real examples first and mathematics second. (SGS is Spirtes, Glymour, and Scheines; CG is Glymour.)

### 1. The Real Examples

*a) Yule (my essay pp. 114–18); cervical cancer (pp. 120–21).*
   Spirtes and Scheines do not comment on these examples.

*b) Rindfuss et al. (pp. 121–26, 130–32).*
   Spirtes and Scheines say "in cases in which investigators have given some care to the treatment and explanation of their data, the algorithm reproduces or nearly reproduces the published accounts of causal relations"; they cite SGS (pp. 132–33). Apparently, the model of fertility and education in Rindfuss et al. was offered in that spirit.
   Spirtes and Scheines do concede (p. 175) that they got the data a little bit wrong. But they simply ignore the logical problems in the model (pp. 124–26). They also ignore the paradoxical empirical results (my table 4.3, p. 131). Finally, as well they might wish to, they ignore the actual output of their own algorithm (my figure 4.5, p. 132). The output tells us that religion causes region of residence, and binary variables have normal distributions (p. 131). This is not a success story for the algorithm; this is a disaster.

*c) The Armed Forces Qualification Test ( pp. 132–34, 151–53).*

According to Spirtes and Scheines ( p. 173), "The point of this example was to show that *given the same background knowledge,* regression techniques could not find the components of the AFQT score, but the PC and FCI algorithms can." This is an afterthought. On p. 133 of my essay, I quote the claim made by SGS ( pp. 243–44) and CG (p. 217). In particular, regression techniques are said to be incapable of solving the problem—background information or no background information—because all the regression coefficients are said to be significant. That is simply false: see my table 4.4, p.133. SGS and CG have misunderstood the data and the statistical logic.

My figure 4.6 (p. 135) shows the actual output from the SGS algorithm, which includes a causal cycle among the AFQT subtests:

$$MC \rightarrow AR \rightarrow WK \rightarrow GS \rightarrow MC.$$

My table 4.4 indicates that conventional regression techniques can solve the AFQT problem, without introducing these cycles. Spirtes and Scheines feel it is unfair for me to use a little information about psychological measurement. They hint darkly that the causal cycle points to some non-linearity in the relationships among the subtest scores ( p. 173–74). None of this will do. Test scores generally look more or less multivariate normal, which should come as no surprise; psychometricians often design their test instruments to achieve just that look. Furthermore, SGS have assumed multivariate normality in applying their algorithms (Spirtes and Scheines, p. 166; Spirtes, Scheines, Glymour, and Meek, pp. 38, 42, 71). The assumption is even repeated in the computer printout of statistical significance levels, headed "Prob.":

> Prob.: Probability that the absolute value of the sample (partial) correlation exceeds the observed value, on the assumption of zero (partial) correlation in the population, assuming a multinormal distribution.

The multivariate normality assumption entails linearity of relationships among scores—up to additive, random errors. That is what "linearity" means in the present context. Linearity is needed to derive the conclusions that SGS and CG trumpet so vigorously. The assumption implies not only that the AFQT is linearly related to the subtests, as Spirtes and Scheines insist; but also that the subtest scores are linearly related among themselves, as Spirtes and Scheines now affect not to believe. Non-linearity can be a great argument; in the present example, it is a red herring. The SGS algorithms are supposed to find acyclic graphs. The cycle in the output cannot be explained by non-linearity. This is logical meltdown.

*d) Timberlake and Williams ( pp. 134–37, 149).*

SGS ( pp. 248–50) criticize this model and reach a definite conclusion ( p. 249): the inferences drawn from the model are "unwarranted." I agree with the bottom line but reject the reasoning, which is based on indefensible statistical assumptions ( pp. 136–37) and an equally unjustified demand for faithfulness ( pp. 147–49). Now Spirtes and Scheines tell us ( p. 173) that the sample size is too small for them to do business, and they don't much like the assumptions either. Fine. Then don't use the example.

*e) Smoking ( pp. 129–30).*

Spirtes and Scheines say ( p. 174) "Freedman disagrees with some of our discussion of the evidence. The relevance of any of this to the question of the reliability of the algorithms we propose escapes us." Freedman actually thinks that the SGS discussion of smoking is wildly off the mark. In the end, they have no substantive findings. Their methodological conclusion is that the statisticians and epidemiologists (including Doll, Peto, Cornfield, and Fisher) didn't understand the issues. Furthermore, the SGS discussion doesn't use any of their algorithms; all the real-looking diagrams they present by way of illustration (for instance, figure 8.16 on p. 232 of CG; figure 27 on p. 237 of SGS) turn out to be hypotheticals. Spirtes and Scheines say they couldn't run the algorithms because they didn't have the data. Next time they publish on an important topic—and dismiss all previous work as incompetent—they might want to analyze some data and look at the literature.

Spirtes and Scheines ( pp. 171–72) criticize my use of a quote from SGS, as "taken out of context," edits that "completely change the meaning of the quotation" ( p. 164). These reproaches are unmerited. We are discussing the use of regression and its variants (including the SGS algorithms) to draw causal inferences from correlational data. There are three major obstacles to this enterprise: ( i) confounding (as SGS acknowledge in the clause I omitted from the beginning of the quote); (ii) specifying the right functional forms of relationships; and (iii) specifying the right stochastic assumptions. If you cannot handle these problems, then you are only pretending to make causal inferences by regression. That was the message of my essay. Spirtes and Scheines are eager to spot potential confounders in epidemiology, as the quote—and surrounding discussion—indicate. However, when it comes to their own work, they just do not see the issues created by confounding and model specification. Despite the rhetoric, SGS simply *assume* there are no omitted variables in the AFQT example or Rindfuss et al., and they simply *assume* independent observations from a common multivariate normal distribution; with Timberlake and Williams, SGS make the same distributional assumptions but allow omitted variables. The models resulting from such arbitrary choices cannot justify causal inferences from correlational data.

*f) The other examples in SGS.*

Spirtes and Scheines now suggest (p. 176) that the virtues of the algorithms will, perhaps, be seen in the "other examples in [SGS] not discussed by Freedman," or in future applications. Sure.

## 2. The Mathematics

### 2.1 Examples

Sections 11–12 of my essay discuss some mathematical examples. Spirtes and Scheines (p. 168) react as if there were priority disputes about this material, or I had offered to provide counterexamples to their theorems. They should relax. I am only trying to explain what SGS are up to, where the assumptions come in, and how unreasonable those assumptions really are.

One example is worth discussion here. My figure 4.10 shows how SGS use the faithfulness assumption to control confounding, the point being that causality is assumed into the picture from the beginning. Spirtes and Scheines say (p. 167) that they "*derive* [causality] from the Causal Markov and Faithfulness conditions and causal sufficiency." I'll meet them half way: causality is assumed into the picture from the beginning, with the Causal Markov condition. Actually, CG (p. 206) uses the language pretty much as I do, although there are a few extra flourishes: my figure 4.10 could be described as "a causal structure 'generating' a probability distribution . . . a directed graph whose edges are given a causal interpretation. . . . " In sum, to use the SGS machinery, you have to assume that the data were generated by a causal graph like my figure 4.10, the faithfulness conditions obtains *relative to that graph,* and so forth. The major obstacles to the research program are removed by assumption.

### 2.2 The Bayesian Argument

On p. 168–69, Spirtes and Scheines repeat their defense of the faithfulness condition: the set of parameters where violations occur has Lebesgue measure 0. I thought my section 13 had disposed of this argument. As the immortal Yogi Berra said, it's déjà vu all over again.

### 2.3 Identifiability and Consistency

I explain (pp. 144–45) that SGS do not have identifiability theory for *linear models,* and for that reason among others, they cannot prove consistency theorems. Careful reading of Spirtes and Scheines shows they concede the point. The comeback (pp. 163–71) is that the algorithms are estimating not *linear models,* but *equivalence classes* of linear models. This is a technical distinction with some impact. In the SGS world, if you can estimate a model, you have inferred causation from association. If it is only possible to estimate an equivalence class—what they call a *pattern*—then the direction of the causal arrows cannot be inferred from the data. For example, in Spirtes and Scheines' figure 5.1, you cannot tell whether *Z* causes *W* or *W* causes *Z*.

Now comes a delicate little question. Did SGS claim to be estimating linear models (as Freedman suggests), or did SGS frankly disavow that ability, claiming only to estimate patterns (as Spirtes and Scheines declare)? My witnesses are Spirtes, Scheines, Glymour, and Meek (1993, pp. 1–2), who ask themselves the question, "What Does TETRAD Do?" Their answer: TETRAD gives you, among other things,

> The power to find linear models in a few minutes that would otherwise have to be painfully extracted by other means, and whose construction is often the point of entire research papers. . . .

> The power to elaborate incomplete linear models. . . .

They say *linear models,* they do not say *equivalence classes of linear models.* (TETRAD embodies the SGS algorithms that we are arguing about, and I am quoting from the documentation to this program.)

I concede that TETRAD output can be in the form of patterns. Whether SGS have a well-developed theory of identifiability for these equivalence classes, I do not know. Whether their Theorem 5.1 (SGS p. 405) gives you the whole equivalence class—as Spirtes and Scheines seem to claim—or just part of the equivalence class, as the theorem seems to state, I also do not know. But of this I am certain: SGS do not have anything like the usual consistency theorems, which say that as more and more data come in, estimates get closer and closer to the truth. Instead, SGS are proving much weaker results, like Fisher consistency. I made this point on p. 145. Spirtes and Scheines do not contest it, nor do they concede it; they waltz around it. From my perspective, identifiability and consistency are part of a technical side-show; but these seem to be the high-voltage issues for Spirtes and Scheines, so I answered in some detail. SGS are ambiguous about what they are estimating, and they do not have consistency theorems in the usual sense.

## 3. Conclusion

I have been doing applied statistics now for many years and have learned some painful lessons. On reflection, I can extract a couple of general principles from this experience, which may be of some use to the reader:

a) *The law of two numbers.* If you get two different numbers that are supposed to be the same, at least one of them is wrong.

b) *The law of conservation of rabbits.* If you want to pull a rabbit out of the hat, you have to put a rabbit into the hat.

The SGS research on causality defies the law of conservation of rabbits.

# III

## NEW
## FOUNDATIONS
## FOR
## CAUSAL
## INFERENCE?

# An Introduction
# to Causal Inference

## *Richard Scheines*

In *Causation, Prediction, and Search* (*CPS* hereafter), Peter Spirtes, Clark Glymour, and I developed a theory of statistical causal inference. In his presentation at the Notre Dame conference (and in his essay, this volume), Glymour discussed the assumptions on which this theory is built, traced some of the mathematical consequences of the assumptions, and pointed to situations in which the assumptions might fail. Nevertheless, many at the conference found the theory difficult to understand and/or assess. As a result I was asked to provide a more intuitive introduction to the theory. In what follows I shun almost all formality and avoid the numerous and complicated qualifiers that typically accompany definitions of important philosophical concepts. They can all be found in Glymour's essay or in *CPS*, which are clear although sometimes dense. Here I attempt to fix intuitions by highlighting a few of the essential ideas and by providing extremely simple examples throughout.

The route I take is a response to the core concern of many with whom I spoke at the Notre Dame conference. Our techniques operate on statistical data and output sets of directed graphs. Most saw how that worked, but could not easily assess the additional assumptions necessary to give such output a causal interpretation, that is, an interpretation that would inform us about how systems would *respond to interventions*. I will try to present in the simplest terms the assumptions that allow us to move from probabilistic independence relations to the kind of causal relations that involve counterfactuals about manipulations and interventions. I first separate out the various parts of the theory: directed graphs, probability, and causality, and then clarify the assumptions that connect causal structure to probability. Finally, I discuss the additional assumptions needed to make inferences from statistical data to causal structure.

## 1. DAGs and d-separation

The theory we developed unites two pieces of mathematics and one piece of philosophy. The mathematical pieces are directed acyclic graphs (DAGs) and probability theory (with the focus on conditional independence), and the philosophy involves causation among variables.

A DAG is a set of vertices and a set of edges (arrows) that connect pairs of these vertices. For example, we might have a set of three vertices: $\{X_1, X_2, X_3\}$, and a set of two edges among these vertices: $\{X_1 \to X_2, X_2 \to X_3\}$. We almost always represent DAGs with a picture, or path diagram, e.g., this DAG looks like:

$$X_1 \to X_2 \to X_3.$$

Prior to any interpretation, a DAG is a completely abstract mathematical object. In our theory, DAGs are given two distinct functions. In the first they represent sets of probability distributions and in the second they represent causal structures. The way they represent probability distributions is given by the Markov condition, which (in DAGs) turns out to be equivalent to a more generally useful graphical relation: d-separation (Pearl 1988).[1] D-separation is a relation between three disjoint sets of vertices in a directed graph. Although too complicated to explain or define here,[2] the basic idea involves checking whether a set of vertices Z blocks all connections of a certain type between X and Y in a graph G. If so, then X and Y are d-separated by Z in G. In the DAG on the left side of figure 7.1, for example, $X_2$ blocks the only directed path connecting $X_1$ and $X_3$, so $X_1$ and $X_3$ are d-separated by $X_2$ in this DAG. By choosing d-separation to connect DAGs to probability distributions, we assume that in all of the distributions P a DAG G can represent, if sets of vertices X and Y are d-separated by a set Z in the DAG G, then X and Y are independent conditional on Z in P. For example, applying d-separation to the DAG in figure 7.1 gives us: $X_1$ and $X_3$ are d-separated by $X_2$. We then assume that in all distributions this DAG can represent, $X_1$ is independent of $X_3$ conditional on $X_2$. We use a notation for independence introduced by Phil Dawid (1979); $X_1 \amalg X_3 \mid X_2$ means: $X_1$ and $X_3$ are independent conditional on $X_2$.

It should be stressed that as long as we remain agnostic and give no interpretation to DAGs, then they are just mathematical objects which we can connect to probability distributions in any way we like. We could just as easily define and then use e-separation, or f-separation, or any graphical relation we please, as long as it produced consistent sets of independencies. When we give DAGs a causal interpretation, it then becomes necessary to argue that d-separation is the correct connection between a causal DAG and probability distributions. Let us put off that task for a few more pages, however.

There are often many distinct DAGs that represent exactly the same set of independence relations, and thus the same set of distributions. And just as one might want a procedure that computes d-separation for any graph, one might

**FIGURE 7.1**

**FIGURE 7.2**

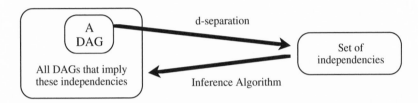

want an algorithm that computes all the DAGs that represent a given set of independence relations ( figure 7.2).

We have developed several such algorithms, one of which is called the PC algorithm and is computed by the TETRAD II program.[3] Its input is a set of independence relations over a set of variables and its output is a set of DAGs over these variables that are d-separation, or Markov, equivalent.[4] Applying the PC algorithm to the same set of independence relations shown on the right side of figure 7.1, you can see (in figure 7.3) that there are two other DAGs that are d-separation equivalent to the DAG in figure 7.1. PC is known to be complete in the sense that its output contains all and only those DAGs that are d-separation equivalent.

## 2. Causal Graphs

If taken no further, d-separation and the Markov condition are just mathematics connecting DAGs and probability distributions and need not involve causation at all.[5] One might be content to use this mathematical theory solely to produce compact and elegant representations of independence structures,[6] or one might take a further step by assuming that when *DAGs are interpreted causally* the Markov condition and d-separation are in fact the *correct* connection between causal structure and probabilistic independence. We call the latter assumption the Causal Markov condition, and it is a stronger assumption than the Markov condition.

DAGs that are interpreted causally are called *causal graphs*. There is an arrow from X to Y in a causal graph involving a set of variables **V** just in case X is a direct cause of Y relative to **V.** For example, if S is a variable that codes for

**FIGURE 7.3**

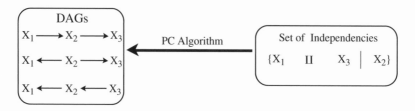

**FIGURE 7.4**

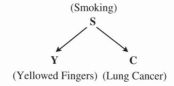

smoking behavior, Y a variable that codes for yellowed, or nicotine-stained, fingers, and C a variable that codes for the presence of lung cancer, then the following causal graph ( figure 7.4) represents what I believe to be the causal structure among these variables.

Causal graphs are assumed to be complete in one sense and not in another. They are incomplete in that they do not necessarily include *all* of the causes of each variable in the system. Thus many of the causes of lung cancer have been left out, e.g., asbestos inhalation, genetic factors, etc. They also leave out many variables that might lie in between a specified cause and its effect, e.g., cillia trauma in the bronchial lining might lie on the "true" causal pathway from smoking to lung cancer. But a causal graph is assumed to be complete in the sense that all of the *common causes* of specified variables have been included. For example, if there is some variable that is a cause of both smoking behavior and lung cancer, e.g., a genetic factor, then the causal graph above is not an accurate depiction of the causal structure among these three variables. The causal graph is also assumed to be complete in the sense that *all* of the causal relations among the specified variables are included in the graph. For example, the graph in figure 7.4 has no edge from Y to S, so it is only accurate if the level of nicotine stains does not in any way cause smoking behavior.

The semantics of a causal graph involve ideal manipulations and the changes in the probability distribution that follow such manipulations. Such an account is circular, because to manipulate *is* to cause. Our purpose, however, is not to provide a reductive definition of causation, but rather to connect it to probability in a way that accords with scientific practice and allows a systematic investigation of causal inference.

To manipulate a variable ideally is to change it in a way that, at least for the moment, leaves every other variable undisturbed. Such a manipulation must directly change only its target and leave changes in the other variables to be produced by these targets or not at all. For example, suppose we are attempting to experimentally test hypotheses concerning the causal relations between athletic performance and confidence. Suppose we intervene to inhibit athletic performance by administering a drug that blocks nutrient uptake in muscle cells, but that this drug also imitates the chemical structure of neurotransmitters that inhibit feelings of insecurity and anxiety, thus serving to directly increase anxiety and lower confidence. This intervention provides little help in trying to reason about the sort of causal relation that exists between athletic performance and confidence, because it directly alters both variables. It is an intervention with a "fat hand."[7] Ideal interventions are perfectly selective in the variables they *directly* change.

The causal graph tells us, for any ideal manipulation we might consider, which other variables we would expect to change in some way and which we would not. Put simply, the only variables we can hope to change must be causally "downstream" of the variables we manipulate. Although we can make inferences upstream, that is from effects to their causes, we cannot manipulate an effect and hope to change its other causes. In figure 7.4, for example, after an ideal manipulation of the level of nicotine stains, nothing at all would happen to the probabilities of smoking and lung cancer. They would take on the same values they would have if we had done no manipulation at all. If we could manipulate the lung cancer level of an individual without directly perturbing his or her smoking behavior or finger stains, then again, we would not expect to change the probability of smoking or of finger stains. If, however, we could manipulate smoking behavior in a way that did not directly perturb any other variable, then (for at least some of these manipulations) we would perturb the probability of the other variables through the direct causal route from smoking to the other variables.

If the causal graph changes, so does the set of counterfactuals about ideal manipulations. If, for example, the causal graph is as I picture it in figure 7.5 (absurd as it may seem), then only the statement concerning manipulations of lung cancer remains unchanged. Any ideal manipulation of smoking will result in no change in Y, but some will result in a change in C's probability, and (some) manipulations of Y will result in changes in S's probability.

Ignoring all sorts of subtleties, the point should be clear: the sort of causation we are after involves the response of a system to interventions.

**FIGURE 7.5**

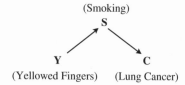

(Smoking)

**S**

**Y**            **C**

(Yellowed Fingers)     (Lung Cancer)

### 3. The Causal Markov Condition

The Causal Markov assumption can be stated simply:

> A variable X is independent of every other variable (except X's effects) conditional on all of its direct causes.

Applying this to each variable in the causal graph in figure 7.4 yields the following independence relations:[8]

> For Y: Y is independent of C conditional on S
> For S: All of the other variables are S's effects, so the condition is vacuous
> For C: C is independent of Y conditional on S

By probability theory, the first and last of these independences are equivalent, so this causal graph entails one independence by the Causal Markov assumption. You can see that figure 7.5 implies the same independence relations as does figure 7.4, even though it is different causally and thus entails different counterfactuals about interventions.

The independence relations entailed by applying the Causal Markov assumption to a causal graph are the same as those obtained from applying d-separation to a causal graph, but it is simpler to justify the connection between causal graphs and probability when stated in a Markov form. The intuition behind the Causal Markov assumption is simple: ignoring a variable's effects, all the relevant probabilistic information about a variable that can be obtained from a system is contained in its direct causes. In a Markov process, knowing a system's current state is relevant to its future, but knowing how it got to its current state is completely irrelevant. Hans Reichenbach (1956) was the first philosopher to explicitly discuss the Markov properties of causal systems, but variants have been discussed by Nancy Cartwright (1989), Wesley Salmon (1984), Brian Skyrms (1980), Patrick Suppes (1970), and many other philosophers.

How does such an assumption capture the asymmetry of causation? For systems of two variables it cannot. The two causal graphs in figure 7.6 imply the same independencies by the Causal Markov condition, and are thus indistin-

**FIGURE 7.6**

$$X \longrightarrow Y \qquad\qquad X \longleftarrow Y$$

**FIGURE 7.7**

$$X \longrightarrow Y \longleftarrow Z \qquad\qquad X \longleftarrow Y \longleftarrow Z$$

Independence Relations Entailed by d-separation

$$X \amalg Z \qquad\qquad\qquad X \amalg Z \mid Y$$

guishable solely on the basis of probabilistic independence. But leaping from this simple indistinguishability to the conclusion that probabilities can never give us information about causal structure is patently fallacious. As Hausman (1984) and Papineau (1985) realized, adding a third variable changes the story entirely. The two graphs in figure 7.7 are *not* Markov or d-separation equivalent, and the difference between their independence implications underlies the connection between independence and causal priority more generally.

We see three main lines of justification for the Causal Markov assumption, although surely there are others. First, versions of the assumption are used, perhaps implicitly, in making causal inferences from controlled experiments. Second, philosophical treatments of probabilistic causality embrace it (Suppes, 1970; Reichenbach, 1956), and third, structural equation models (Bollen, 1989), which are perhaps the most widely used class of statistical causal models in social science are Causally Markov. Elaborating on the first two lines of support are beyond the scope of this essay; they are covered in *CPS* or in Glymour's essay. Here I will try to make the connection between structural equation models and the Causal Markov assumption a little more explicit.

In a structural equation model, each variable is equal to a linear function of its direct causes plus an "error" term. Thus the causal graph in figure 7.4 would translate into the following structural equation model:

$$Y = \beta_1 S + \varepsilon_y$$
$$C = \beta_2 S + \varepsilon_c$$

where $\beta_1$ and $\beta_2$ are real valued coefficients and $\varepsilon_c$ and $\varepsilon_y$ are error terms with strictly positive variance. If the system in figure 7.4 is complete as specified, that is, its causal graph is complete with respect to common causes, then a structural equation modeller would assume that $\varepsilon_c$ and $\varepsilon_y$ are independent of each other and of S. Indeed, in structural equation models in which all of the common causes are included the error terms are assumed to be independent. It turns out that such models *necessarily satisfy* the Causal Markov assumption (Kiiveri and Speed 1982). Spirtes (1994) has generalized the result to models in which each effect is an *arbitrary* function of its immediate causes and an independent error.[9] The nature of the function connecting cause and effect is not so important as the independence of the error terms.

The connection between structural equation models and causation (as it involves the response of a system to interventions) arises through the connection between independent error terms and ideal manipulations. Although ideal manipulations provide the semantic ground for causal claims, such manipulations are sometimes only ideal and cannot be practically realized. For example, although poverty may cause crime, we cannot ethically intervene to impoverish people. In such situations we resort to collecting data passively. Since experimental science specializes in creating arrangements in which ideal manipulations exist and are subject to our will, it is no surprise that critics of causal inference from statistical data insist that experiments are the only means of establishing causal claims. But besides "I can't imagine how else it could be done," what is their argument?

In the first place, there might well be ideally selective sources of variation that exist in nature but which we cannot now or ever hope to control. For example, the moon's position exerts a direct effect on the gravitational field over the oceans, which causes the tides. But though the moon is a source that we cannot control, at least we can measure it.

In other systems, such ideal sources of variation might exist, but be both beyond our control and unobservable. In fact, a natural interpretation of the error terms in structural equation models gives them precisely these properties. There is a unique error term $\varepsilon_x$ for each specified variable X, and in systems which include all the common causes, $\varepsilon_x$ is assumed to be a source of X's variation that *directly* affects only X. And although we cannot measure them or control them, structural equation modellers assume that such error terms exist. It is this assumption that makes such systems Causally Markov.[10]

It is tempting to think that even if we do interpret error terms as representing unobservable but ideal manipulations, we are stopped dead for purposes of causal inference just because we cannot observe them. But this is a fallacy. It is true that we cannot learn as much about the causal structure of such systems as we could if the error terms were observable (the experimenters' world), but by no means does it follow that we can learn nothing about them. In fact this is precisely where causal inference starts, with systems that are assumed to be Causally Markov.

## 4. Inference

Accepting the Causal Markov assumption, I now turn to the subject of inference: moving from statistical data to conclusions about causal structure. Beginning

**FIGURE 7.8**

with statistical data and background knowledge, we want to find all the possible causal structures that might have generated these data. The fewer assumptions we make constraining the class of possible causal structures, the weaker our inferential purchase. I should note, however, that it is not the job of our theory to dictate which assumptions a practicing investigator should endorse, but only to characterize what can and cannot be learned about the world given the particular assumptions chosen.

In this section I discuss a few of the assumptions that we have studied. There are many others that we are now studying or that would be interesting to study. An enormous class of problems I will not deal with at all involves statistical inferences about independence: inferring the set of independence relations in a population from a sample. In what follows I assume that the data are statistically ideal and that in effect the population lies before us, so that any probabilistic independence claim can be decided with perfect reliability.

*Faithfulness*

The first assumption I will discuss is *Faithfulness*. By assuming that a causal graph is Causally Markov, we assume that any population produced by this causal graph has the independence relations obtained by applying d-separation to it. It does not follow, however, that the population has exactly these and no additional independencies. For example, suppose figure 7.8 is a causal graph that truly describes the relations among exercise, smoking, and health, where the + and – signs indicate positive and inhibitory relations respectively.[11]

In this case the Causal Markov assumption alone puts no constraints on the distributions that this structure could produce, because we obtain no independencies whatsoever from applying d-separation or the Markov condition to its DAG. But in some of the distributions that this structure could produce, Smoking might be independent of Health "by coincidence." If Smoking has a negative direct effect on Health, but Smoking has a positive effect on Exercise (absurd as this may seem) and Exercise has a positive effect on Health, then Smoking serves to directly inhibit Health and to indirectly improve it. If the two effects happen to exactly balance and thus cancel, then there might be no association at all between

Smoking and Health. In such a case we say that the population is *unfaithful* to the causal graph that generated it.

If there are any independence relations in the population that are not a consequence of the Causal Markov condition (or d-separation), then the population is unfaithful. By assuming Faithfulness we eliminate all such cases from consideration. Although at first this seems like a hefty assumption, it really isn't. Assuming that a population is Faithful is to assume that whatever independencies occur in it arise not from incredible coincidence but rather from structure. Some form of this assumption is used in every science. When a theory cannot explain an empirical regularity save by invoking a special parameterization, then scientists are uneasy with the theory and look for an alternative that could explain the same regularity by an appeal to structure rather than luck. In the causal modeling case, the regularities are (conditional) independence relations, and the Faithfulness assumption is just one very clear codification of a preference for models that explain these regularities by invoking structure and not by invoking luck. By no means is it a guarantee; nature might indeed be capricious. But the existence of cases in which a procedure that assumes Faithfulness fails seems an awfully weak argument against the possibility of causal inference. Nevertheless, critics continue to create unfaithful cases and display them (see, for example, David Freedman's essay in this volume).

Assuming Faithfulness seems reasonable and is widely embraced by practicing scientists. The inferential advantage gained from the assumption in causal inference is enormous. Without it, all we can say on the basis of independence data is that whatever causal structure generated the data, it cannot imply any independence relations by d-separation that are not present in the population. With it, we can say that whatever structure generated the data, it implies by d-separation *exactly* the independence relations that are present in the population. For example, suppose we have a population involving three variables $X_1$, $X_2$, $X_3$, and suppose the independence relations in this population are as in table 7.1.

Even if we assume that all the Causally Markov graphs that might have produced data with these independencies involve only $X_1$, $X_2$, and $X_3$, then there are still nine such graphs. Their only shared feature is that each has some direct connection between $X_1$ and $X_3$ and between $X_2$ and $X_3$. Adding Faithfulness reduces the set of nine to a singleton ( figure 7.9).

## Causal Sufficiency

In this example we have managed to infer that both $X_1$ and $X_2$ are direct causes of $X_3$ from a single marginal independence between $X_1$ and $X_2$. This gives many people pause, as it should. We have achieved such enormous inferential leverage in this case not only by assuming Faithfulness, but also by assuming Causal Suf-

**FIGURE 7.9**

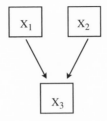

**TABLE 7.1**

| All Possible [12] Independencies among $X_1, X_2, X_3$ | In population | Not in population |
|---|:---:|:---:|
| $X_1 \amalg X_2$ | √ | |
| $X_1 \amalg X_3$ | | √ |
| $X_2 \amalg X_3$ | | √ |
| $X_1 \amalg X_2 \mid X_3$ | | √ |
| $X_1 \amalg X_3 \mid X_2$ | | √ |
| $X_2 \amalg X_3 \mid X_1$ | | √ |

ficiency, which I noted above by writing: "all the Causally Markov graphs that ... involve only $X_1$, $X_2$, and $X_3$."

The assumption of Causal Sufficiency is satisfied if we have *measured* all the common causes of the measured variables. Although this sounds quite similar to the assumptions about the completeness of causal graphs, it is not exactly the same thing. When we assume that a causal graph is complete with respect to common causes, it is in service of being clear about what sorts of systems we are representing with such graphs. In the inferential case we are making two assumptions: one involves the existence of some causal graph that is complete with respect to common causes and that is Causally Markov, and the other is an assumption about the variables we have measured as opposed to those we have not. For example, we might build a model in which we specify a variable called Intelligence which we cannot directly or perfectly measure, but four of whose effects we can measure, say test scores $X_1$–$X_4$ ( figure 7.10). Supposing that the causal graph among Intelligence and $X_1$–$X_4$ is complete with respect to common causes, and that it is Causally Markov and Faithful to whatever population it produces over {Intelligence, $X_1$–$X_4$}, then the following list of independence relations will hold in this population.

$X_1 \amalg X_2 \mid$ Intelligence
$X_1 \amalg X_3 \mid$ Intelligence
$X_1 \amalg X_4 \mid$ Intelligence
$X_2 \amalg X_3 \mid$ Intelligence
$X_2 \amalg X_4 \mid$ Intelligence
$X_3 \amalg X_4 \mid$ Intelligence

Since Intelligence is unmeasured, however, our data will only include independencies that do not involve it, which in this case is the empty set. Thus the causal graph involving Intelligence and $X_1 - X_4$ is complete with respect to common causes, but the measured variables $X_1 - X_4$ are not Causally Sufficient. To summarize, the Causal Markov assumption, although it involves a representational form of causal sufficiency, is an assumption about the way causation and probability are connected, while Causal Sufficiency is an assumption about what we have managed to measure. I have so far discussed three different assumptions:

1) the Causal Markov assumption: upon accurately specifying a causal graph G among some set of variables V (in which V includes all the common causes of pairs in V), *at least* the independence relations obtained by applying d-separation to G hold in the population probability distribution over V.

2) the Faithfulness assumption: *exactly* the independence relations obtained by applying d-separation to G hold in the population probability distribution over V.

3) the Causal Sufficiency assumption: the set of *measured* variables M include all of the common causes of pairs in M.

In the example concerning Faithfulness, we managed to infer the unique causal structure in figure 7.9 from the single marginal independence $X_1 \amalg X_2$ by making all three assumptions. It is still possible to make inferences about the structure(s) underlying the data without the Causal Sufficiency assumption, but of course we cannot learn as much.

**FIGURE 7.10**

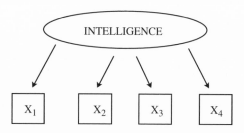

**FIGURE 7.11**

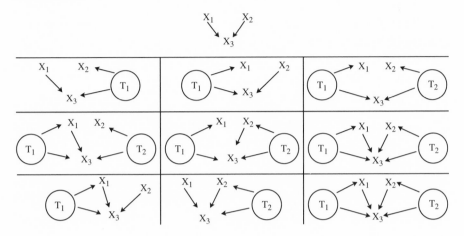

When we do not assume Causal Sufficiency, we still assume that there *is* *some* structure involving the measured variables (and perhaps other variables) that is complete with respect to common causes and that satisfies the Causal Markov assumption, but we must acknowledge that we might not have measured all the common causes. So whatever algorithm we use to move from independence relations to all the causal graphs that might have produced these independence relations, the set of graphs must include members that have common causes we have not measured. In the example in figure 7.9, we have measured $X_1$–$X_3$, and observed a single independence: $X_1 \perp\!\!\!\perp X_2$. If we assume Causal Markov and Faithfulness, but not Causal Sufficiency, the set of ten causal graphs that would produce exactly this independence appears above ( figure 7.11), where the T variables in circles are the common causes that we might not have measured.

In fact this set is still too small, for where we have specified a single unmeasured common cause of two variables (such as $X_1$ and $X_3$) and named it "$T_1$," in actuality there might be any number of distinct unmeasured common causes of $X_1$ and $X_3$. So wherever $T_1$ appears as a common cause (of say $X_1$ and $X_3$) this is really an abbreviation for: there exists some unmeasured common cause(s) of $X_1$ and $X_3$.

Although dropping the assumption of Causal Sufficiency has reduced our inferential power considerably, it has not completely eliminated it. Notice that in none of the structures in figure 7.11 is $X_3$ a cause of any other variable. So we have learned something about what causal relations do not exist: $X_3$ is not a cause of $X_1$ or of $X_2$, even though it is associated with both. In other words, we have inferred from independence data the following: if we were to ideally manipulate $X_3$, then we would do nothing to alter $X_1$ or $X_2$.

Can we ever gain knowledge about what causal relations do exist without assuming Causal Sufficiency? Yes, but not unless we either measure at least four variables or make additional assumptions. For example, if the following independencies are observed among $X_1 - X_4$, then assuming Causal Markov and Faithfulness we can conclude that in every graph that could possibly have generated this data, $X_3$ is a cause of $X_4$.

$$X_1 \amalg X_2$$
$$X_1 \amalg X_4 \mid X_3$$
$$X_2 \amalg X_4 \mid X_3$$

That is: if Causal Markov and Faithfulness are satisfied, then from these independence relations we can conclude that a manipulation of $X_3$ would change the probability of $X_4$.

Adding other sorts of knowledge often improves the situation, e.g., knowledge about the time order of the variables. In the following case from James Robins, for example, we can obtain knowledge that one variable is a cause of another when we have only measured three variables, and we do so by assuming Causal Markov, Faithfulness, but not Causal Sufficiency. If we know that $X_1$ occurs before $X_2$ and $X_2$ before $X_3$, and we know that in the population $X_1 \amalg X_3 \mid X_2$, then under these assumptions we can conclude that $X_2$ is a cause of $X_3$. We can also conclude that there is no unmeasured common cause of $X_2$ and $X_3$.

## 5. Conclusion

Contrary to what some take to be our purpose, we do not intend to magically pull causal rabbits out of a statistical hat. Our theory of causal inference investigates what can and cannot be learned about causal structure from a set of assumptions that seem to be made commonly in scientific practice. It is thus a theory about the inferential effect of a variety of assumptions far more than it is an endorsement of particular assumptions. There are situations in which it is unreasonable to endorse the Causal Markov assumption (e.g., in quantum mechanical settings), Causal Sufficiency rarely seems reasonable, and there are certain situations where one might not want to assume Faithfulness (e.g., if some variables are completely determined by others). In the Robins case immediately above, for example, we inferred that there was no unmeasured common cause, or "confounder," of $X_2$ and $X_3$. Robins, however, believes that in epidemiological contexts there are always unmeasured confounders, and thus makes an informal

Bayesian argument in which he decides that his degrees of belief favor giving up Faithfulness before accepting the conclusion that in this case it forced.

If our work has any good effects on practice, it will be as much to cast doubt on pet theories by making it easy to show that reasonable equivalent alternatives exist as it will be to extract causal conclusions from statistical data. If it succeeds in clarifying the scientific rationale that underlies causal inference, which is our real goal, then its most important effect will be to change the way studies are designed and data is collected.

### NOTES

I thank Martha Harty and Anne Boomsma for commenting on drafts of this essay.

1. If directed graphs have cycles, or chains of arrows that lead from a variable back to itself, then this equivalence breaks down.

2. We try to explain it in *CPS,* pp. 71–74.

3. For those interested in getting acquainted with the ideas as they are embodied in the program that computes many of the discovery algorithms presented in *CPS,* the TETRAD II program is available from Lawrence Erlbaum Associates, Hillsdale, N.J.

4. Sometimes there are no DAGs that can represent a given set of independence relations.

5. In fact, Judea Pearl originally developed the theory connecting DAGs and probability in order to afford robots or other AI agents an efficient way to store and use probability distributions which represented the agent's uncertainty over states of the world.

6. In fact we have often heard just such a purpose endorsed explicitly in public by able statisticians, but in almost every case these same people over beer later confess their heresy by concurring that their real ambitions are causal and their public agnosticism is a prophylactic against the abuse of statistics by their clients or less careful practitioners.

7. Kevin Kelly suggested this nomenclature.

8. A variable X is always independent of Y conditional on Y, so in this list I do not include the trivial independences between each variable and its direct causes when we condition on these direct causes.

9. In fact it is only necessary for the proof that the error term is a function of the variable for which it is an error and that variable's immediate causes.

10. In certain contexts the detrimental effect on causal inference of violating this assumption is well understood. For example a regression model, in which some of the regressors are correlated with the error term, results in a bias in estimating the causal effect of these regressors.

11. This example is from Cartwright (1983).

12. All possible non-trivial independencies, that is.

# A REVIEW OF RECENT WORK
# ON THE FOUNDATIONS OF
# CAUSAL INFERENCE

## *Clark Glymour*

I, like others, want us to understand the causal processes behind phenomena for the sake of that understanding itself, and so that we can improve the future through interventions the outcomes of which we can correctly predict. These goals require us to make arrangements to obtain data; they require us to make inferences from data to hypotheses about causal structure; and they require us to use those hypotheses to make predictions about the outcomes of interventions. Methodologies and methodologists ought to aim to provide the means to make such arrangements, inferences, and predictions as reliably, informatively, and economically as possible, and to do so in a fashion consistent with ethical constraints. The first question for methodologists ought therefore to be: just what *is* possible and what is not? A great deal has been learned on this score in the last decade. Very little of the news has reached practitioners in the social sciences, epidemiology, or elsewhere, and for reasons I will explain later, may not find a welcome in those quarters. I will try to provide a more or less systematic review, although many important details are unavoidably omitted. Most of this essay summarizes results in Spirtes, Glymour, and Scheines (1993); some of the presentation is different, and I have included some very recent results obtained since the publication of that book.

Investigations into the connections between causality and probability can be instructed by the history of investigations in probability. Whether or not one thinks, as I do, that every attempt to offer a substantive definition of probability is unsatisfactory, certainly none have gained a general consensus. Progress in the theory of probability and statistical inference was obtained nonetheless through the examination of kinds of cases, from gambling to astronomical measurement, and through the development of informal proofs. Attempts at axiomatization led eventually to the Kolmogoroff axioms, and variants, which permit the explication of historical cases and the reconstruction of informal proofs, and which continue to provide a rich theory. In the same way, there is no settled definition of causation, nor is it likely to be fruitful to insist on one before other questions are addressed. We know that causal relations are different from simple statistical associations and that the difference has to do with the fact that causal claims entail something subjunctive, something about what would happen under ideal interventions or manipulations. We have a wealth of cases from experimental design and from forms of causal explanation of nonexperimental data. And, most recently, we have an array of plausible proposals for axiomatizing the relations between causal and stochastic claims. Progress in understanding lies in exploring the mathematical consequences of various combinations of such axioms and in seeing what they imply for familiar cases and methods.[1]

The content of this essay is as follows:

1. The Markov Condition, relating causal structure and stochastic independence, is implicit in Fisher's experimental design, Rubin's experimental design, in factor analysis and elsewhere. Limitations of the condition are described.
2. The faithfulness condition, asserting that the Markov condition captures all conditional independencies, is stated and some of its limitations are described.
3. Mathematical consequences of the Markov and faithfulness conditions for graphical model equivalence and model specification search are given and illustrated using empirical and simulated cases with and without unmeasured common causes of measured variables.
4. Prediction is analyzed with the Markov condition and the "No Side Effects Principle." An analysis with these assumptions is given of Lindley and Novick's Simpson's paradox cases and of Rubin's discussion of a confounded experimental design. Connections with the Rubin framework are described. The elements are given of the theory of prediction when causal knowledge is incomplete.
5. Implications are drawn for the design of empirical studies.
6. Some other axiomatic assumptions are discussed: directed independence graphs, directed chain graphs, deterministic systems, and recent results on feedback systems represented by directed cyclic graphs.

I can no longer read as fast as relevant new results appear, and I have therefore left out a good deal from this review, including recent work (that has appeared since the first draft of this essay) on Bayesian techniques for model discovery, combinations of Bayesian and classical search methods, work on the sample bias, prediction, etc. by my associates at Carnegie Mellon, by Judea Pearl at UCLA, by David Heckerman, Dan Geiger, and others at Microsoft, by Stefan

Lauritzen and Nanny Wermuth in Denmark and Germany, and by Gregory Cooper at the University of Pittsburgh.

## 1. The Markov Condition

Fisher begins *The Design of Experiments* by considering an experiment to decide whether a particular lady can determine from the taste of a cup of tea with milk, whether the milk was poured into the cup before the tea or after. Fisher proposes an experiment in which four cups are chosen at random to have the tea first, four chosen to have the milk first.[2] In that case, Fisher claims, it follows that a definite statistical hypothesis is associated with the claim that the order in which the tea and milk are poured has no influence on the lady's judgment of the order, namely the "null hypothesis" that the order of pouring for any trial is independent of the lady's judgment on that trial. Here is one reconstruction of Fisher's inference. Let the first eight positive integers identify the cups. Represent an independent random device by R. A value r for R is a subset of four of the first eight positive integers, which identifies the cups that will have tea poured first; suppose we let F represent the order of pouring, milk first or tea first, a deterministic function of the value of R. Represent the lady's judgment by a variable J, and various other causes of the lady's judgment by a variable C. If we represent the claim that the value of one variable X influences the value of another Y by X → Y, then if order of pouring does influence the lady's judgment the causal relations in the randomized experiment are assumed to be as in figure i of 8.1, and if order of pouring does not influence her judgment the causal relations are assumed to be as in ii.

**FIGURE 8.1**

Now assume:

> *Fisher's Condition*: For any two variables, F, J in such a set-up, if F does not cause J, J does not cause F, and F and J have no common causes, then F ⊔ J.[3]

Fisher's null hypothesis thus follows from (1) the assumption that one of these graphs describes the causal relations among the variables in the set-up of the experiment, (2) the hypothesis that which liquid is poured first has no effect on the lady's judgment, and (3) the condition just stated.

My reading of two Bayesian critics of Fisher's argument, Kadane and Seidenfeld (1990), is that they agree with Fisher's assumption about what the experimental set-up and the alternative causal hypotheses entail for independence, but see no need for R to be random. R could be any means of determining pouring order, so long as it has no influence on J except possibly through F.

Consider another piece of work in statistical experimental design. Rubin (1974) considers an educational experiment to compare the effects of alternative school reading programs on reading skills as measured by a post-test, $(Y)$. The experiment assigns students to reading programs, $(T)$ on the basis of each student's pre-test score $(X)$ and a random factor that helps ensure that students with every pretest score will be represented in every reading program. $U$ represents other unmeasured causes of $Y$. Rubin wishes to predict the average difference $\tau$ in $Y$ values if all students in the population were given treatment $T = 1$ as against if all students were given treatment $T = 2$. In such a design one worries that the pre-test score and the post-test score may share common unmeasured causes $V$. The situation in the experiment is represented in figure 8.2.

**FIGURE 8.2**

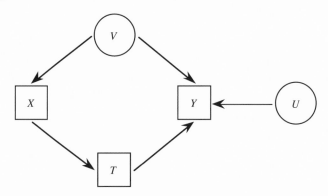

I will return later to Rubin's discussion of this case, but suffice for now that Rubin solves the problem by estimating the dependency of $Y$ on $T$, *conditional* on $X$. The inference assumes:

> *Rubin's Condition:* If $T$ has no influence on $Y$, $Y$ has no influence on $T$, and all common causes of $T$, $Y$ influence $T$ only through $X$, then in this set-up $T$, $Y$ are independent conditional on $X$.

Related assumptions connecting causal hypotheses with conditional independence relations are implicit in statistical methodology outside of experimental design. Consider factor analysis. Thurstone (1935) introduced factor analysis under an equivocal interpretation, claiming that his factor models were nothing but simplifications of data, which suggests his book ought to have been called something like *Vector Methods for Data Simplification, with Psychometric Applications* rather than the juicer title he in fact gave it, *The Vectors of Mind*. Of course, Thurstone's latent factors were immediately and almost universally interpreted as hypothetical causes. Some more recent writers on factor analysis (Bartholomew 1987) are less coy than Thurstone. Factor models assume that measured variables have no direct effects on one another, and that neither do measured variables influence latent, unmeasured factors. But one of the purposes of factor analytic algorithms is to find a collection of latent variables and "factor loadings" such that the measured variables are independent conditional on the factors. Understood as causal hypotheses, factor models assume something like this:

> *Factor Condition:* In a population of like systems in which $X$ and $Y$ themselves each influence no other variables, $X$ and $Y$ are distributed independently conditional on the set of all of their common causes.

Some treatments of factor analysis, such as Bartholomew's, assume the factor condition without further foundation. Others derive it, or special cases of it, from stochastic models of individuals in which for each individual and each measured variable $X$, there is a probability distribution over that individual's responses to $X$, and given the individual, the distributions of $X$, $Y$ etc. are independent (Ellis and van den Wollenberg 1993).

Or consider the parallel literatures on "structural equation models" (sociology and many other social sciences), "path models" (biology), and "simultaneous equation models" (economics), so many names for much the same thing. Simon (1954), and following him, Blalock (1961), emphasized that these linear models imply various sets of vanishing partial correlations, which for normal variates are conditional independence hypotheses. Blalock's examples illustrate that there is some intricate connection between the *graphical structure* of a set of linear equations and the vanishing partial correlations required by the corresponding statistical model with uncorrelated errors, but Blalock himself did not characterize the connection. That characterization appeared twenty years later, in work by Kiiveri and Speed (1982). They consider "recursive" linear statistical models (those in which no two variables are functions of one another) and error terms are distributed independently; they associate each model in the obvious way with a directed acyclic graph. They propose that the conditional independence relations associated with such a model[4] are characterized by what they call the Markov condition, one version of which is as follows:

> *Markov Condition:* Consider the vertices of a directed acyclic graph G as random variables. Let **Y** be any set of variables none of which are de-

scendants of X—there is no directed path from X to any member of **Y**. Let **Parents** (X) be the set of all variables connected to X by an edge directed into X. Then **Y** is independent of X conditional on **Parents** (X).

Applied to the graph in figure 8.2, for example, the Markov condition implies that X ⊔ Y | {V, T}. In structural equation models, the hypothetical linear dependencies are typically, if vaguely, understood as causal claims, and so in the graphical representation the directed edges bear the same causal interpretation, whatever it is. The idea of the graphical representation is roughly that a directed edge X → Y in a graph G indicates that there is a possible (I don't know in exactly what sense of possible) intervention that would change the value of X, and possibly hold constant some descendents of X, and that through and only through the change in X the intervention would also change the value of (or the probability of some value of) Y and would do so even if all other variables that are not descendants of X were to remain unchanged.

Fisher's condition, Rubin's condition, and the Factor condition are each special cases of the Markov condition applied to graphs describing hypothetical causal relations in experimental and nonexperimental settings. These and many other examples indicate to me that where causal inference is involved, statistical practice in experimental design and elsewhere is premised on a causal interpretation of the Markov Condition. Allowing the rather vague notion of a causal structure "generating" a probability distribution (whether through individual propensities, i.i.d. sampling from a population of like systems, or by constraining degrees of belief conditional on the hypothesis of the causal structure), and understanding a "causal graph" to be a directed graph whose edges are given a causal interpretation, whatever that is, we could formulate the idea this way:

> *Causal Markov Condition (Frequency Version):* Let *G* describe the causal relations among a set **V** of variables, where every common cause of variation of two or more variables in **V** is itself in **V**, and let P be a population of units all sharing the same causal relations *G*. Let *P* be the frequency distribution of variables in **V**. Then <G, P> satisfies the Markov condition.

> *Causal Markov Condition (Subjective Version):* Let *G* describe the causal relations among a set **V** of variables, where every common cause of variation of two or more variables in **V** is itself in **V**, and let *P* be the distribution of degrees of belief over **V** conditional on *G*. Then <*G, P*> ought to satisfy the Markov condition.

The Markov condition can fail to describe a population of causal systems for any number of reasons:

1. The functional dependencies and corresponding directed graph may be cyclic, as in certain feedback processes. Consider a unit u, a set **X** of variables $X_i$: $u \to R$ , and any system of linear functional dependencies among the $X_i$ that determines a directed *cyclic* graph, G, for example

$$X_3 = g(X_1, X_4, \varepsilon_3)$$
$$X_4 = f(X_2, X_3, \varepsilon_4)$$

corresponding to the directed graph in figure 8.3 (with the error terms not represented):

**FIGURE 8.3**

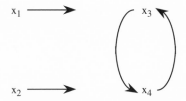

Assume the system is linear and the linear coefficients are consistent with finite correlations among the variables, which are jointly normally distributed in a population **U** of such like systems, with $X_1$ and $X_2$ and the $\varepsilon_i$ jointly independent. Then $<G, PU>$ does not satisfy the Markov condition. For example, $X_2$ is not a descendant of $X_3$, whose parents are $X_1$ and $X_4$, but $X_2$ and $X_3$ are not independent conditional on $X_1$ and $X_4$. Moreover, $PU$ need not be the marginal of any distribution whose conditional independencies are given by applying the Markov condition to some directed acyclic graph.[5]

2. Each unit may have the same directed graph, but the particular functions may not be the same in all units: in one unit $X_4$ may depend linearly on $X_1$ and $X_2$, while in another unit quadratically, and so on. Except under special conditions, the Markov condition will not be satisfied in such cases even when the exogenous variables are independently distributed.

In this case the Markov condition can always be resurrected by introducing an extra variable T, whose values identify relevantly different types of units in the population, and revising the functional dependencies and the directed graph appropriately. In general, the mixed structure may form a tree of special cases. Whether the classification variable T is regarded as a cause or as a fiction will depend on context and, no doubt, on the regarder.

3. There may be no directed graph common to all units. Certain cases of this sort may be reconciled similarly to 2. For example if there exists a directed acyclic graph G such that the directed graph of each unit is a subgraph of G, we

can introduce a variable T and an extension of G to a graph G* that includes T, in such a way that special values of T yield the functional dependencies and corresponding directed graphs appropriate to each unit. Consider, by contrast, a case in which some units have one of the graphs in figure 8.4 and some the other:

**FIGURE 8.4**

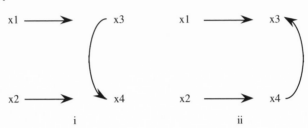

In the simplest case, let $X_1$, $X_2$ have the same distributions in the two subpopulations. The resulting population distribution will not satisfy the Markov condition for either graph i or ii, nor for the combined graph formed from the edges of both. When, however, no graphs in the population have edges in opposite directions between the same vertices, the distribution can again always be represented by the directed acyclic graph containing all edges occurring in any graph and also having a latent variable.

4. The units may be homogenous in the sense relevant here, and the directed graph may be acyclic, but the exogenous variables are statistically dependent in the population. Again, for correlations (or normal distributions) we can introduce a classification variable T (or variables, as need be), extending the graph, and in the enlarged system satisfying the Markov condition. This is essentially what happens when "correlated errors" are interpreted as produced by common latent factors. I know of no algorithm, however, for choosing the latent structure appropriately.

5. The population may be defined by conditioning on some variable Z that is a common effect of two or more variables under study that have no mutual influence. In that case the variables of interest will be statistically dependent.

There may be other reasons why the Markov condition might fail, but I haven't thought of them. Of course, if the relation among variables is intrinsically stochastic, there can be no derivation of the Markov condition, although it may be satisfied.[6]

## 2. Faithfulness

Another connection between causal hypotheses and conditional independence relations lies behind a lot of applied statistical practice. In describing Kiiveri and

Speed's claim, I said they say the Markov condition "characterizes" the conditional independence relations associated with a recursive linear model without correlated errors. That isn't quite what they say, and it isn't quite right either. Strictly, for such a model M with a set Θ of zero linear coefficients and a set Φ specifying for each variable V which other variables occur in the equation for V, the Markov condition characterizes the conditional independence relations that hold in common among *all* such linear models that agree with M on Θ and Φ. In a linear recursive model correlations can vanish either because of the graphical structure among the variables—that is, because of the Markov condition—or because of special parameter values. For example consider the model:

$$X = 12\,Y + 3\,Z + \varepsilon_X$$
$$Z = -4\,Y + \varepsilon_Z$$

with the error terms uncorrelated with each other and with variables on the same side of the equation in which they appear. The corresponding directed acyclic graph yields no conditional independence relations, but X and Y are uncorrelated. But notice that the slightest variation in the value of any one of the linear coefficients will produce a model in which X and Y are correlated. Informally, the vanishing correlation of X and Y is unstable; formally, if we use the natural measure on the space of parameters of the model, the vanishing correlation is a measure zero condition, which doesn't, of course, mean such things can't happen.

I will say that a joint distribution *P* on a set of variables is *faithful* to a directed acyclic graph *G* if every conditional independence in *P* follows from the Markov condition applied to *G*. The formulation is due to Pearl (1988). Faithfulness requires, for example, that if an acyclic graph contains a directed edge X → Y, that X and Y are not independent or conditionally independent in the associated distribution.

Kiiveri and Speed in effect assume they are dealing with marginal distributions over the "non-error" variables that are faithful to the corresponding graphs with error variables omitted. So do Simon and Blalock, again without making the assumption explicit. In fact, in the 1960s various criticisms of Simon and Blalock's ideas turned on (made-up) examples with unfaithful distributions, forming one of those too frequent intellectual disputes in which the real issue is never joined. Almost every recursive structural equation model with uncorrelated errors that I have come across describes a graph and a faithful distribution. Fisher argues from the assumption that a treatment does not affect an outcome, and has no common causes with the outcome, to independence and the null statistical hypothesis; he makes it clear that he has no objection to accepting and rejecting the null hypothesis, and the hypothesis of no causal efficacy, so long as the acceptance or rejection is tentative and reversible and the probabilistic history is not lost. But any inference from the assertion of the null hypothesis—that is from the independence of treatment and outcome—to the denial of a causal link from treatment to outcome, is a special case of the faithfulness condition. Bartholomew's text on factor models explicitly assumes that statistical dependencies among measured variables will be due to common latent causes, which is a special case of the faithfulness condition. In an analysis of prediction in experimental and nonexperimental contexts, Pratt and Schlaifer (1988) produce conditions which they claim are sufficient and "almost necessary" to enable the prediction of the distribution of a variable that will result from an intervention that alters the value of another variable. Their analysis takes a bit of deciphering, but what turns out to be required for the necessity of their condition is faithfulness. And so on.

So faithfulness is a common assumption, but it is not necessarily true. It can fail because of special parameter values, and it can also fail if some variable in the graph is a deterministic function of other variables in the graph, and in certain marginal distributions of deterministic systems.[7] We can only expect it to hold for a collection of observed variables if, for each observed variable, some cause of variation of that variable is unobserved.

## 3. Consequences

We have two very general assumptions implicit in social statistics (almost) wherever causal inferences are made, both in experimental and nonexperimental settings. Either or both can be false. One could, I suppose, stop with that reflection and reject inferences that depend on those assumptions, although I can't think of any criticisms of causal inference in the social sciences that rely on that kind of doubt. By parity of skepticism, one would have, I think, to reject virtually all work in epidemiology, and even causal inferences from randomized clinical trials, no matter how large the sample and no matter how well conducted the experiment. Skepticism is a lazy and uninformative game once we know assumptions are not necessarily true. In view of the practice, I think the first serious thing to do is to investigate the consequences of the following:

> *Working Assumption:* The distribution of observed variables in a population studied empirically, whether experimentally or otherwise, is the marginal of a distribution satisfying the Markov and faithfulness conditions for a directed acyclic graph describing the causal (or causal and classificational) relations among the observed variables, and possibly also unobserved variables, in each unit of the population.[8]

Once we have figured out the consequences of this assumption, we can turn to the consequences of some of the ways we have seen it can fail.

The first problem is that while the Markov condition implicitly characterizes conditional independence, it is not in the least obvious how to determine whether for disjoint sets of vertices **X, Y, Z** in a directed acyclic graph, **X** is independent of **Y** given **Z**. That problem was solved by Pearl, Verma and Geiger (Pearl 1988):

> *Theorem 1 (Pearl, Verma, Geiger):* The Markov condition for directed acyclic graph G implies that **X** is independent of **Y** given **Z** if there does not exist a sequence $<U_1 \ldots U_k>$ of vertices in G, such that (i) all $U_i$ are distinct, (ii) $U_1$ is in **X**, $U_k$ is in **Y**, and each $U_i$, $i < k$, is adjacent in G to $U_{i+1}$, (iii) if edges between $U_{i-1}$ and $U_i$ and between $U_i$ and $U_{i+1}$ are not both into $U_i$, then $U_i$ is not in **Z**, and (iv) if edges between $U_{i-1}$ and $U_i$ and between $U_i$ and $U_{i+1}$ are both into $U_i$, then there is a directed path in G from $U_i$ to some member of **Z**.

The condition is complicated but fully algorithmic and fully general. There is even an intuition to it, which I will not detail here. Two particular vertices X, Y satisfying the three conditions with respect to a set **Z** are said to be *d-connected* given **Z**, and any sequence $<U_i \ldots U_k>$ satisfying the conditions is said to be a *d-connecting path*. If the conditions are not satisfied, then X, Y are said to be *d-separated* by **Z**.

A further essential question is equivalence or indistinguishability: when do two directed acyclic graphs admit exactly the same family of distributions satisfying the Markov condition? Two definitions are required: two edges both directed into a vertex Y are said to *collide* at Y. If edges from X and Z collide at Y, and X, Z are not adjacent, the pair of edges is said to be an *unshielded collider*.

> *Theorem 2 (Frydenberg 1990; Verma 1990).* The set of all distributions satisfying the Markov condition for directed acyclic graph $G_1$ is the set of all distributions satisfying the Markov condition for directed graph $G_2$ if and only if (i) the two graphs have the same vertex set; (ii) the two graphs have the same adjacencies; (iii) the two graphs have the same unshielded colliders.
>
> The set of all distributions satisfying the Markov and faithfulness conditions for directed acyclic graph $G_1$ is the set of all distributions satisfying the Markov and faithfulness condition for directed graph $G_2$ if and only if (i) the two graphs have the same vertex set; (ii) the two graphs have the same adjacencies; (iii) the two graphs have the same unshielded colliders.[9]

It follows that the question of the equivalence of two directed graphs under Markov or Markov and faithfulness can be solved in time $O(n^3)$ where n is the number of vertices.

We now have the theoretical resources to ask a question that is logically central to understanding causal inference in the context of the Working Assumption. Through d-separation, the Markov and faithfulness conditions map any directed graph to a unique set of conditional independence relations. Call the map I. Through the indistinguishability relations, each directed acyclic graph is mapped to a unique equivalence class of graphs. Call that map E. The inverse of I composed with E is a map from sets of conditional independence relations to an indistinguishability class of graphs; under the Working Assumption, E I⁻¹ specifies everything that conditional independence and dependence facts can tell us about causal structure when it is assumed that there are no unmeasured common causes of observed variables. If we had a feasible way to compute E I⁻¹ and a feasible way of making the appropriate statistical decisions, we would have a discovery procedure which would be reliable in all cases in which the Working Assumption is true, the statistical decisions correctly made, and there are no latent common causes. The following theorem shows there is a computationally *infeasible* method:

> *Theorem 3 (Verma, Pearl*[10]*)* Distribution $P$ satisfies the Markov and faithfulness conditions for directed acyclic graph $G$ if and only if (i) any two vertices are adjacent in $G$ if and only if they are statistically dependent conditional on every subset of vertices of $G$ not containing them, and (ii) $X \rightarrow Y \leftarrow Z$ is an unshielded collider in $G$, then X, Z are not independent conditional on Y.

Theorem 3 suggests several discovery procedures, independently pursued by Pearl and Verma (1991) and by Spirtes, Glymour, and Scheines, for example:

> *SGS Algorithm (Spirtes, Glymour, Scheines 1990)* Start with a complete undirected graph. Choose a pair of variables, X, Y, and for each subset **Z** of vertices not containing X or Y, decide if X is independent of Y conditional on **Z** until such a subset is found or every **Z** is tried. If such a **Z** is found, remove the X, Y edge; otherwise proceed to a new pair of variables in place of X,Y, and repeat the procedure until all pairs of variables are tested. When all pairs have been tested a reduced undirected graph $C$ is the result. For each triple of variables X, Y, Z with X adjacent to Y, Y adjacent to Z and X, Z not adjacent in $C$, if X is not independent of Z conditional on Y, orient the edges to collide at Y. Finally, orient any further edges as required to avoid further unshielded colliders or directed cycles.

The algorithm is correct in the sense that given an oracle that provides a set of conditional independencies for a distribution satisfying the Markov and faithfulness conditions for a directed acyclic graph whose vertices are all measured, the algorithm returns a characterization of the equivalence class of the true graph. It is infeasible for two reasons: When two edges are in fact adjacent, the procedure tests every subset for independence, so that except in the case of completely disconnected graphs, more than $2^{(n-2)}$ tests are always required. Further, tests of conditional independence become less reliable the larger the conditioning set; in tests for vanishing partial correlations in normal distributions, for example, we must effectively reduce the sample size by 1 for every conditioning variable, and so reduce the power of the statistical decision. Loss of power as a function of conditioning set is worse with discrete variables, and further complicated by the fact that when conditioning on several variables, there may be few or even no in-stances in the sample for some combinations of values of the conditioning variables.

A further theorem offers a way around the problems of the SGS algorithm. I don't know who first proved it; probably Pearl or someone in his group.

*Theorem 4:* Any two distinct, non-adjacent vertices, X, Y in a directed acyclic graph are d-separated by the set of parents of X or the set of parents of Y.

Only variables adjacent to X can be parents of X, a fact that can be used in the following way: start with a complete graph and for each pair of variables, test whether they are independent; if so remove the edge between them. Continue until all pairs have been tested for independence. The result is an undirected graph, often with fewer edges. Then, for each pair of variables X, Y that remain adjacent, check whether X, Y are independent conditional on some one variable adjacent to X or some one variable adjacent to Y; if such a variable is found remove the X, Y edge. Continue in this way for all adjacent pairs. The result is an undirected graph which may be sparser than at the previous stage. For pairs of variables still adjacent, repeat this procedure for pairs of variables adjacent to one or adjacent to the other, then for triples, and so on. Stop when conditioning on all adjacent variables removes no edges. Now orient the edges in the same way as in the SGS procedure. This procedure is called the PC algorithm, and it has the following properties:

*Theorem 5 (Spirtes, Glymour):* For independence facts from distributions satisfying the Markov and faithfulness conditions for some directed acyclic graph G, the output of the PC algorithm is the same as the output of the SGS algorithm. If G is of maximal degree k, PC requires tests conditional on no more than $k + 1$ variables. The number of tests required is bounded by $n^2(n - 1)^{k-1}/(k - 1)!$

There are alternative procedures for making the same inferences. Wedelin (1993), for example, has described a very different procedure for inference with discrete variables using minimum description length criteria and interleaving estimates of the joint probability distribution with the construction of graphs.[11] Spirtes and Meek (1995) have described slower Bayesian procedures that in simulations are more reliable than PC.

### 3.1 Illustrations

Although these results may seem abstract they can be immediately applied to social scientific data and practice.[12] A recent paper by Rodgers and Maranto (1989) considers hypotheses about the causes of academic productivity drawn from sociology, economics, and psychology, and produces a combined "theoretically based" model. Their data were obtained from solicitations and questionnaires were sent to 932 members of the American Psychological Association who obtained doctoral degrees between 1966 and 1976 and were currently working academic psychologists. Equal numbers of male and female psychologists were sampled, and after deleting respondents who did not have degrees in psychology, did not take their first job in psychology, etc., a sample of 86 men and 76 women was obtained. The authors give a very elaborate explanation of causal theories suggested by the pieces of sociological, economic, and psychological literature. Rodgers and Maranto estimate no fewer than six different sets of structural equations and corresponding causal theories. None of the structural equation systems based on these models save the phenomena. But combining all of the edges in the "theoretical" models, adding two more that seem plausible, and then throwing out statistically insignificant (at .05) dependencies leads Rodgers and Maranto instead to propose a different causal structure which fits the data quite well. It would appear that the tour through "theory" was nearly useless, but Rodgers and Maranto say otherwise:

> Causal models based solely on the pattern of observed correlations are highly suspect. Any data can be fitted by several alternative models. The construction of the best-fit model was thus guided by theory-based expectations.[13]

If the Rodgers and Maranto theory were completely correct, the undirected graph underlying their directed graph would be uniquely determined by the conditional independence relations, and the orientation would be almost uniquely determined. When the PC algorithm is applied to their correlations with the commonsense time order using a significance level of .1 for tests of zero partial correlations, the output is the graph on the left side of figure 8.5, which we show alongside the Rodgers and Maranto model.

**FIGURE 8.5**

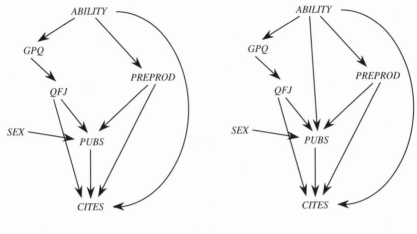

PC Output                    Rodgers and Maranto Graph

All but one of the edges in the Rodgers and Maranto model is produced instantaneously from the data and commonsense knowledge of the domain—the time order of the variables.

Rindfuss, Bumpass and St. John (1980) were interested in the mutual influence in married women of education at time of marriage (*ED*) and age at which a first child is born (*AGE*). On theoretical grounds they argue at length for the model on the left in figure 8.6, where the regressors from top to bottom are as follows:

$$
\begin{aligned}
DADSO &= \text{father's occupation} \\
RACE &= \text{race} \\
NOSIB &= \text{absence of siblings} \\
FARM &= \text{farm background} \\
REGN &= \text{region of the United States} \\
ADOLF &= \text{presence of two adults in the subject's childhood family} \\
REL &= \text{religion} \\
YCIG &= \text{cigarette smoking} \\
FEC &= \text{whether the subject had a miscarriage.}
\end{aligned}
$$

Regressors are correlated. The sample size is 1766. Apparently to their surprise, the investigators found on estimating coefficients that the *AGE* → *ED* parameter is zero. Given the prior information that *ED* and *AGE* are not causes of the other variables, the PC algorithm (using the .05 significance level for tests) directly

**FIGURE 8.6**

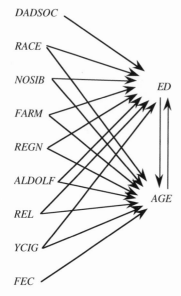

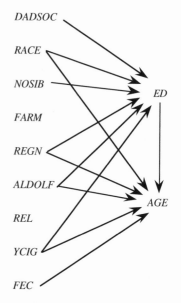

Rindfuss, et al. theoretical model; AGE → ED coefficient not statistically significant

TETRAD II model

finds the model on the right in figure 8.6, where connections among the regressors are not pictured.

I think it is likely that the PC algorithm captures the essential features of published models because in both cases, despite what Rindfuss et al. suggest, the models were developed to fit the conditional independence relations suggested by the sample. My guess is that various heuristics to fit independence constraints are commonly used throughout the social sciences without much public acknowledgment.

The PC algorithm doesn't always give the models social scientists advocate, especially not when the models are "theory driven" and the theory is unlikely in view of the sample statistics. *Causation, Prediction and Search* gives a number of examples of this kind. In these cases to be sure, no one knows which theory is correct, that produced by the social scientists and psychologists or the alternatives produced by the program, but at the very least the automated procedures offer concrete cautions to data analyses driven by theoretical commitments.

A more interesting sort of case involves independent verification or refutation of computer-generated predictions. Here is one example: The *AFQT* is a test battery used by the United States armed forces. It has a number of component tests, including:

Arithmetical Reasoning (*AR*)
Numerical Operations (*NO*)
Word Knowledge (*WK*)

In addition a number of other tests, including those listed below, are not part of the *AFQT* but are correlated with it and with its components:

Mathematical Knowledge (*MK*)
Electronics Information (*EI*)
General Science (*GS*)
Mechanical Comprehension (*MC*)

Given scores for these eight measures on 6224 armed forces personnel, a linear multiple regression of *AFQ*T on the other seven variables gives significant regression coefficients to only five. Two of the variables have a significant *negative* coefficients, which could result from scaling or from an unmeasured common cause.

> Given the prior information that *AFQT* is not a cause of any of the other variables, an experimental version of the PC algorithm in TETRAD II correctly picked out {*AR*, *NO*, *WK*} as the only variables adjacent to *AFQT*, and hence the only variables that can possibly be components of *AFQT*. (Spirtes, Glymour, and Scheines 1993)[14]

### 3.2 Consequences Resumed

The Working Assumption is that observed distributions are the *marginals* of a distribution satisfying the Markov and faithfulness conditions. The two previous subsections assume further that there are no unrecorded common causes of measured variables, an assumption I now drop. Consider a directed acyclic graph *G* representing a causal process, and any associated probability distribution *P*, where <*G*, *P*> satisfy the Markov condition. Suppose that only a proper subset **O** of variables in the graph are measured or recorded. What conditional independence relation among variables in **O** is required by the Markov condition applied to *G*? What graphical object represents those marginal conditional independence relations and also represents information about *G*? A nice answer to both questions is given in Pearl and Verma's (1991) notion of the *inducing path graph* for *G*.

An undirected path *U* between *X* and *Y* is an inducing path over **O** in *G* if and only if (i) every member of **O** on *U* (except the endpoints) occurs at the collisions of two arrowheads on the path, and (ii) for every vertex *V* on *U* where two arrowheads collide, there is a directed path from $V$ to $X$ or from $V$ to $Y$. It has been shown that there is an inducing path between $X$ and $Y$ in $G$ over $\mathbf{O}$ if and only if $X$ and $Y$ are not independent conditional on any subset of $\mathbf{O}\backslash\{X,Y\}$. For variables $X$, $Y$ in $\mathbf{O}$, in the inducing path graph $H$ over $\mathbf{O}$, $X \leftrightarrow Y$ in $H$ if and only if there is an inducing path between $X$ and $Y$ over $\mathbf{O}$ in $G$ that is into $X$ and into $Y$; there is an edge $X \rightarrow Y$ in $G$ if and only if there is no edge $X \leftrightarrow Y$ in $H$, and there is an inducing path between $X$ and $Y$ over $\mathbf{O}$ in $G$ that is out of $X$ and into $Y$. (It is easy to show that there are no inducing paths in $G$ over $\mathbf{O}$ that are out of $X$ and out of $Y$.) The two kinds of edges in an inducing path graph $H$ have a straightforward causal interpretation: a directed edge $X \rightarrow Y$ occurs in $H$ only if there is a directed path from $X$ to $Y$ in $G$, i.e., $X$ is a cause of $Y$; a double headed edge $X \leftrightarrow Y$ occurs in $H$ only if there is an unmeasured $T$ and a directed path from $T$ to $X$ and a directed path from $T$ to $Y$, the two paths intersecting only at $T$, i.e., only if $X$ and $Y$ have an unmeasured common cause.

A POIPG (partially oriented inducing path graph) represents the set of all inducing path graphs with the same adjacencies as the POIPG, but in which some of the orientations of the edges have not been specified. A directed edge in a POIPG indicates that all inducing path graphs in the class have that edge; a bidirected edge indicates that all inducing path graphs in the class have that bidirected edge. However, POIPGs can have edges ending in a mark, an "o" as in $X \circ\!\!\rightarrow Y$, allowing some of the inducing path graphs represented to have $X \leftrightarrow Y$ and some to have $X \rightarrow Y$. Similarly a POIPG may contain an edge $X \circ\!\!-\!\!\circ Y$. Two edges sharing a vertex, each with a mark at that vertex, can be underlined, as in $-\!\!\underline{\circ\ X\ \circ}\!\!-$, indicating that the two "o" marks cannot simultaneously be arrowheads in any inducing path graph it represents. POIPG $A$ is at least as informative as POIPG $B$ if and only if they have the same adjacencies and $A$ has more orientations than $B$, i.e., every "o" in $A$ is a "o" in $B$, and every arrowhead in $A$ is either an arrowhead or a "o" in $B$. For systems without reciprocal causation or feedback, when for all one knows at the outset latent variables may be acting, informative POIPGs are therefore natural objects with which to analyze causal inference founded on conditional independence relations.

Here are some simple examples of POIPGs for conditional independence relations taken from empirical examples given by Cox and Wermuth (1993).

    (i) $Y \amalg W \mid V$ and $X \amalg V \mid W$ (represented by the POIPG in i).
    (ii) $Y \amalg W$ and $X\ V$: (represented by the POIPG in ii)
    (iii) $Y \amalg W$ and $X \amalg V$ and $V \amalg W$. (represented by the POIPG in iii).[15]

We say that a path is a POIPG is *directed* if every edge has a single arrowhead, no edges have any "o" mark, and all the arrowheads point in the same direction. We say a path is semi-directed if there are no arrowheads on it pointing in oppo-

**FIGURE 8.7**

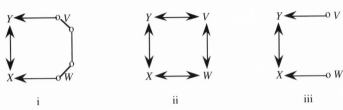

i             ii             iii

site directions. We say a directed acyclic graph *G* is *for* a POIPG Π if Π is a POIPG of *G* over a subset of the vertices in *G*.

> *Theorem 6 (Spirtes, Glymour):* If a POIPG Π has a directed path from X to Y, then every directed acyclic graph for Π has a directed path from X to Y. If every directed path from X to Y in Π contains a vertex Z, then in every graph for Π, every directed path from X to Y contains Z. If P contains no semi-directed path from X to Y, then no graph for Π contains a directed path from X to Y. If X, Y are connected by a double-headed arrow in Π then in every graph G for Π there is a vertex U not in Π and directed paths from U to X and from U to Y, each path containing only one variable in Π.

For many cases, a simple modification of the PC algorithm with the same complexity bounds as before gives the correct output even when there are latent variables. Reflection on the PC algorithm shows that it allows edges to have double arrows, and that in many cases with latent variables, when the PC algorithm is run on the observed marginal, double-headed arrows appear in the full graph between variables that have a common unobserved cause. PC can be modified to add "o"s at the ends of each edge in the original complete undirected graph, and to remove an "o" if the orienting procedure says there definitely cannot be an arrowhead at the corresponding edge end (Spirtes, Glymour, and Scheines 1993). Unfortunately, for a complete and correct algorithm a much more intricate procedure is required. In general, to determine adjacency of X, Y in the inducing path graph it is not sufficient to test, as the PC algorithm does, for independence conditional only on variables adjacent to X or Y. Spirtes (1992) first described a correct, complete algorithm (the FCI algorithm) for constructing a POIPG from conditional independence relations among observed variables and optional background knowledge. The algorithm is intricate and I will not describe it here (see Spirtes 1992; Spirtes, Glymour, and Scheines 1993).

**FIGURE 8.8**

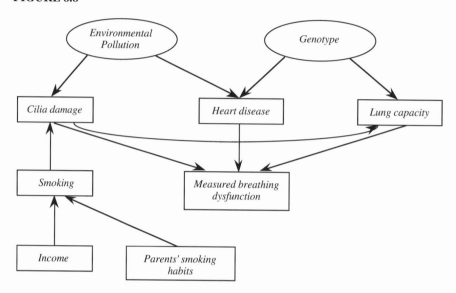

POIPGs can be more or less informative depending on how many of the edges in the inducing path graph they orient implicitly or explicitly. Consider, for example, the graph in figure 8.8, devised by Christopher Meek, where the variables in ovals are unobserved.

The figure 8.9 shows an uninformative POIPG produced by the modified PC algorithm from input consisting of relevant marginal conditional independence facts relating the observed variables, and the maximally informative POIPG from the FCI algorithm produced from analogous input.

The output of the FCI algorithm represents the indistinguishability class of POIPGs for a given set of measured variables and their conditional independence relations:

*Theorem 7 (Spirtes, Verma 1992):* Distributions satisfying the Markov and faithfulness conditions for directed acyclic graph $G_1$ and distributions satisfying those conditions for directed acyclic graph $G_2$ have the same conditional independence relations over a subset **O** of vertices common to the two graphs if and only if the POIPG produced by the FCI algorithm for every distribution faithful to $G_1$ is the POIPG produced by the FCI algorithm for every distribution faithful to $G_2$.

A corollary of the Spirtes and Verma proof is that equivalence of directed acyclic graphs in the sense of Theorem 7 can be decided in $O(n^6 + m^8)$ where n

**FIGURE 8.9**

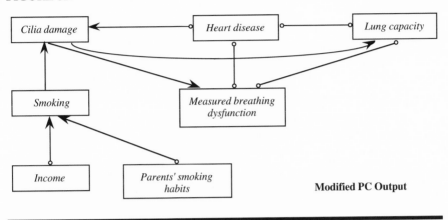

Modified PC Output

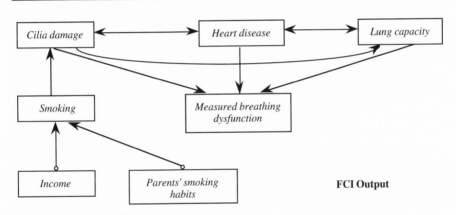

FCI Output

is the maximum of the number of vertices in $G_1$ or the number of vertices in $G_2$, and m is the number of observed vertices. But Thomas Richardson has shown that to decide adjacency the FCI procedure is exponential in the number of vertices even when the degree of the graphs is held constant.

### 3.3 Latent Variable Illustrations

I have no social scientific or psychometric examples in which these inference procedures for latent variables correctly identify the presence of a latent variable whose presence is independently known. There are striking examples with large samples of simulated data, illustrated by the FCI output in the previous figure. The most immediate empirical illustrations provide reasons to doubt some causal inferences in the social scientific literature.

Blau and Duncan's (1967) study of the American occupational structure has been praised by the National Academy of Sciences as an exemplary piece of social research and criticized by one statistician (Freedman 1983b) as an abuse of science. Using a sample of 20,700 subjects, Blau and Duncan offered a preliminary theory of the role of education ($ED$), first job ($J_1$), father's education ($FE$), and father's occupation ($FO$) in determining one's occupation ($OCC$) in 1962. They present their theory in the following graph, in which the undirected edge represents an unexplained correlation:

**FIGURE 8.10**

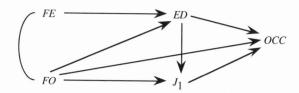

Blau and Duncan argue that the dependencies are linear. Their salient conclusions are that father's education affects occupation and first job only through the father's occupation and the subject's education. Blau and Duncan's theory was criticized by Freedman as arbitrary, unjustified, and statistically inadequate (Freedman 1983b). Indeed, if the theory is subjected to the asymptotic $\chi^2$ likelihood ratio test of the EQS (Bentler 1985) or LISREL (Joreskog and Sorbom 1984) programs, the model is decisively rejected ($p < .001$), and Freedman reports it is also rejected by a bootstrap test.

If the conventional .05 significance level is used to test for vanishing partial correlations, given a commonsense ordering of the variables by time, from Blau and Duncan's covariances the PC algorithm produces the pattern in figure 8.11:

**FIGURE 8.11**

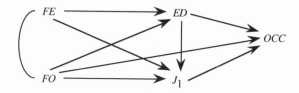

The model shown passes the same likelihood ratio test with $p > .3$. Freedman argues that in the American population we should expect that the influences among these variables differ from family to family, and therefore that the assumption that all units in the population have the same structural coefficients is unwarranted. The same possibility is raised by the data itself, if we apply the FCI or modified PC algorithms. If a population consists of a mixture of subpopulations of linear systems with the same causal structure but different variances and linear coefficients, then unless the coefficients are independently distributed or the mixture is in special proportions, the population correlations will be different from those of any of the subpopulations, and variables independent in each subpopulation may be correlated in the whole. When subpopulations with distinct linear structures are mixed and these special conditions do not obtain, the directed graph found from the correlations will typically be complete. We see that in order to fit Blau and Duncan's data we need a graph that is only one edge short of being complete. Mixtures can be thought of as produced by latent variables, and the Blau and Duncan data with the modified PC algorithm produces a POIPG in which every directed edge in figure 8.14 has a small "o" at the tail. In other words, the data are consistent with a model in which all of the observed dependencies are due to unmeasured factors.

The same moral is if anything more vivid in another linear model built from the same empirical study by Duncan, Featherman, and Duncan (1972). They developed the following model of socioeconomic background and occupational achievement, where *FE* signifies father's education, *FO* father's occupational status, *SIB* the number of the respondent's siblings, *ED* the respondent's education, *OCC* the respondent's occupational status and *INC* the respondent's income.

In this case the double-headed arrows merely indicate a residual correlation. The model has four degrees of freedom, and entirely fails the EQS likelihood ratio test ($\chi^2$ is 165). When the correlation matrix is given to the TETRAD II program (Spirtes, Glymour, and Scheines 1993) along with an obvious time ordering of the variables, the modified PC algorithm produces a complete graph: all the dependencies may be due to unmeasured factors. This data set and its analysis have served as a model for quantitative sociology; unfortunately, I think the only reasonable conclusion is that it tells us almost nothing about which of these variables influences others, and how much.

A more positive application concerns Needleman's study of the influence of low level lead on children's I.Q. Needleman, Geiger, and Frank (1985) estimated the effect of lead exposure on I.Q. by regressing I.Q. scores for 221 children on six variables, including measured lead concentration in baby teeth and found a significant negative Beta coefficient for lead exposure. S. Klepper assumed reasonably that all variables were measured with error and using an errors in variables model and interval bounding techniques concluded that the data

**FIGURE 8.12**

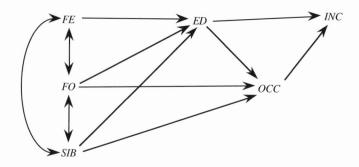

were consistent with no lead effect. Scheines (1996) analyzed the data with the TETRAD II program and found that three of Needleman's regressors have no influence on I.Q., that the negative influence of lead is increased in a regression on the remaining set of variables, and that for any reasonable priors on measurement error, the posterior probability of lead effect is large.

## 4. Prediction and Causation

For practical purposes, the chief reason we are interested in inferences to causal structure at all is in aid of predicting the results of interventions for an individual or a population, and many writers have regarded the connections between causal claims and claims about the results of ideal interventions as analytic. One influential view about causation goes something like this: If in a system S with a set of variables V, variable X is a direct cause of variable Y then an ideal intervention that changes X sufficiently (i) changes no variable that is not an effect of X, and (ii) changes Y. Consider again deterministic relations in three examples.

The graphs describing the functional dependencies that result when a value of T is forced from outside the system are given on the right-hand side of figure 8.13 in each case. Variables that may be altered by such an intervention are denoted in the equations with a subscript t.

The equations in case i are: and with the intervention:
$$Y = f(U, V, T)$$ $$Y_t = f(U, V, T_t)$$
$$X = g(V)$$ $$X = g(V)$$
$$T = h(X)$$ $$T = t$$

**FIGURE 8.13**

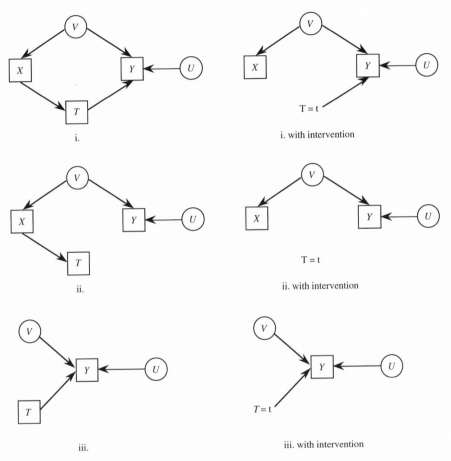

i.

i. with intervention

ii.

ii. with intervention

iii.

iii. with intervention

Notice that the value of $Y_t$ may depend on the value of X because $Y_t = f(U, V, t)$ and X is a function of V. It may be that for some pair of values $x_1$, $x_2$ of X, and for some $v_1 \in g^{-1}(x_1)$ and some $v_2 \in g^{-1}(x_2)$, and for some u and t, $f(u, v_1, t) \neq f(u, v_2, t)$.

In cases iii, the intervention makes no change in the equations or in the directed graph. In general, if we consider ideal interventions to fix the value of a variable T in a system of equations, then in the corresponding directed graph all edges directed into T are removed. Of itself, the Working Assumption seems to provide little information about the results of interventions. If we consider an intervention that fixes a value for T, then in systems so treated T is independent of all other variables, and so all edges into and out of T should be removed. A further assumption suggests itself, however:

*No Side Effects principle:* In a population of systems whose causal structure is described by a directed graph *G*, an ideal intervention that directly manipulates T may alter the distribution of T conditional on the parents of T, but the distribution of every other variable conditional on its parents is unchanged.

The No Side Effects principle can be expressed as a consequence of the Markov condition:

*Theorem 8 (Glymour, Spirtes, Fienberg):* Let an ideal manipulation of a variable T in a system represented by a graph G be represented by extending *G* to $G^*$ by adding a variable M and an edge directed into T and no other edges. Let the original distribution *P* on *G* be the distribution $P^*$ on $G^*$ conditional on $M = m_1$ and let the distribution that results from the intervention be the distribution conditional on $M = m_2$. If $<P^*, G^*>$ satisfies the Markov condition then the No Side Effects principle is necessarily satisfied.

The theorem holds as well for simultaneous interventions that directly manipulate several variables. Define a *trek* between variables X, Y as a directed path from X to Y or a directed path from Y to X or a pair of directed paths from some third variable, S, respectively to X and to Y and intersecting only at S.

*Theorem 9 (Spirtes, Glymour):* If **Z** is a set of variables containing one variable from every trek between T and Y, and M is a direct manipulation of T satisfying the conditions of Theorem 8, then for every value t of T that has positive probability conditional both on $M = m_1$ (no intervention) and $M = m_2$ (intervention), $P^* (Y \mid T = t, M = m_1, Z) = P^* (Y \mid T = t, M = m_2, Z)$.

Theorems 8 and 9 are the basis for predictions from a causal hypothesis subject to the Markov half of the Working Assumption and interventions satisfying the No Side Effects principle. The theorems are illustrated, without being stated, in the statistical literature. I will give two examples, Lindley and Novick's (1981) discussion of two cases of Simpson's paradox, and Rubin's (1974) discussion of the confounded experimental design previously mentioned.

Lindley and Novick consider the two following Simpson-like cases:

Case 1. We obtain data on recoveries for samples of males and females who have received a treatment (t) and a control (c):

| Males | R = 1 | R = 0 |
|-------|-------|-------|
| T = t | 18 | 12 |
| T = c | 7 | 3 |

| Females | R = 1 | R = 0 |
|---------|-------|-------|
| T = t | 2 | 8 |
| T = c | 9 | 21 |

| Combined | R = 1 | R = 0 |
|----------|-------|-------|
| T = t | 20 | 20 |
| T = c | 16 | 24 |

The sample is understood to be very, very large. We are not given information as to how persons were chosen for the sample or for treatment value. Note that the recovery rate is higher for T = c for both males and females but the recovery rate is higher for T = t for the combined group.

Question 1: In view of this information, if we are presented with a new subject whose gender is unknown, which treatment should we prefer, t or c?

Lindley and Novick say we should prefer T = c.

Case 2: We obtain analogous data on yields and heights for samples of black and white plants.

| Tall | Y = 1 | Y = 0 |
|------|-------|-------|
| C = w | 18 | 12 |
| C = b | 7 | 3 |

| Short | Y = 1 | Y = 0 |
|-------|-------|-------|
| C = w | 2 | 8 |
| C = b | 9 | 21 |

| Combined | Y = 1 | Y = 0 |
|----------|-------|-------|
| C = w | 20 | 20 |
| C = b | 16 | 24 |

Question 2: This time we must decide whether to plant a white (C = w) or a black variety of plant, in ignorance of the height the plant will grow to.

Lindley and Novick say we should prefer C = w.

Lindley and Novick say the issue is whether the new patient is "exchangeable" with the patients in the sample and whether the new plant is "exchangeable" with the plants in the sample. I find this an unilluminating way to say that the question is whether we should let the frequency determined from the samples equal our probability for R if we impose treatment T and, in parallel, equal our probability for Y if we choose a plant of color C. Lindley and Novick make it quite plain that the difference they see in the two cases is causal, but say they prefer not to talk about that. So much the worse for their discussion, since the very issue is the interaction of causal constraints and probabilities.

Their idea, which I think is correct, is that in case 1 in the sample data, T and gender (which they denote by M) are statistically dependent, but since one does not know the gender of the new subject, the new subject's gender can have no "effect" (their word), no influence, on one's choice for the value of T for the new subject, and therefore (?) for the new subject one should treat T and G as statistically independent. By contrast, in case 2 the decision as to which variety of plant to grow does nothing to alter the causal processes that produce in the sample the statistical dependency between plant height and color, and therefore (?) one's probability distribution for the new subject should treat plant height and color as statistically dependent just as they are in the sample. So in the second case the sample conditional probability of Y on C should equal the probability of Y given that variety C is planted. They use without argument the assumption that in case 1 the distribution for the new subject should be exactly as before save that T and M should be independent.

Theorem 8 gives a smooth account of what is going on in Lindley and Novick's case 1. We don't need to know the actual graph, only that gender (M) is not an effect (a descendant) of treatment T, and we find that when we intervene to impose a treatment on the new subject, T and M must be independent, and the result is immediate. In case 2, the decision to plant one variety or another does not interfere in the causal processes (e.g., the genetic features) that produce the association between height and color in the sample, and whatever the causal structure may be, no processes are altered that terminate in color and connect to height and yield, and we have Lindley and Novick's solution.

For a use of Theorem 9, consider Rubin's treatment of the example given earlier. In an educational experiment in which reading program assignments $T$ are assigned on the basis of a randomly sampled value of some pre-test variable $X$ which shares one or more unmeasured common causes, $V$, with $Y$, the score on a post-test, we wish to predict the average difference $\tau$ in $Y$ values if all students in the population were given treatment $T = 1$ as against if all students were given treatment $T = 2$. The situation in the experiment is represented in figure 8.14.

**FIGURE 8.14**

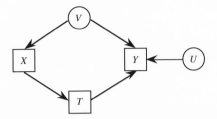

Provided the experimental sample is sufficiently representative, Rubin says that an unbiased estimate of $\tau$ can be obtained as follows: Let $k$ range over values of $X$, from 1 to $K$, let $\overline{Y1k}$ be the average value of $Y$ conditional on $T = 1$ and $X = k$, and analogously for $\overline{Y2k}$. Let $n1k$ be the number of units in the sample with $T = 1$ and $X = k$, and analogously for $n2k$. The numbers $n1$ and $n2$ represent the total number of units in the sample with $T = 1$ and $T = 2$ respectively.

Let $\bar{Y}_{Tf=1}$ = expected value of $Y$ if treatment 1 is forced on all units. According to Rubin, estimate

$$\bar{Y}_{Tf=1}$$

by:

$$\sum_{k=1}^{K} \frac{n1k + n2k}{n1 + n2} \, \overline{Y1k}$$

and estimate $\tau$ by:

$$\sum_{k=1}^{K} \frac{n1k + n2k}{n1 + n2} \, \overline{Y1k} - \overline{Y2k}$$

The basis for this choice may not be apparent. If we look at the hypothetical population in which every unit is forced to have $T = 1$, then it is clear from Rubin's tacit independence assumptions that he treats the manipulated population as if it had the causal structure shown in figure 8.15, as the following derivation shows.[16]

$$\bar{Y}_{Tf=1} = \sum_{Y}^{\rightarrow} Y \times P(Y_{Tf=1}) =$$

$$\sum_{Y}^{\rightarrow} Y \times \sum_{k=1}^{K} P(Y_{Tf=1}|X_{Tf=1} = k, T_{Tf=1} = 1)P(X_{Tf=1} = k|T_{Tf=1} = 1)P(T_{Tf=1} = 1) =$$

$$\sum_{Y}^{\rightarrow} Y \times \sum_{k=1}^{K} P(Y_{Tf=1}|X_{Tf=1} = k, T_{Tf=1} = 1)P(X_{Tf=1} = k)$$

**FIGURE 8.15**

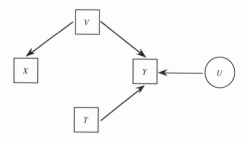

The second equality in the above equations holds because $P(T_{Tf=1} = 1) = 1$, and $X_{Tf=1}$ and $T_{Tf=1}$ are independent according to the causal graph shown in figure 8.15. By Theorem 9, both $P(Y_{Tf=1}|X_{Tf=1}, T_{Tf=1})$ and $P(X_{Tf=1})$ are invariant under direct manipulation of $T$ in the graph of figure 8.15. This entails the following equation.

$$\bar{Y}_{Tf=1} = \overrightarrow{\sum_{Y}} Y \times \sum_{k=1}^{K} P(Y_{Tf=1}|X_{Tf=1} = k, T_{Tf=1} = 1)P(X_{Tf=1} = k) =$$

$$\sum_{k=1}^{K} P(X = k) \times \overrightarrow{\sum_{Y}} Y \times P(Y|X = k, T=1) = \frac{n1k + n2k}{n1 + n2} \times \overline{Y1k}$$

Note that $X$ and $T$, unlike $X_{Tf=1}$ and $T_{Tf=1}$ are *not* independent. Rubin's assumption that $X_{Tf=1}$ and $T_{Tf=1}$ are independent indicates, as I see things, that he is implicitly assuming that the causal graph of the manipulated population is the graph of figure 8.15, not the graph of figure 8.14, which is the causal structure of the unmanipulated population. $\bar{Y}_{Tf=2}$ can be derived in an analogous fashion.

If there is no trek between T and Y containing an edge directed into T, then the distribution of Y after an ideal intervention that directly manipulates T is the same no matter what value T has before the intervention. In an informal sense, $Y_t$ is independent of T, although it is not obvious that there is a common sample space on which $Y_t$ and T are random variables.

A recent tradition in statistics developed from Rubin's work has tried to capture the import of Theorem 9, without the aid of directed graphical representations, by assuming that there *is* in fact a sample space for $Y_t$ and T. The formalism of the "Rubin framework," as it has been developed by Rubin, by Rosenbaum, by Holland (1986), and by Pratt and Schlaifer (1988) and others, has been one of the bright spots in statisticians' discussions of causality in the last twenty years, and it seems to have influenced writing in some substantive areas, especially epidemiology.

The Rubin framework treats every unit in a population as having not only variables whose values a unit exhibits but also dispositional variables, whose values may never be exhibited. Thus a student not only has a post-test score Y, but also has the unobserved post-test score $Y_t$ that she *would have exhibited* if she had been given reading program t. So far this is nothing unusual (at least to philosophers). But T and $\{Y_t\}$, for all values t, are also regarded as random variables with a joint distribution. Holland gives an argument that under a condition he calls "strong ignorability" the difference between the expected values of $Y_{t1}$ and $Y_{t2}$, for any values t1, t2 of T, can be computed from the expected value of the difference of the expected values of Y conditional on T = t1 and Y conditional on T = t2. The chief result Pratt and Schlaifer claim is that under strong ignorability, $Y_t$ has the same distribution as does Y conditional on T = t. No proof is given. All of their examples accord with the No Side Effects principle.

The accounts just sketched, that is Theorems 8 and 9 and the corresponding aspects of the Rubin framework, are for the easy case when the relevant parts of the causal structure are completely known. With nonexperimental data the real problem is when and how to predict when the causal structure *isn't* known. Without experimental control or independent knowledge constraining the possibilities, the best one can expect to have on the basis of sample data is a POIPG or finite collection of alternative POIPGs. Spirtes has described a Prediction Algorithm that provides a sufficient condition for predicting the distribution of a variable Y, given a POIPG, an intervention satisfying the No Side Effects principle, and an observed distribution. Rather than giving the algorithm, which is intricate, I'll describe the idea and repeat an illustration from Spirtes, Glymour, and Scheines (1993). Suppose we measure *Genotype (G)*, *Smoking (S)*, *Income (I)*, *Parents' smoking habits (PSH)* and *Lung cancer (L)*. Suppose the unmanipulated distribution is faithful to the unmanipulated graph that has the partially oriented inducing path graph shown in figure 8.16.

The partially oriented inducing path graph does not tell us whether *Income* and *Smoking* have a common unmeasured cause, or *Parents' smoking habits* and *Smoking* have a common unmeasured cause, and so on. If we directly manipulate *Smoking* so that *Income* and *Parents' smoking habits* are not parents of *Smoking* in the manipulated graph, then no matter which graph produced the marginal distribution, the partially oriented inducing path graph and the Manipulation Theorem tell us that if *Smoking* is directly manipulated that in the manipulated population the resulting causal graph will be consistent with the POIPG shown in figure 8.17.

In this case, we can determine the distribution of *Lung cancer* given a direct manipulation of *Smoking*. Three steps are involved. First, from the partially oriented inducing path graph we find a way to factor the joint distribution in the manipulated graph. Let $P_{Unman}$ be the distribution on the measured variables and let $P_{Man}$ be the distribution that results from a direct manipulation of *Smoking*. It can be determined from the partially oriented inducing path graph that

$$P_{Man}(I, PSH, S, G, L) = P_{Man}(I) \times P_{Man}(PSH) \times P_{Man}(S) \times P_{Man}(G) \times P_{Man}(L|G, S)$$

where $I = Income$, $PSH = Parents'$ $smoking$ $habits$, $S = Smoking$, $G = Genotype$, and $L = Lung$ $cancer$. This is the factorization of $P_{Man}$ corresponding to the immediately preceding graph that represents the result of a direct manipulation of $Smoking$.

Second, we can determine from the partially oriented inducing path graph which factors in the expression just given for the joint distribution are needed to calculate $P_{Man}(L)$. In this case $P_{Man}(I)$ and $P_{Man}(PSH)$ prove irrelevant and we have:

$$P_{Man}(L) = \overset{\longrightarrow}{\underset{G, S}{\sum}} P_{Man}(S) \times P_{Man}(G) \times P_{Man}(L|G, S)$$

**FIGURE 8.16**

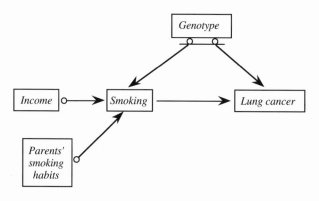

**FIGURE 8.17**

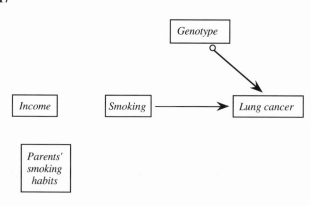

Third, we can determine from the partially oriented inducing path graph that $P_{Man}(G)$ and $P_{Man}(L|G,S)$ are equal respectively to the corresponding unmanipulated probabilities, $P_{Unman}(G)$ and $P_{Unman}(L|G,S)$. Furthermore, $P_{Man}(S)$ is assumed to be known, since it is the quantity being manipulated. Hence, all three factors in the expression for $P_{Man}(L)$ are known, and $P_{Man}(L)$ can be calculated.

Note that $P_{Man}(L)$ can be predicted even though $P(L)$ is most definitely not invariant under a direct manipulation of $S$.

## 5. Implications for the Design of Empirical Studies

The results of the Working Assumption and the No Side Effects principle contradict some standard methodology in social science and statistics, both theoretical and applied. For example, they reveal new defects, as if more were needed, in regression as a technique for causal inference. I will focus instead on what they imply about the design of empirical studies.

Arguing against epidemiologists' inferences that smoking causes cancer, Fisher (1959) claimed again and again that observational evidence can never suffice to determine whether a statistical dependency among two or more variables is due entirely to an unobserved common cause. His argument is that a correlation or other dependency between X, Y will appear if X is a function of Y (and other stuff, of course), Y of X , or both are functions of some other variable Z. The argument is invalid: it does nothing to show that relations among X, Y and other observed variables are never sufficient to decide whether X, Y have an unobserved common cause. And the conclusion is false: under the Working Assumption, there are purely statistical sufficient conditions for a dependency between two measured variables from a larger set of measured variables to be entirely due to unmeasured common causes; in parallel, there are sufficient conditions for a dependency to be due entirely to the influences of measured variables; there are sufficient conditions for the dependency of X and Y to be due to an influence of X on Y, and finally, sufficient conditions for there to be no influence of X on Y.

Fisher's argument is a piece of a larger view that holds causal inference from experimental and nonexperimental data to rely on quite different principles, and holds the latter to be unsound in principle; I think it is fair to say that those who hold such views have given no rigorous systematic account of the supposed differences. Under the Working Assumption there are intricate connections between the capacities of experimental and nonexperimental methods, but the basic message is simple: in experimental designs one knows something that may be true in nonexperimental designs but, if true, has to be discovered.

Consider three alternative causal structures, and let us suppose for the moment that they exhaust the possibilities and are mutually exclusive: (i) *A* causes

$C$, (ii) some third variable $B$ causes both $A$ and $C$, or (iii) $C$ causes $A$. If by experimental manipulation we can produce a known distribution on $A$ not caused by $B$ or $C$, and if we can produce a known distribution on $C$ not caused by $A$ or $B$, we can distinguish these causal structures. It might be, for example, that the population is divided randomly into two groups, and each group is given a different fixed value of $A$. In that case values of $A$ are assigned on the basis of the output of some random device which we can think of as a further variable $U$. In the experiment, all of the edges into $A$ in the causal graph of the nonexperimental population are broken, and replaced by an edge from $U$ to $A$; furthermore there is no non-empty undirected path between $U$ and any other variable in the graph that does not contain the edge from $U$ to $A$. Any procedure in which $A$ is caused only by a variable $U$ with these properties I will call a *randomized experiment*. In a randomized experiment we know three useful facts about $U$: $U$ causes $A$, there is no common cause of $U$ and $C$, and if $U$ causes $C$ it does so by a mechanism that is blocked if $A$ is held constant (i.e., in the causal graph if there is a directed path from $U$ to $C$ it contains $A$.).

The randomized experimental setups for the three alternative causal structures are shown in figure 8.18, where an $A$-experiment represents a manipulation of $A$ breaking the edges into $A$, and a $C$-experiment represents a manipulation of $C$ breaking edges into $C$. If we do an $A$-experiment and find partially oriented inducing path graph (ia*) over $\{A,C\}$ then we know that $A$ causes $C$ because we know that we have broken all edges into $A$. Similarly, if we perform a $C$-experiment and find partially oriented inducing path graph (iiic*) then we know that $C$ causes $A$. If we perform an $A$-experiment and get (iia*) and a $C$-experiment and get (iic*) then we know that there is a latent common cause of $A$ and $C$ (assuming that $A$ and $C$ are dependent in the nonexperimental population).

Now suppose that in the nonexperimental population there are variables $U$ and $V$ *known* to bear the same relations to $A$ and $C$ respectively as in the experimental setup. (We assume in the nonexperimental population that $A$ is not a deterministic function of $U$, and $C$ is not a deterministic function of $V$.) That is, $U$ causes $A$, there is no common cause of $U$ and $C$, and if there is any directed path from $U$ to $C$ it contains $A$; also, $V$ causes $C$, there is no common cause of $V$ and $A$, and if there is any directed path from $V$ to $A$ it contains $C$. Can we still distinguish (i), (ii), and (iii) from each other without an experiment? The answer is yes. In figure 8.19 (io*), (iio*) and (iiio*) are the partially oriented inducing path graphs corresponding to (io), (iio), and (iiio) respectively. Suppose the FCI algorithm constructs (io*). If it is known that $U$ causes $A$, then from the fact that the edge between $U$ and $A$ and the edge between $A$ and $C$ do not collide, we can conclude that the edge between $A$ and $C$ is oriented as $A \rightarrow C$ in the inducing path graph. It follows that $A$ causes $C$. Similarly, if the FCI algorithm constructs (iiio*) ideally

**FIGURE 8.18**

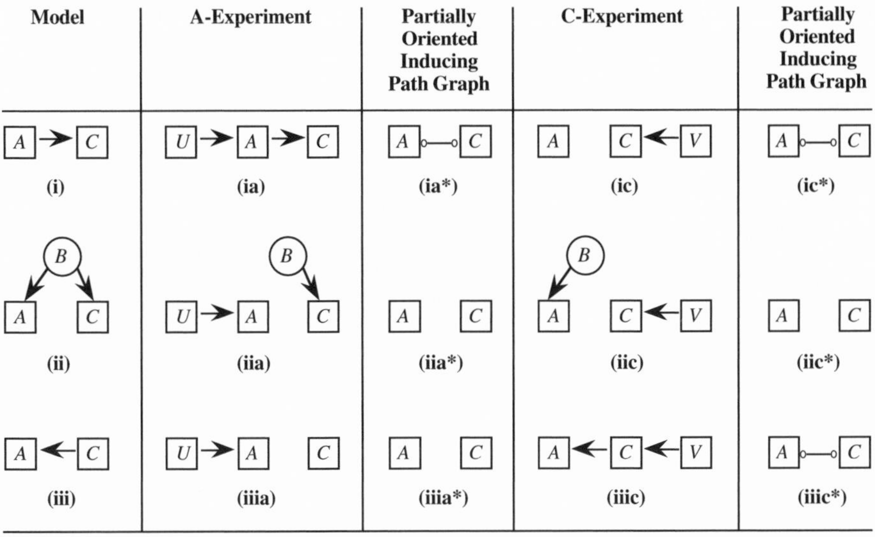

| Model | A-Experiment | Partially Oriented Inducing Path Graph | C-Experiment | Partially Oriented Inducing Path Graph |
|---|---|---|---|---|
| $A \rightarrow C$ <br> (i) | $U \rightarrow A \rightarrow C$ <br> (ia) | $A \circ\!\!-\!\!\circ C$ <br> (ia*) | $A \quad C \leftarrow V$ <br> (ic) | $A \circ\!\!-\!\!\circ C$ <br> (ic*) |
| $B$ <br> $A \quad C$ <br> (ii) | $B$ <br> $U \rightarrow A \quad C$ <br> (iia) | $A \quad C$ <br> (iia*) | $B$ <br> $A \quad C \leftarrow V$ <br> (iic) | $A \quad C$ <br> (iic*) |
| $A \leftarrow C$ <br> (iii) | $U \rightarrow A \quad C$ <br> (iiia) | $A \quad C$ <br> (iiia*) | $A \leftarrow C \leftarrow V$ <br> (iiic) | $A \circ\!\!-\!\!\circ C$ <br> (iiic*) |

we can conclude that $C$ causes $A$. The partially oriented inducing path graph in (iio*) indicates that there is a latent common cause of $A$ and $C$, and that $A$ does not cause $C$, and $C$ does not cause $A$.

Note that if we had measured variables such as $W$, $U$, $V$, and $X$ in figure 8.20 then the corresponding partially oriented inducing path graphs would enable us to distinguish (i), (ii), and (iii) without experimentation and without the use of any prior knowledge about the causal relations among the variables.

In the more complex cases in which the possibilities are (i) $A$ causes $C$ and there is a latent common cause $B$ of $A$ and $C$, (ii) there is a latent common cause $B$ of $A$ and $C$, and (iii) $C$ causes $A$ and there is a latent common cause $B$ of $A$ and $C$, a parallel analysis holds. Each of the structures (i), (ii) and (iii) can be distinguished from the others by experimental manipulations in which for a sample of systems we break the edges into $A$ and impose a distribution on $A$ and for another sample we break the edges into $C$ and impose a distribution on $C$. The analysis of the corresponding nonexperimental case is more complicated. Assume that there is a variable $U$ and it is known that $U$ causes $A$, there is no common cause of $U$ and $A$, and if there is any directed path from $U$ to $C$ it contains $A$, and that there is a variable $V$ and it is known that $V$ causes $C$, there is no common cause of $V$ and $A$, and if there is any directed path from $V$ to $A$ it contains $C$. The directed acyclic graphs and their corresponding partially oriented inducing path graphs are shown

**FIGURE 8.19**

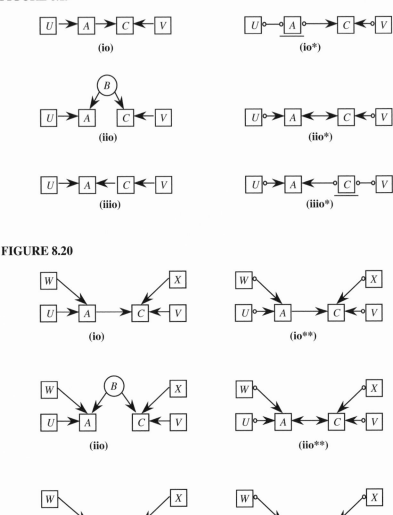

**FIGURE 8.20**

in figure 8.21. Now suppose that the directed acyclic graphs are true of an observed nonexperimental population. We can still distinguish (i), (ii), and (iii) from each other:

For example, suppose that an application of the FCI algorithm produces (io*). The existence of the $U \circ\!\!\to C$ edge entails that either there is a common cause of $U$ and $C$ or a directed path from $U$ to $C$. By assumption, there is no common cause of $U$ and $C$, so there is a directed path from $U$ to $C$. Also by as-

**FIGURE 8.21**

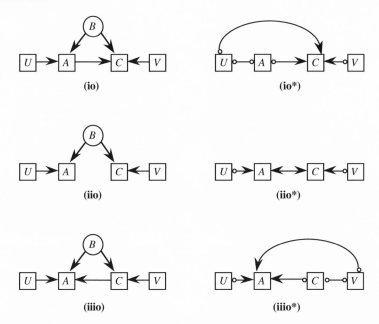

sumption, all directed paths from *U* to *C* contain *A*, so there is a directed path from *A* to *C*. Given that there is an edge between *U* and *C* in the partially oriented inducing path graph, and the same background knowledge, it also follows that there is a latent common cause of *A* and *C*. Similarly, if we obtain (iiio*) in figure 8.21 then we know that *C* causes *A* and there is a latent common cause of *A* and *C*. If we obtain (iio*) then we know that *A* and *C* have a latent common cause but that *A* does not cause *C* and *C* does not cause *A*.

It is also possible to distinguish (i), (ii), and (iii) from each other without any prior knowledge of particular causal relations, but it requires a more complex pattern of measured variables, as shown in figure 8.22. If we obtain (io**) then we know without using any such prior knowledge about the causal relationships between the variables that *A* causes *C* and that there is a latent common cause of *A* and *C*, and similarly for (iio**) and (iiio**).

The example illustrates one important advantage to experimentation over passive observation. By performing an experiment we can make a quantitative prediction about the consequences of manipulating *A* in (i), (ii), and (iii). But if (i) is the correct causal model, we cannot make a quantitative prediction of the effects of manipulating *A*. (In the linear case, a prediction could be made because *U* serves as an "instrumental variable.")

Suppose finally that we want to know whether there are two causal pathways that lead from *A* to *C*. More specifically, suppose we want to distinguish

**FIGURE 8.22**

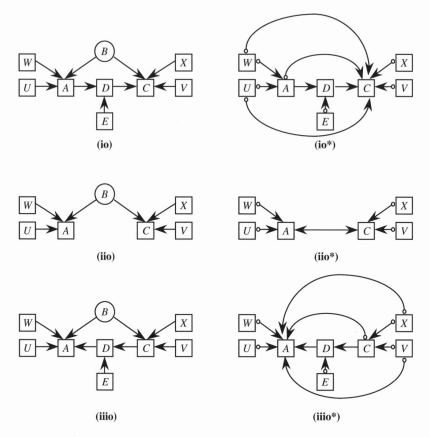

which of (i), (ii) and (iii) in figure 8.23 obtains, remembering again that $B$ is unmeasured.

By experimental manipulation that breaks the edges directing into $A$ and imposes a distribution on $A$, independence relations distinguish structures (i) and (iii) from structure (ii) but not from one another.

Assume once again that in a nonexperimental population it is known that $U$ causes $A$, that there is no common cause of $U$ and $C$, and that if there is any path from $U$ to $C$ it contains $A$. It is also known that $V$ causes $C$, and that there is no common cause of $V$ and $A$, and that if there is any path from $V$ to $A$ it contains $C$. The directed acyclic graphs and their corresponding partially oriented inducing path graphs are shown in figure 8.24.

Unlike the randomized experimental case, where (i) and (iii) cannot be distinguished, in the nonexperimental case they *can* be distinguished. Suppose we obtain (iiio*). We know from the background knowledge that $U$ causes $A$, and

**FIGURE 8.23**

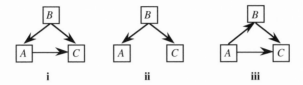

i          ii          iii

**FIGURE 8.24**

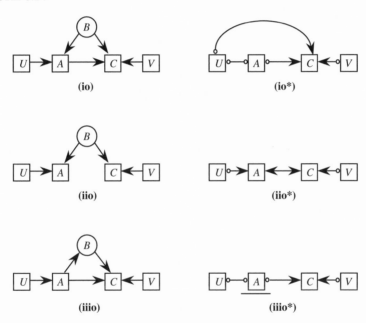

from (iiio*) that the edge between *U* and *A* does not collide with the edge between *A* and *C*. Hence in the corresponding inducing path graph there is an edge from *A* to *C* and in the corresponding directed acyclic graph there is a path from *A* to *C*. (Of course we cannot tell how many paths from *A* to *C* there are; (iiio*) is compatible with a graph like (iiio) but in which the <*A,B,C*> path does not exist.) We also know that there is no latent common cause of *A* and *C* because (iiio*) together with our background knowledge entails that there is no path in the inducing path graph between *A* and *C* that is into *A*. Suppose on the other hand that we obtain (io*). Recall that the background knowledge together with the partially oriented inducing path graph entail that *A* is a cause of *C* and that there is a latent common cause of *A* and *C*.

**FIGURE 8.25**

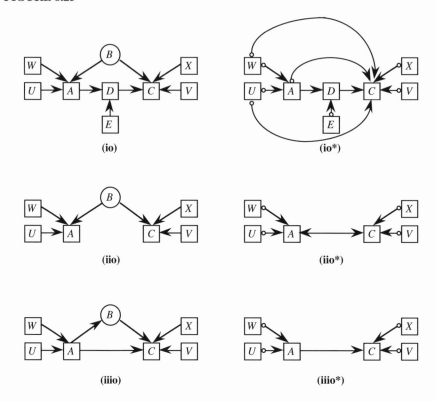

(io)                    (io*)

(iio)                    (iio*)

(iiio)                   (iiio*)

Once again if more variables are measured, it is also possible to distinguish these three cases without any background knowledge about the causal relationships among the variables, as shown in figure 8.25.

It may seem extraordinary to claim that structure (i) in figure 8.23 cannot be distinguished from structure (iii) by a randomized experiment, but can be distinguished without experimental control if the structure is appropriately embedded in a larger structure whose variables are measured. It runs against common sense to claim that when $A$ causes $C$, a randomized experiment cannot distinguish $A$ and $C$ also having an unmeasured common cause from $A$ also having a second mechanism through which it effects $B$, but that observation without experiment sometimes can distinguish these situations. But randomized experimental manipulation that forces a distribution on $A$ breaks the dependency (in the experimental sample) of $A$ on $B$ in structure (i), and thus information that is essential to distinguish the two structures is lost.

While a randomized experiment alone cannot distinguish (i) from (iii) in figure 8.23, the combination of a simple observational study and randomized experimentation can do so. We can determine from an $A$-experiment that there is a path from $A$ to $C$, and hence no path from $C$ to $A$. We know if $P(C|A)$ is not invariant under manipulation of $A$ then there is a trek between $C$ and $A$ that is into $A$. Hence if $P(C|A)$ is different in the nonexperimental population and the $A$-experimental population we can conclude that there is a common cause of $A$ and $C$. If $P(C|A)$ is invariant under manipulation of $A$ then we know that either there is no common cause of $A$ and $C$ or the particular parameter values of the model "coincidentally" produce the invariance. By combining information from an observational study and an experimental study it is sometimes possible to infer causal relations that cannot be inferred from either alone. This is often done in an informal way.

These examples show, I hope, that the relations between experimental and observational procedures for identifying causal relations are considerably more intricate than acknowledged, and that Fisher's casual dismissals of causal inference from observational data, by now a standard textbook paragraph, are far too casual.

I want finally to note the implications of the Working Assumption and the No Side Effects principle for a fundamental problem in the design of ethical clinical trials. Where treatments of unknown safety and efficacy are to be compared experimentally, a respect for the autonomy of individuals argues that subjects in the experiment should be free to choose their own treatment. If, however, some variable feature of patients that influences treatment outcome also influences individual treatment choices, the experimental outcome will be "confounded," that is, one won't know how much of any association between treatment and outcome is due to a specific effect of treatment and how much to the common influence on patient choice and outcome. So, according to conventional wisdom, it is not possible to reconcile good ethics and good method.

If in a particular study we had sufficient reason to believe no factor influences both patient's preferences about treatment and the outcome of treatment, then there would be no objection to self-selection of treatment except that it might lead to some unrepresented treatment possibilities. If then also patient treatment preference did not entirely determine treatment, but instead biased a random assignment so that patients were more likely to get their preferred treatment, the usual methodological objections to self-assignment would not apply. The question is then whether it is in principle possible to identify circumstances in which there is no confounding. Extensions of the examples in this section show that the answer is affirmative. Details are given in Spirtes, Glymour and Scheines (1993).

## 6. Without the Working Assumption

### *Directed Independence Graphs*

Each way of abandoning the Working Assumption determines a new class of problems, some better motivated than others. Assumptions weaker than faithfulness have often been imposed. Wermuth and Lauritzen (1983) introduced the idea of a recursive diagram or directed independence graph, which is just a directed acyclic graph paired with a distribution such that there is a linear ordering of the vertices in the graph, and an edge $X \to Y$ occurs if and only if Y is later than X and X and Y are dependent conditional on the set of all vertices, except X, that precede Y in the ordering. If the distributions must be positive, this condition is equivalent to a condition introduced by Pearl, called minimality, which requires that G, P satisfy the Markov condition but for any proper subgraph G' of G, G', P do not satisfy the Markov condition. The edges in the graphs can be given the same sort of causal interpretation as has been suggested for edges in graphs satisfying the Markov and faithfulness conditions, and given a graph the same analysis of prediction under interventions can be given.

The problem of the equivalence of directed independence graphs was solved by Spirtes:

> *Theorem 10 (Spirtes):* Two directed acyclic graphs, G, G' admit the same family of probability distributions satisfying the conditions for directed independence graphs if and only if G and G' have the same vertex set, the same adjacencies, and the same colliders.

The very definition of directed independence graph suggests a discovery procedure if a linear ordering of the variables, as by time, is known. But the obvious procedure based on Wermuth and Lauritzen's definition is not feasible in practice, since it requires conditional independence tests controlling for an ever-increasing set of variables. Gregory Cooper and his collaborators have developed a heuristic Bayesian procedure, the K2 algorithm, for categorical variables—the procedure is not guaranteed to generate the indistinguishability class corresponding to the ordering it uses and the actual probability distribution. Nonetheless, the procedure works extraordinarily reliably and rapidly in simulation studies; when variables are categorical and prior knowledge allows only a few (hundred) linear orderings of the variables, and one knows there are no unmeasured common causes, the procedure can be applied in practice.

I know of no work concerning independence constraints induced on the marginal over a subset of measured variables by a directed independence graph with latent variables. There is neither a theory of indistinguishability nor a theory of inference for directed independence graphs with unobserved variables.

## Directed Chain Graphs

From a mathematical point of view, it seems natural to weaken the linear ordering required in directed independence graphs to a partial ordering. Wermuth and Lauritzen introduced the notion of a directed chain graph, which partitions the variables and linearly orders the sets in the partition. A directed edge X → Y appears if X is a member of a partition set prior to the set to which Y belongs, and X, Y are dependent conditional on the set of all variables, excluding Y, not later than Y in the ordering. An undirected edge X — Y occurs if X, Y are members of the same partition set and are dependent conditional on the set of all variables (excluding X and Y) not later than either of them. Frydenberg (1990) characterized indistinguishability for such graphs.

Some controversy has arisen over the interpretation of the undirected edges in directed chain graphs. Because they are undirected they do not lend themselves to an interpretation related to interventions. Only in special cases can the conditional independence relations in observed marginals of directed acyclic graphs satisfying the Markov condition be correctly represented by directed chain graphs. Whittaker (1990) and Lauritzen, Thiesson, and Spiegelhalter (1993), have claimed that the undirected edges represent statistical dependencies that result at equilibrium from feedback processes, but that edges in the chain graphs do not describe causal relations in the feedback process itself. That interpretation requires showing that some interesting class of time series have limiting distributions that are represented by the conditional independence relations encoded in directed independence graphs. Lauritzen has suggested (in conversation) that the Gibbs sampler has just such properties.

## Determinism

One of the interesting ways faithfulness can fail is if some observed variables are deterministic functions of other observed variables. In that case extra independence relations generally arise. Dan Geiger (1990) gave an algorithm for calculating the independencies from a directed acyclic graph in which some children are deterministic functions of their parents; Peter Spirtes generalized the procedure to other cases of determinism, but there is no proof that Spirtes' algorithm is complete. No one has worked out an indistinguishability characterization or inference procedures for structures with deterministic relations, but if Spirtes' algorithm is complete the theory should be reasonably straightforward.

## Feedback and Directed Cyclic Graphs

Perhaps the most interesting way to deny the Markov condition is to consider feedback processes whose causal interpretation seems to require a cyclic graph.

In the form of linear equations such systems are not unusual in econometrics and in engineering. Nonlinear forms are frequently encountered in control theory, and the most biologically realistic form of neural net is a nonlinear feedback system. Although the basis for representing correlations in linear feedback systems was described by Haavelmo fifty years ago, and a purely graphical rule for determining correlations can be derived from Mason's work in the 1950s, until very recently virtually nothing was known about the relations between directed cyclic graphs and vanishing partial correlations in distributions given by corresponding linear dependencies with independent noise terms. Glymour et al. (1987) obtained a purely graphical necessary and sufficient condition for a directed cyclic graph to linearly entail a vanishing partial correlation controlling for a single variable. This result, and a number of examples of vanishing second-order partial correlations, led Spirtes, Glymour, and Scheines (1993) to conjecture that a directed cyclic graph linearly entails a vanishing partial correlation (of any order) if and only if the correlated variables are d-separated by the set of variables controlled for. An odd generalization of the result for first order partials can be derived from a technique of Whittaker (1990): in a directed cyclic graph, X, Y are d-separated by a set $\mathbf{Z}$ of variables if and only if the graph linearly implies $\rho XY.\mathbf{Z} = 0$, provided every variable that is a parent in the graph of any two or more members of $\{X,Y\} \cup \mathbf{Z}$ is also in the set $\{X,Y\} \cup \mathbf{Z}$.

A completely general result has recently been obtained by Spirtes (1993) and, independently, by Koster (1996) for the linear case.

*Theorem 11 (Spirtes 1993, Koster):* A directed graph linearly entails that vertex X is independent of vertex Y conditional on a set $\mathbf{Z}$ of vertices if and only if X, Y are d-separated by $\mathbf{Z}$.

The result has a number of obvious applications to recent discussions; for example, it gives the correct account of the independence relations in a simultaneous equation model that Whittaker (1990) claims is uninterpretable—in fact the very example used at the beginning of this essay—and it shows that patterns of conditional independence relations that Cox and Wermuth (1993) find especially puzzling can be generated by cyclic graphs. Thus a linear model with the graph of figure 8.4 can be described by linear simultaneous equations:

$$X_3 = a_3 X_1 + b_3 X_4 + \varepsilon_3$$
$$X_4 = a_4 X_2 + b_4 X_3 + \varepsilon_4$$

or by a process in which $x_{3t}$ depends on $x_{1t-1}$, $x_{4t-1}$, and $\varepsilon_{3t-1}$, while $x_{4t}$ depends on $x_{3t-1}$, $x_{2t-1}$, and $\varepsilon_{4t-1}$, with the same linear coefficients as in the simultaneous equations representation:

$$X_{3t} = a_3 X_{1t-1} + b_3 X_{4t-1} + \varepsilon_{3t-1}$$
$$X_{4t} = a_4 X_{2t-1} + b_4 X_{3t-1} + \varepsilon_{4t-1}$$

where all error terms $\varepsilon_3$ have the same value, and all error terms $\varepsilon_4$ have the same values. For values of the linear coefficients such that the simultaneous equation system has a solution, the simultaneous equations are satisfied by the limiting values of the X variables as t increases without bound. More generally, the conditional independence relations in linear cyclic models correspond to those in time series with constant errors.

Thomas Richardson (1995) has provided conditions for two directed graphs, of whatever cyclicity, to have the same d-separation properties. Richardson's conditions are intricate and I will not give them here, but they lead to a polynomial time decision procedure for equivalence, and they yield an algorithm for generating a representation of the equivalence class from a faithful set of independence and conditional independence relations.

Spirtes has shown that if we consider "non-recursive" *non-linear* systems with independently distributed error terms, d-separation fails to give the correct independence and conditional independence relations. He provides, however, an algorithm for computing the conditional independencies implied assuming each error variable is a function of the substantive variable it influences and of that variable's parents. Pearl and Dechter (1996) have shown that d-separation implies conditional independence in cyclic graphs representing systems of discrete variables for which values of all variables are uniquely determined by values of exogenous, independent discrete error variables. However, d-separation does not in general give all conditional independencies implied by such discrete models, and in most discrete cyclic systems the values of error terms do not determine unique values for all variables.

## 7. Concluding Remarks

Statisticians and social scientists often dismiss automated search procedures out of hand. I find their reasons untenable, but there are some concerns that deserve comment. One is that automated search procedures don't come with analytic error probabilities. I can't tell you, for example, as a function of the sample size alone, the probabilities of various sorts of graphical errors in the output of the PC algorithm or the FCI algorithm or the K2 algorithm, even when the family of distributions is correctly assumed. Why this should be a reason for dismissing the procedures, however, is a mystery to me. No one can tell me the probabilities of various sorts of errors in the models a particular sociologist or econometrician conjectures or publishes. The automated procedures have at least two advantages in this regard. First, we do know that under various assumptions connecting distributions and graphs, the automated procedures are asymptotically correct; we don't know any such thing about the output of the social scientist. Second, we can study empirically the error probabilities of automated procedures in a wide range of circumstances by running the procedures on simulated data from known structures; extensive studies of this kind have been carried out for the procedures I have mentioned in this essay. In principle but not in practice, the same thing could be done with the econometrician or the sociologist or the statistician. It never is. I think the general case for automated search is overwhelming: anything the human researcher actually knows before considering an actual data sample can be elicited and told to a computer; anything further the researcher can infer from examining a data sample can be inferred by a computer, and the computer can otherwise avoid the inconsistencies and the biases of humans. Practitioners would benefit even from the necessity, forced by the TETRAD II program, of making clear which of their claims are prior assumptions forced on the data and which are conjectures extracted from the data.

The methods I have described have made few inroads into social scientific practice, partly because they have only recently appeared in literatures social scientists don't read, and partly because they contradict the dogmas in which almost every student of statistics or social science is tutored. (They have, however, already found a number of applications in medicine, psychiatry, biology, and physics). But I think the case of Blau and Duncan, and many others I could give, illustrate another, ultimately graver, reason why computerized discovery methods may be resisted in the social sciences, no matter how strong the argument may be for the reliability of such methods. With social science data sets automated methods, using provably reliable methods based on assumptions the investigator tacitly makes, often deliver the news least wanted: the data are uninformative about matters of most interest.

## NOTES

Research for this paper was supported in part by the National Science Foundation, and by the Office of Naval Research and the Navy Personnel Research and Development Center. I am indebted to Chris Meek, Thomas Richardson, Peter Spirtes, Richard Scheines, and Teddy Seidenfeld. This essay was prepared for the Notre Dame Conference on Causation and the Social Sciences, in 1993; an abbreviated version was delivered to the Institute for Mathematical Statistics, Cleveland, 1994.

1. The value of clear, axiomatic foundations is illustrated by some of the recent statistical literature on causal inference that proceeds otherwise. Thus Richard Stone

("The assumptions on which causal inferences rest," *Journal of the Royal Statistical Society B* 55 [1963]: 455–466) offers an analysis of the probabilistic conditions for x to influence y. Stone's analysis, which involves a set z of measured covariables and a set u of unmeasured covariates, is false unless we exclude cases in which x influences y only through members of z or of u, and also exclude cases in which members of z or u are influenced by both x and y, and also exclude cases in which y influences x. What remains is trivial.

2. Fisher's own words are unclear as to whether the order of presentation of the cups is also to be randomized. The wording of two passages in chapter 2 of the book implies that the sequence with which the cups are presented to the lady is to be randomized; a third passage suggests that the cups that receive first tea then milk are chosen at random.

3. The notation **X ⊔ Y | Z** indicates that the set X of variables is independent of the set Y of variables conditional on the set Z of variables.

4. They further note but do not prove that an analogue of the Markov condition, with vanishing partial correlations in place of conditional independencies, holds for linear models with uncorrelated errors regardless of any normality assumption.

5. Cox and Wermuth (1993) claim that there is no possible causal interpretation of such graphs and simultaneous equations and their distributions. They give no reasons. I expect their discovery would be news to economists, electrical engineers, and control theorists.

6. Frank Arntzenius (1992) has given a number of other kinds of interesting purported counterexamples to the Markov condition. Some depend on the quantum theory, while the rest seem not to be counterexamples either because they involve logical rather than causal relations or because they import extra conditions for causal dependency, such as that the causes increase the probability of their effects, which are not part of the Markov condition and in my own view are not correct.

7. For example, if X is a deterministic function of S and Y, and S, Y are each deterministic functions of Z (i.e., Z simply codes the values of S and Y), and Y is a function of X and $e_Y$ and S is a function of W and $e_S$, then if we marginalize out S, Y, and the error variables, X and W are independent and remain independent conditional on Z.

8. Of course this doesn't mean that when we do nonexperimental studies the sampling distribution satisfies the Markov condition for the causal structure of the units; that depends on how we sample.

9. It may seem surprising that while the set of distributions satisfying the Markov condition for any incomplete graph is larger than the set of distributions satisfying the Markov and faithfulness condition for the same graph, the graphical equivalence condition is the same for the Markov condition alone as for the Markov and faithfulness conditions jointly. A little reflection on the faithfulness condition should convince the reader that this is so.

10. The actual genesis of this result is unclear to me. A result that easily entails it was stated by Pearl and Geiger.

11. Any distribution satisfying the Markov condition for a directed acyclic graph G can be "factorized," that is, written as a product of the conditional probability of each

variable on its parents. The factorization formula associated with such a graph permits a direct maximum likelihood estimate for multinomially distributed variables.

12. Immediately, that is, with some hundreds of hours of programming and program testing. The PC program was implemented by Peter Spirtes; Richard Scheines aided in tests and applications.

13. They continue: "By using the two measures of productivity, *PUBS* and *CITES*, and the five causal antecedents, we initially estimated a composite model with all of the paths identified by the six theories. This model produced a large positive deviation between the observed and predicted correlation of *ABILITY* with *PREPROD*, suggesting that we omitted one or more important paths. Reexamination of our initial interpretation of the six theories led us to conclude that two paths had been overlooked. One such path is from *ABILITY* to *PREPROD*. . . . The other previously unspecified path is from *ABILITY* to *PUBS*. These two paths were added and all nonsignificant paths were deleted from the composite model to arrive at the best-fit model." Thus is data-mining disguised as theory.

15. In fact, we were inadvertently misinformed that all seven tests are components of *AFQT* and we first discovered otherwise with the SGS algorithm.

16. The POIPG actually represents these independence relations only under the assumption of composition, i.e. that for any four disjoint sets of random variables, **X**, **Y**, **Z**, and **W**, $\mathbf{X} \amalg \mathbf{Y} \mid \mathbf{Z}$ and $\mathbf{X} \amalg \mathbf{W} \mid \mathbf{Z}$ entail $\mathbf{X} \amalg (\mathbf{Y},\mathbf{W}) \amalg \mathbf{Z}$. Composition holds for any positive distribution.

17. The arrow over the summation over Y indicates that the sum is taken over values of Y.

# A Critical Appraisal of Causal Discovery Algorithms

## *Paul Humphreys*

The research program described in detail in Spirtes, Glymour, and Scheines (1993) and outlined by Clark Glymour in the essay published in this volume has more than one goal. One of these goals is to provide a connection between causal structure and probabilities by using an explicit axiomatization that links directed graphs with probability distributions over the vertices of those graphs. In using an axiomatic approach, these authors approach the project of connecting probability and causation from a different direction than most other authors, for in the philosophical literature connections between probability and causation have usually taken the form of explicit definitions. Sometimes these definitions attempt to reduce causation to probabilistic concepts (e.g., Suppes 1970), in other approaches the procedure is to build up complex causal relations from simple causal relations taken as primitive, using probabilistic invariance as the defining characteristic of causal relations (e.g., Humphreys 1989).

The axiomatic approach has, in contrast to these philosophical efforts, been used to good effect in Judea Pearl's earlier work on probabilistic inference in artificial intelligence (Pearl 1988). Spirtes, Glymour, and Scheines' (hereafter referred to as SGS) interest is also in making inferences, the idea being that on the basis of statistical independence and dependence relations between variables, perhaps supplemented by auxiliary information, we can infer the presence or absence of causal relations between those variables and represent the causal relations by edges in a directed graph. I shall address later the important question of the exact sense in which these graphs are actually representations of causal relations, rather than of probabilistic dependency relations, but for now it bears emphasizing that Glymour's account is not a probabilistic theory of causation in

---

*This essay was written as a set of comments on a paper by Clark Glymour that was presented at the 1993 Notre Dame conference. The version of Glymour's essay that appears in this volume was not made available to me before I submitted my paper, and differs from his earlier essay in a number of crucial respects. My more recent views on these issues can be found in Humphreys and Freedman (1996).

the sense usually understood by philosophers—the relations between variables are deterministic for Glymour, and increases in probability play no role for him in identifying causal relations.[1]

A second goal is to provide computationally feasible algorithms based on those axioms. These algorithms are designed to discover at least some of the causal relations between specified variables of scientific interest and, importantly, SGS claim to be able to sometimes make these discoveries without using theoretical background knowledge about which variables can and cannot be causally connected. In doing this, these algorithms are said to have an advantage over existing methods of inferring causal structure, such as structural equation models, which do explicitly have to rely on non-statistical assumptions about what are plausible and implausible candidates for causal relations between variables. Such assumptions are often called, very misleadingly in my view, "*a priori*" assumptions.[2] It is better to view them as essentially using background knowledge about causal relations, where this background knowledge does not come from the data. This background knowledge may be based on explicit theory or it may be tacit practitioner's knowledge or it may be just "common sense" but it will rarely be *a priori* in the traditional philosophical sense of knowable independently of any particular empirical experience.

The extent to which the SGS algorithms require background knowledge varies, and their presentation of their case sometimes makes the data-driven nature of their enterprise seem purer than it actually is. Sometimes, a time ordering on the vertices, together with other assumptions, is used to direct edges (SGS 1993, 133: more on this in section II below) and indeed all of their algorithms employ that information (ibid., 127); in other examples background causal knowledge is incorporated into the programs (ibid., 127). Yet they also claim that their programs recovered almost all of a thirty-seven variable model of an emergency medical system without any prior information about the ordering of the variables (ibid., 11, 146). It would be more accurate, I think, to portray their primary interest as an investigation of to what extent nonexperimental methods can arrive at causal conclusions, and under what assumptions (e.g. ibid., vii, 3, 10, 22).

A third goal is to develop methods for predicting the results of interventions in existing causal structures. I shall not have much to say here about this third goal, for it has been discussed in admirable detail by James Woodward and David Freedman in their contributions to this volume. Fourthly, the research program suggests that most existing methods of representing causal structure, including regression, factor analysis, structural equation models and so on—essentially the whole spectrum of contemporary econometric and sociometric causal methods—are seriously inadequate to their tasks. Finally, and perhaps most surprisingly, SGS's approach does all this in a way that is supposedly free from the need to say explicitly what causation is or to provide a definition of causation. To a lesser extent SGS are inexplicit about what is the appropriate interpretation of probability to use in their approach. This ability to work with (explicitly) undefined terms within an axiomatic framework is, of course, one of the great methodological advantages of the axiomatic approach but it also has its well-known drawbacks, the worst being that it results in too broad a class of structures that satisfy the axioms. We shall see that this is a defect that is clearly present in SGS's framework, and it undermines their claim to have given algorithms that discover causal structure; to eliminate noncausal epistemic relations, the supplementary information used in the discovery process plays a crucial role.

By building on the results of Judea Pearl and his research group, SGS have provided a constructive approach to some complex and difficult problems. There is no doubt that their contributions have moved discussions of causal structures a step forward. And I agree with the spirit of their remarks that an unhealthy strain of skepticism on the part of philosophers is often an excuse for avoiding hard constructive thinking about what can be done in this area. It must be said, though, that this skepticism is by no means confined to philosophers, for statisticians have tended to be at least as skeptical about these matters. (See, e.g., Freedman 1983b; this volume). Yet I find their belief that an intuitive understanding of causation can carry us a long way in using their methods quite wrong. In fact, when we come to closely examine the SGS methods, we shall find that not only is it hard to put a consistent interpretation on the axioms, but it is by no means obvious that they have any real causal content at all. As I shall argue, the most plausible construal of the graphs is that they are not representations of causal relations understood in some robustly realistic sense of causation, but are devices for representing epistemic dependency relations.

## I.

Let us begin with the relation between graphs and probability, which will be the principal focus of my remarks. Some straightforward and obvious comments about the assumptions that SGS use are in order because it is easy for elementary points to get lost once one becomes immersed in the complex details of the SGS program. In essence, SGS's idea is to get at the notoriously elusive concept of causation by showing that certain causal relations, represented in graph-theoretic form, entail certain probabilistic relations, primarily stochastic independence and dependence. Because there is a sense that direct contact between probabilistic relations and empirical data is possible through relative frequency values or subjective probability values in a way that direct contact between causal relations and empirical data is not, the SGS methodology hypothesizes that certain causal relations entail specific probabilistic dependency relations. Using those entailment relations, one can at the very least *eliminate* the possibility of a causal connection between two variables by virtue of certain observed statistical independencies. There are many questions that arise from this apparatus. Here I shall focus on the following three:

1. Do the assumptions on which the representational apparatus is based contain or presuppose specifically causal content?
2. What is the correct interpretation of causation to use with these methods?
3. What is the status of the representational apparatus that SGS use to capture causal relations?

I begin with the first of these but I want to stress that these three questions are intimately linked and one cannot properly answer the first without clear answers to the second and third. Some of the assumptions on which the algorithms are based are explicitly cited in Glymour's essay in this volume, others are to be found in SGS (1993).

1. *The Markov Assumption.* This has a number of different formulations but one of the most general is:

   Given a directed acyclic graph (DAG) G over a set of vertices $\underline{V}$ representing random variables, and a joint probability distribution $P(\underline{V})$ over $\underline{V}$, then $<G,P>$ satisfies the Markov condition iff for every vertex W in $\underline{V}$, then given Parents (W), W is probabilistically independent of any other vertex $X \in V - \{Descendants (W)\} \cup \{Parents (W)\}$.

2. *The Faithfulness Assumption.* The probability distribution P that is associated with a DAG G representing the causal structure of a system is such that every conditional probabilistic independence relation in P follows from the Markov condition applied to G. Roughly, there are no "accidental" independencies.

3. *Working Assumption.* The distribution of observed variables in a population is the marginal of a distribution satisfying the Markov and faithfulness conditions.

4. *Causal Sufficiency.* The set of vertices of the DAG used to represent causal structure is closed under common causes; i.e., every common cause of measured variables is explicitly included in the model.

5. *No Side Effects Assumption.* An ideal intervention that directly manipulates T may alter the distribution of T conditional on the parents of T, but the distribution of every other variable conditional on its parents is unchanged.

I have here noted only those assumptions used in determining the causal structure when the model is correctly specified—i.e., when all the relevant variables have been identified and included in the model and the goal is to discover which variables are causes of which others. One exception here is that error variables are not explicitly included in the model, but are always assumed to be present for *every* variable in the system. The reason for this is that if one variable is a (deterministic) function of some others then the discovery procedure can give the wrong results.[3] So to avoid this, some source of purely exogenous stochastic variation is attached to every variable in the model, and these error variables are mutually independent. When this condition is satisfied the graph is called *pseudo-indeterministic*. We add this as an explicit assumption.

6. *Pseudo-Indeterminism.* Each vertex of the DAG has an unrepresented single parent (an 'error variable') which is independent of every other such error variable and which has a non-degenerate distribution. All DAGs considered are pseudo-indeterministic.

(I note here that the independence of the error terms follows from the Causal Sufficiency assumption, and that the Causal Sufficiency assumption is dropped for partially oriented inducing path graphs.)

In addition to these assumptions, there is one definition that lies at the core of the SGS algorithms. This is the definition of *d-connectedness*, and it is here that I find SGS's position on causation essentially opaque.

*Definition.* Two vertices X, Y in a DAG G, where $(X \neq Y)$ and X, Y do not belong to Z, are *d-connected* by a set of vertices Z if and only if there is an undirected path U between X and Y of the form (1) every collider in U has a descendant in Z (including the null descendant) and (2) every other vertex on U is outside Z. X, Y are *d-separated* by Z just in case they are not d-connected by Z.

The concept of d-connectedness is not easy to grasp, but what it means is this: Suppose we have the case in figure 9.1.

**FIGURE 9.1**

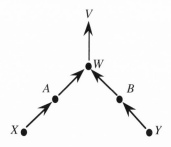

Then X,Y are d-connected by {V} but d-separated by {V,A}. Thus, if we knew V, then from a knowledge of X, we could reliably infer what value Y would have.

This concept of d-separation plays a key role in both the SGS algorithm and the PC algorithm used by SGS because it is the principal device by means of which directions are given to edges in a DAG on the basis of statistical dependency relations. As SGS put it "d-separation in fact characterizes all and only the conditional independence relations that follow from satisfying the Markov condition for a directed acyclic graph" (1993, 72).[4] The definition itself is in purely graph theoretic terms—there are no causal concepts involved. The first question I want to raise here is this: Is there a consistent causal interpretation of d-separation that can motivate its use in the algorithms used by SGS to arrive at causal structures?

Before I discuss this, let's be clear about what the perceived difficulty is. At the beginning of his original essay, Glymour asserted, "While I will rely on some intuitions about relevant features of the notion of causation, I avoid throughout any jejune attempt at definition." That approach is recognizable, and fits well with the axiomatically based apparatus he has developed. The method seems to be this: We begin with a preformal set of intuitions about causation, and about the relations between causal connectedness and probabilistic dependence. Then the Working Assumption (number 3 above) is used as the basis for justifying the claim that every observed statistical independence relation is a result of the causal relations embodied in the DAG and the Markov condition. That is, the claim is that whenever one has a purported example of statistical independence that does *not* result from a causal connection, it falls into one of a set of recognizable misapplications of the method. These include, for example, nonhomogeneity of the population to which the model is applied or underspecification of node variables.

But SGS also claim (1993, 41) in their chapter on axioms and their interpretations: "We advocate no definition of causation, but in this chapter we try to make our usage systematic, and to make explicit our assumptions connecting causal structure with probability, counterfactuals, and manipulations. With suitable metaphysical gyrations the assumptions could be endorsed from any of these points of view, perhaps including even the [view that prefers not to talk of causation at all]." I do not think that SGS have been successful in keeping their use systematic, and I shall show that only an epistemic interpretation fits their apparatus as presented.

When Judea Pearl introduced the concept of d-connectedness, it was within a framework where *inferential* connections were the primary focus. In fact, one of the principal reasons for using directed graphs rather than undirected graphs is because of the latter's inability to correctly represent cases of common effects. Pearl uses an example in which two coins are independently flipped and a bell rings if and only if both coins are the same. When we know that the bell rang (and when we know it did not) we can infer with certainty from the outcome of one coin toss what the other coin outcome was. This *induced dependency* between the coin outcomes is clearly epistemic—it is certainly not causal. As Pearl says (1988, 93) "This weakness in the expressive power of undirected graphs severely limits their ability to represent informational dependencies." Within what Pearl calls a Bayesian network, a distinction is made between a joint effect of two causes (a collider) on the one hand, and a common cause or an intermediate cause between X and Y on the other. If you examine the definition of d-connectedness, clause (1) requires that on a d-connecting path all such colliders be in the conditioning set or a descendant of a collider be in it. On an epistemic interpretation, this makes sense. As we saw in the bell example, learning about the common effect gives us information about the causes, and, by an inference that goes against the causal direction, a descendant of the common effect can do the same thing (suppose a dog salivates if and only if the bell rings, for example). Clause (2) requires that one cannot condition on a common cause or an intermediate cause. So with d-connectedness, by conditioning on Z, we can learn something about events further back in the causal network.[5] It is this asymmetry between common effects and common causes or intermediate causes that allows the SGS and PC algorithms to discover colliders in the case where conditioning on the middle element Z in a sequence X-Z-Y fails to render X and Y independent. Thus far, we have discussed d-connectedness in terms of information. But as we have seen, SGS want a much wider class of interpretations for causal networks than that.

In his essay in this volume, Glymour is not very forthcoming about what interpretation to give to d-separatedness. He says, "There is even an intuition to [d-connectedness] which I will not detail here" (p. 211). An intuition is in fact detailed in the SGS book on pages 72–73, but it is not consistent. We are first given an analogy to causal flow in a pipe, where a collider corresponds to a closed valve but a common cause and an intermediate cause correspond to open valves. This causal flow analogy makes sense along the lines of an interpretation of causation involving transfer of energy or transmission of a mark or manipulability, for example. Next, we are told that conditioning on an open (active) node converts it to a closed (inactive) node because we know that keeping a variable of that type fixed renders X and Y independent. This makes sense on a manipulability criterion, and we know that this conversion from one status to another is a familiar fact about common causes and "screening off." It makes sense on an inferential view too: once we know the common cause, knowledge of one effect does not provide further information about the other, and knowledge of proximate causes renders knowledge of remote causes irrelevant. But when it comes to common effects (colliders), SGS (1993, 72) say merely "that conditioning on a collider makes it active was noted in section 3.5.2 above."

When we turn to section 3.5.2 we find two types of cases where this can happen. One involves a "Bayesian example" drawn from Pearl which involves inference from the known facts that your car will not start and that your battery is not dead to the fact that the fuel tank is empty. The other involves a case related to Simpson's paradox whereby factors independent in the whole population become dependent when conditioned upon some factor that subdivides the population frequencies. Yet the second of these gives us no insight into how causal flow can be directed through a collider, and since SGS are opposed to specifically probabilistic accounts of causation, either of the type that relies solely on statistics to characterize causal relations or one that uses probability increases and decreases as the key to causation, the onus must lie on some other interpretation of the graph. Certainly a manipulability view of causation will not account for why a collider becomes active when it or a descendant is conditionalized upon; nor does this make sense on a causal flow view. The only view on which it makes sense is an inferential view where it is as legitimate to infer from an effect to a cause as vice versa, and, in certain circumstances, from one joint cause to another, conditional on their common effects.

This has two consequences. If the only consistent interpretation that can be laid on the DAGs is one of *inference* and not of causation itself, then the liberality of interpretation suggested by SGS is misplaced. Given the centrality of d-separation in their discovery procedure, there is one and only one thoroughgoing interpretation of these "causal" graphs and that is an epistemic one. Second, when SGS turn to prediction, they focus almost exclusively on manipulation as the basis of the No Side Effects Assumption and its consequences. This must mean that we have to switch causal horses when moving from one part of the research program to another.

## II.

It might be said that there is no need to *have* an interpretation of d-separation. It could be viewed as an instrumental device that in some rather hard to understand way gets us the right causal graphs in conjunction with the other assumptions. All that is important is arriving at the correct causal graphs and the evidence SGS provide in terms of comparative performance against existing studies does that. This is a reasonable response, and one that revolves around two different uses of the axiomatic method: one where all the axioms are interpreted and truth flows from those axioms to the theorems and the structures that result from applying the theorems, and the other where all the truth lies in the consequences and the axioms pick up whatever interpretation, if any, is consistent with those consequences. Suppose we adopted this second view (which I am not personally in favor of). An immediate question then arises: how do we know that the causal structures which result from applying these axioms are true? Now sometimes

SGS declare a successful application because their result almost conforms to the structure of an existing model arrived at by other means (e.g. the Rodgers and Maranto example, Glymour's essay, pp. 214–15; SGS 1993, 134 ff). But on other occasions, such as with the Rindfuss, Bumpass, St. John example (Glymour, pp. 215–16; SGS 1993, 139–40) the PC algorithm is declared to yield results superior to existing methods, in this case because Rindfuss et al. initially included a causal connection from the variable representing the age at which a first child is born to the variable representing the education level of women at time of marriage, but they later found the regression coefficient to be zero, whereas the PC algorithms deleted that edge in the course of the discovery process and it was not included in the final model. Now since causal connections are not directly observable, if we are employing the second axiomatic strategy where content flows from the consequences to the axioms, we cannot rely on directly knowing the truth of the axioms but instead we must rely on our intuitive or theoretical judgments about causation in particular cases to validate specific models. (The same point holds with other cases, such as the Blau and Duncan example, but I have not discussed them because they employ partially oriented inducing path graphs rather than the basic apparatus discussed here). This, of course, is one of the original questions we raised about these methods: What kind of causal knowledge, if any, must we have in order to use the SGS methods? There are (at least) two kinds of causal knowledge that might be involved. The first is knowledge required to understand the assumptions—this will be in general knowledge about what kind of causation and probability is involved. The other is knowledge required to apply the assumptions in particular cases. Consider in this regard the Markov assumption, upon which much of the SGS method is built. Within the SGS approach, the Markov assumption is taken as an axiom, and there are two versions of it. One version (SGS 1993, 33), which we earlier listed as the first assumption, is a purely formal criterion linking graph structures with probabilistic relations, and has no causal content. The Causal Markov condition (ibid., 54) is slightly less formal, but not much. It simply asserts: Let G be a causal graph with vertex set $\underline{V}$ and P be a probability distribution over the vertices in $\underline{V}$ generated by the causal structure represented by G. G and P satisfy the Causal Markov condition if and only if for every W in $\underline{V}$, W is independent of $\underline{V}$ / (Descendants (W) $\cup$ Parents (W)) given Parents (W).

But this reference to the causal interpretation of the graph seems unmotivated, because the (formal) Markov condition is a consequence of a completely acausal condition. To see this, compare the SGS approach with the one taken by Pearl, within which the axiomatic formulation concerned qualitative probabilistic independence and independency maps. Take the following definition:

*Definition.* A DAG G is a **directed independence graph** of P(**V**) for an ordering > of the vertices of G if and only if A → B occurs in G if and only if ¬ (A ⊔ B | **K**(B)), where **K**(B) is the set of all vertices V such that V ≠ A and V > B, where A ⊔ B | Z means that the joint probability density of A and B, given Z, factors into the conditional densities of A given Z and of B given Z.

Pearl then proved this result:

If P(**V**) is a positive distribution, then for any ordering of the variables in **V**, P satisfies the Markov and Minimality conditions for the directed independence graph of P(**V**) for that ordering.[6]

So, my point is this: since, (a) given some ordering of the vertices (which could be merely a temporal ordering), a strictly positive distribution over the vertices guarantees that the formal Markov condition will be satisfied, (b) the fact that we can check whether there are any null probabilistic dependencies in a purely numerical way, and (c) the fact that SGS often eschew prior causal knowledge in favor of temporal ordering or no ordering at all (SGS 1993, 112) in arriving at their graphs, what role does the causal interpretation of the Causal Markov condition play in their algorithmic procedures? This question is particularly pressing in the case of DAGs that are also directed independence graphs. To put the point another way, we could eliminate the formal Markov condition for those DAGs that are also directed independence graphs if we had a prior ordering on the vertices and a strictly positive distribution over those vertices. SGS state (1993, 111) that, aside from the computational inefficiencies of algorithms based on directed independence graphs (which is not a causal issue), they want to eliminate the need for a prior ordering on the variables, presumably because at least in some cases, a time ordering is inappropriate or unavailable. But then if the causal interpretation of the graphs in the Causal Markov condition is not being used to impose a prior ordering on the vertices, what role does it play?

### III.

I turn now to the question of whether the assumptions of the SGS approach require specific causal information in order to be correctly applied. SGS are quite explicit that in general the SGS and PC algorithms give an equivalence class of causal graphs rather than a unique causal graph. The algorithms first give an undirected graph that is usually less than complete (i.e., it is derived from a completely connected graph by deleting edges) and then they identify unshielded colliders.[7] (Unshielded colliders have the form in figure 9.2 where A is not directly connected to B and vice versa.) This partially orients the undirected graph,

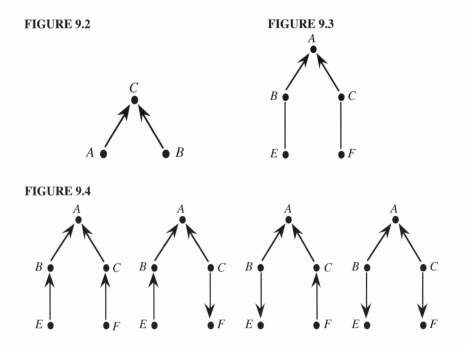

**FIGURE 9.2**

**FIGURE 9.3**

**FIGURE 9.4**

but leaves open a number of different ways to complete the DAG. For example, the graph in figure 9.3 could be completed in any of the following ways, as shown in figure 9.4.

Again, with reference to the Rodgers and Maranto example, the graph in figure 8.6 (Glymour, p. 216) is but one of a number consistent with the initial output of the PC algorithm. In the case of that example, an appeal to time order of the variables is used to select the final graph and to eliminate other members of the statistically indistinguishable equivalence class. But this strategy is not always possible. To see why, consider another case that SGS describe.

Because the assumptions used are graph theoretic and probabilistic, we can ask whether these methods can model relationships that are not causal in kind. And so they can. In the example involving the AFQT test (Glymour, pp. 216–17), the relations involved are classificational rather than causal. The variables AR (Arithmetical Reasoning), NO (Numerical Operations), and WK (Work Knowledge) are, as SGS note, components of the whole AFQT test and the PC algorithm picks them out as variables adjacent to AFQT. But AR, NO, WK no more cause AFQT than the numbers 2, 3, 4 cause the number 9. They are constitutive components of it. (Note that if there is at least one other component of AFQT that is not included in the graph, and this is independent of AR, NO, WK, then the system can be pseudo-indeterministic.) It will not do merely to extend our concept of causation to include constitutive relations, for a standard objection to counterfactual analyses of causation carries over to the manipulability view too. It is true that if X's score on the arithmetical reasoning test had been different, X's AFQT score would have been different too (assuming the No Side Effects principle and no fortuitous cancelling by the other components) but the AR score is not a cause of X's AFQT, any more than an atom's having two electrons is a cause of its being a helium atom—it is simply part of the defining conditions for what it is to be a helium atom.[8]

Here we can perhaps draw a sharp distinction between the use of these methods as prediction and manipulation devices, and their use for causal purposes. There is no doubt that from the AR scores we can partially predict the AFQT score and that by manipulating the AR score we can change the AFQT score. But neither should be confused with discovering what the causes of the AFQT score are. So if we wish to isolate the causal, as opposed to the classificational, connections in a graph, we will have to bring in additional information, and that information will not always be available on the basis of common sense or of obvious temporal order. For example, if an individual took the AR test first and the NO and WK tests later, the AFQT score would only be available later than the AR score. And we might need a significant amount of theoretical knowledge rather than mere common sense to know which variables are components of others. (I note here that the Working Assumption explicitly allows that classificational relations can be included in a DAG.)

## IV.

Finally, there is an issue not discussed by Glymour in his contribution to this volume that is important and bears bringing out explicitly. The SGS discovery procedure is designed to find *independence* relations and to eliminate corresponding edges in the graph, leaving directed edges as representations of causal connections. But we ought to remember that there are two kinds of dependencies, positive and negative—in the linear case these will be represented by positive and negative correlation coefficients. This difference is obviously important when one is interested in interventions, as SGS sometimes are. Unless we know whether Y is positively or negatively connected with X we do not know whether increasing X will increase Y or decrease it, and without that knowledge, interventions can be counterproductive, decreasing SAT scores by paying school administrators higher salaries for example. Of course, the SGS algorithms are supplemented by estimation procedures to arrive at numerical values for parameters in the graph, and this will provide the signs of the coefficients. The point I want to bring out here is this: SGS discuss the kind of situations within which the

Faithfulness condition can fail (1993, 64–69), focusing on contexts within which Simpson's paradox occurs. Correctly noting that Simpson's original example involved cases in which positive associations within two subpopulations became independent when the subpopulations were combined, SGS prove a powerful theorem[9] ( for the continuous case). They show that in linear models the set of parameter values for which the Faithfulness condition fails has measure zero, and hence for pseudo-indeterministic linear models, we can proceed comfortably with the knowledge that the Faithfulness condition will fail only under probabilistically extraordinary conditions. But Simpson's original paradox is not all SGS need to worry about. It will be as bad for them if their algorithms, supplemented by standard statistical methods, erroneously tell us that a pair of variables are negatively causally connected when they are also positively connected in subpopulations. Moreover, these kinds of problems are not limited to Simpson's paradox. There is a related and well-known problem in sociological sampling known as the ecological fallacy,[10] wherein associations at the group level, e.g., between mean educational level and racial composition within counties, are transferred to associations at the level of individuals. In such cases, however, positive associations at the group level can arise from negative associations at the individual level. So it is not enough for SGS to avoid causal structure ambiguities caused by independence relations in the data not being properly represented in the graph—they must also, like other causal modelers, be concerned with selecting the right populations on which to impose their models. That this problem is not peculiar to their method scarcely means that it does not need to be addressed. And then the question is: Can we apply the graph to the correct population without using prior causal knowledge of which type of association, positive or negative, is correct? Or be able to rule out as meaningless signed variables attached to populations that properly belong only to individuals? It is implausible to believe that this could be done using only common sense, and hence the problem of signs is one that seems to require prior causal knowledge to resolve.

## V.

Let us summarize the project we have been discussing. SGS have two aims. One is to provide a computationally feasible algorithm for identifying the correct causal relations between variables in models of social phenomena. The other is to provide an explicit set of assumptions on which correct causal inference can (in principle) be based. This second part of the project is of more philosophical interest than the first, but because the applied aspect is an essential part of the larger project, it is perfectly reasonable to ask whether the assumptions on which these algorithms are based are applicable in real cases rather than in principle only.

It seems clear that some of them are not. The frequency version of the Markov condition requires that the probability distribution arises from a homogeneous population, i.e., one in which all the units share the same causal relations (i.e., they all have the same graph). But in the nonexperimental contexts within which these algorithms are supposed to be applied, this will rarely if ever be the case. Social variation is far too great for that. Second, there is extensive use of "prior information" that some variables are not causes of others. Thus, for example in the use of the PC algorithm (Glymour, p. 215) it is said that ED (the level of education at the time of marriage) is not causally related to the other variables. One of these other variables is REL (the individual's religious affiliation). Why is it supposed to be obvious that the level of education at marriage is not a cause of one's religious affiliation? One knows many women whose agnosticism is a direct causal effect of their high level of education. Similarly, in the United States one's level of education is frequently a cause of where one chooses to live (REGN), another link supposedly excluded by "prior information." The same point could be made with respect to YCIG (cigarette smoking). It might well be that the associations between such variables and educational level are due to some relationship other than a direct causal link in the direction I have suggested, but to rule such links out *a priori* is implausible and methodologically suspect.

## VI.

So, to conclude: It is regrettable that the genuine achievements in the research of Spirtes, Glymour, and Scheines are accompanied by such an inexplicit account of the exact sense in which their graphs are causal. Until this is remedied, I remain skeptical that their algorithms are genuine causal discovery engines. Moreover, movement onto new ground inevitably brings with it some old questions as well as new ones. So here again are the questions that future work on their approach will need to answer:

1. In what sense are DAGs representations of causal relationships, rather than of conditional probabilistic dependency relations?
2. Is there a consistent non-epistemic interpretation of the d-separation condition?
3. (a) How much causal knowledge is actually needed to apply the algorithms?
   (b) In particular, do aggregation problems require background causal knowledge to select the correct level of causal analysis?
4. Can these methods distinguish causal graphs from classificatory graphs?

**NOTES**

I should like to thank David Freedman, Richard Scheines, Glenn Shafer, Peter Spirtes, and James Woodward for helpful conversations and correspondence. Not all of them agree with the positions taken in this essay.

1. Glymour's essay, note 7.
2. An explicit example of this can be found in section 3 of Simon (1954).
3. See Glymour's essay, pp. 209–210.
4. See also clauses B) and C) of the SGS algorithm (SGS 1993, 114), and clauses B) and C) of the PC algorithm (ibid., 117).
5. See the reference in note 1 above.
6. This result is cited on p. 35 of SGS (1993).
7. Glymour's essay, pp. 212–13; SGS (1993), p. 96.
8. A similar error can be found in Sosa (1980), sections 1–4.
9. SGS (1993), pp. 68–69.
10. See, e.g., Robinson (1950).

# Causal Models, Probabilities, and Invariance

## *James Woodward*

I want to begin by saying how much I admire Clark's essay and the book, *Causation, Prediction, and Search* (Spirtes, Glymour, and Scheines 1993)—hereafter SGS—that lies behind it. Unlike so many philosophical discussions of causality, the SGS program is original, constructive, systematic, and highly technically sophisticated. It is informed about and in extremely close contact with related literature in statistics and artificial intelligence and has a very clear practical and methodological import. It is full of provocative ideas about an enormous range of topics. Indeed, I think I have probably learned more from it (and from thinking about the problems it poses) than from any other book I have ever read on the subject of causal inference—one measure of this is that many of the comments that follow simply draw attention to or expand on observations that SGS themselves make. I hope that all of you will read this extraordinary book and will derive as much pleasure and instruction from it as I have.

It is useful to begin by comparing SGS's program to the conventional approach to the problem found in the econometrics literature. Consider a general simultaneous equation model of the form

(1.1)   $Y\Gamma + XB + U = 0$

where the matrix Y represents observations on the endogenous variables, X the exogenous variable, U the error term and B and $\Gamma$ are the coefficient matrices. It is well known that multiplication of the coefficient matrices in (1.1) by any nonsingular matrix will yield a distinct structural model with the same reduced form equation as (1.1) and hence a model which implies exactly the same facts about statistical relationships among measured variables as (1.1). If we accept the view that distinct structural models represent distinct possible systems of causal relationships, it seems hard to avoid the conventional conclusion that one finds in econometrics textbooks: that facts about statistical relationships among measured variables greatly underdetermine the causal facts, even given a fixed stock of variables that we have correctly identified as the causally relevant variables. If we also consider the possibility, discussed in more detail below, that the variables we have identified are not really the causally relevant ones—that they instead are merely correlated with the truly causally relevant variables, either because they are proxy variables that imperfectly measure the relevant variables or for some other reason—or the possibility that the units in the population with which we are dealing are causally heterogeneous, the underdetermination of causal relationships by facts about statistical dependencies looks even more massive.

Conventional approaches in econometrics and causal modeling literature try to solve this underdetermination problem by making strong additional background assumptions which in conjunction with statistical information will exclude large numbers of alternative models. The standard view within this literature is that this required additional background information will take the form of highly concrete and subject matter specific considerations—that it will be information about whether some particular variable can or cannot cause another, about particular causal orderings, or about the distribution of the error terms and so on. In effect, this is information that certain terms in the coefficient matrices in (1.1) are equal to zero, or information about relationships among these coefficients (so-called cross equation restrictions) or information about the covariances of the error terms in (1.1). One uses this information to single out a particular structural model (or a class of such models) from a larger class of observationally equivalent competitors.

A very elementary illustration of such an appeal to subject matter specific background information, in this case in order to justify the inclusion or exclusion of certain variables from a regression equation, is provided by the following example drawn from Tufte's *Data Analysis for Politics and Policy* (1974). Tufte is interested in the causes of variations in automobile fatality rates across states. He regresses a variable representing death rate against a variable representing presence or absence of automobile inspections, a variable representing population density, and variables representing whether or not a state was one of the original thirteen states and has seven or less letters in its name. He obtains a nonzero regression coefficient in each case—death rate is correlated with each of these variables—and then comments that

> While we observe many different associations between the death rate and other characteristics of the state, it is our substantive judgement and not merely the observed association, that tells us density and inspections might have something to do with the death rate and that the number of letters in the name of the state has nothing to do with it. (Tufte 1974, 9)

Even though there is a statistical association both between fatalities and the number of letters in a state's name, and between fatalities and inspections, Tufte holds, on the basis of subject matter specific background considerations (his "substantive judgement" that inspections are the sort of thing that can affect the death rate while the number of letters in the state's name is not), that only the second relationship reflects a direct causal connection.

How does the SGS program compare with this conventional approach to the undetermination problem? As I see it, one of the most distinctive features of Glymour's approach is that it claims to dispense with or at least largely dispense with the need for the large amount of domain specific background information required by a more conventional approach. Instead SGS propose to infer causal conclusions by combining information about statistical dependencies with highly general, subject matter independent, background assumptions connecting causes and probabilities—these are the connections that are embodied in Glymour's working assumption.[1]

A second respect in which SGS's program may or may not differ from more conventional approaches has to do with the interpretation of causation it assumes. As I will attempt to explain in more detail below, the conventional approach adopts what I shall call a control conception of causation and links the notion of causal relationship to the notion of an invariant relationship. It is unclear to me whether SGS adopt a similar conception of causation or whether they attempt to solve the underdetermination problem in part by adopting a different and weaker conception of causation.

The remarks that follow will be structured in the following way. Sections I and II will explore Glymour's working assumption and the role it plays in causal inference. Sections III and IV will then focus on the interpretation of causation assumed in Glymour's program.

Before turning to a more specific look at the working assumption, however, there is a general point which, although perhaps obvious, is nonetheless worth keeping in mind. This is that if the working assumption is to help in solving the underdetermination problem it must be playing something like the role assigned to more substantive background knowledge in conventional approaches. That is, when we confine our attention to systems satisfying the working assumption we must be excluding, on the basis of very general subject matter independent considerations, a large number of alternative models. I mention this because I suspect that many philosophers, at least, may think that the working assumption is relatively weak and uncontroversial. In fact, if the working assumption is to do its job at all, it must be very powerful in what it rules out.

To provide just one illustration of this, the working assumption seems to require that the underlying system from which the observed distribution of variables comes has something like a recursive structure. At the very least, the working assumption seems to rule out a very large class of the non-recursive simultaneous equation models that are postulated in many areas of economics and political science. In particular the working assumption requires that the graph representing the underlying causal relations for the system under investigation be acyclic and that, as noted below, all exogenous variables be uncorrelated. This noncorrelation assumption will not in general be true in non-recursive systems since in such systems the error terms will often be correlated across equations. Moreover, as is well known, if we confine ourselves to recursive systems, and are given a complete ordering of the variables as, for example, by time, and all variables are measured, statistical information will suffice to pick out a *unique* causal graph. Now SGS do not in general assume that all variables are measured or that we know the complete ordering of all variables. Nonetheless, in their treatment of particular examples, SGS often appeal to information about time-order to causally order at least some of the variables, and it is clear that the assumption that the underlying model is recursive (or at least that it is acyclic and all exogenous variables are uncorrelated), particularly when combined with temporal considerations, greatly cuts down on the space of alternative models among which they must distinguish. As we shall see in more detail below, other features of the working assumption—for example, the requirement that the units in the population under investigation be homogenous—are also quite restrictive. My point is not that this is necessarily unreasonable, but rather that we need to remember that the SGS program does not avoid undetermination problems, but rather attempts to solve them by adopting different assumptions than conventional approaches. What we need to do is to inquire into the relative merits of these two strategies.

## I

I turn now to a more detailed look at the working assumption which, you may recall, reads as follows:

> *Working Assumption:* The distribution of observed variables in a population studied empirically, whether experimentally or otherwise, is the marginal of a distribution satisfying the Markov and faithfulness conditions for a directed acyclic graph describing the causal (or causal and classificational) relations among the observed variables, and possibly also unobserved variables, in each unit of the population.

The Markov and faithfulness conditions read as follows:

> *Markov Condition:* Consider the vertices of a directed acyclic graph G as random variables. Let **Y** be any set of variables none of which are de-

scendants of X—there is no directed path from X to any member of **Y**. Let **Parents** (X) be the set of all variables connected to X by an edge directed into X. Then **Y** is independent of X conditional on **Parents** (X).

*Faithfulness Condition:* A joint distribution P on a set of variables is faithful to a directed acyclic graph G if every conditional independence in P follows from the Markov condition applied to G.

This part of my discussion has several related goals. First, I want to consider some of the circumstances under which these conditions will fail and to ask how prevalent such failures are likely to be in typical contexts in which causal modeling schemes are used. Glymour rightly complains about a kind of lazy skepticism that would reject all inference procedures based on these conditions merely because there are some circumstances under which they are violated. Still, it is surely worthwhile to ask how widespread, as an empirical matter, such violations are likely to be. Another related question I want to explore is what the consequences of such violations are likely to be—is there a characteristic bias or mistake that will be produced by relatively atheoretical inference procedures like those of SGS when used in circumstances in which the working assumption is violated? Here I will be pursuing the following line of thought: Roughly speaking, the SGS inference procedures assume that statistical dependencies always reflect causal connections. If, as seems to be the case, violations of the working assumption can produce dependencies that do not reflect causal connections, then one worry is that, when such violations are present, there will be a systematic tendency within the SGS program to find causal connections where none exist, or at least to output conclusions that, in a sense to be explained below, are uninformative. Even if their inference schemes perform as reliably as SGS claim in circumstances in which the working assumption is satisfied, we will need to consider the possibility that other procedures, less biased toward discovering causal connections whenever there are statistical dependencies, will be more reliable when the working assumption is violated.

Relatedly, I want to ask about the implications of violations of the working assumption for the use of causal modeling techniques generally when they are used, as they often are in the social sciences, without extensive substantive background knowledge. Glymour claims that those who construct causal models commonly assume that the conditions described in his working assumption are satisfied. As a general proposition this seems to me to be something of an exaggeration at least for methodologically sophisticated modelers. In fact, I think that one of the main reasons many investigators insist on the importance of substantive theory in causal modeling is precisely because of the worry that in many situations in which causal modeling techniques are used the conditions described by the working assumption may be violated and that in such situations, causal modeling unguided by theory is likely to produce mistaken or at least unhelpful results.

What I suspect *is* true is something like the following: First, while there are perhaps more cases than Glymour supposes in which reliable domain specific background knowledge provides useful guidance, he is surely right in thinking that in many cases in which causal modeling techniques are used, particularly in disciplines like sociology and social psychology, no such knowledge is available. Instead, actual practice often resembles what we see in Rodgers and Maranto's study (1989), as described in Glymour's essay—the investigator ransacks the data, looking for dependence and independence relations and then postulates causal connections on their basis. "Theory" does not really act as an independent constraint on this enterprise, but instead is an after-the-fact rationalization of the results of an investigation that is almost completely data driven. In contexts like this—in which we lack background knowledge or are going to engage in a data-driven search anyway—I find it very plausible that Glymour's procedures will do as well or better than alternative atheoretical procedures. Moreover, when the working assumption fails, it is hard to see how causal modeling without strong guidance from substantive background assumptions could possibly work effectively.

This suggests that we can think of SGS's program as a double-edged sword. On the one hand, one of the ambitions of the program is to improve social research: to provide powerful, computationally efficient procedures for causal inference, even in the absence of much substantive background theory. On the other hand—and I take it from the remarks at the end of his essay this is Glymour's view as well—by thinking about the circumstances in which the working assumption will be violated, one can gain some insight into why relatively atheoretical (by this I simply mean approaches that do not rely very heavily on substantive background theory) causal modeling exercises, unguided by strong, domain specific background knowledge often fail to produce reliable causal knowledge. Indeed—and I intend this not as a criticism but as a very strong compliment—as I read through Glymour's essay and the book that lies behind it, I had the unsettling feeling that I was both being presented with a set of powerful new inferential techniques and, at the same time, with some deep insights into why relatively atheoretical causal modeling techniques, including his own, are unlikely to produce outcomes that are very informative about causal connections when applied to many of the data sets to which social scientists have access.

A final issue I want to explore is whether the conditions spelled out in the working assumption can really stand alone as autonomous principles or whether they in fact require supplementation by domain specific background knowledge of various sorts for their successful application. Glymour claims, and illustrates his claim by reference to the previously mentioned study by Rodgers and Maranto (1989), that substantive domain specific theory is often too weak and unfounded to provide real guidance in the sorts of contexts in which causal modeling techniques are used. He urges as one of the main practical advantages of his approach that it identifies conditions which if satisfied permit reliable causal inference in the absence of such domain specific theory. Clearly this advantage will be at least partly undercut if, as I will suggest, it turns out that, to apply the Markov and faithfulness conditions in a sensible way one must make use of a fair amount of domain specific information.

I turn now to a more detailed look at some of the conditions set out in or implied by the working assumption, beginning with a principle that will be familiar to philosophers—the principle of the common cause. There are many different versions of the common cause principle but for our purposes we can take the principle to consist of two claims. The first claim, which Glymour labels Fisher's condition, is that if two variables X and Y are statistically dependent, then either X causes Y or Y causes X or there is some "common cause" or set of such causes for X and Y. Put informally the claim is, as SGS put it in their book (1993), "statistical dependence is produced by a causal connection." The second claim is that when there is such a set of common causes, it induces certain independence relations—in particular, X and Y are independent conditional on this set, or as it is often put, the common cause or causes "screen off" X from Y. Both claims are consequences of the Markov condition. Another striking consequence of the Markov condition is that all exogenous variables must be statistically independent. If, in a graph representing some causal structure, putatively exogenous variables are not independent, SGS assume that the graph has omitted some relevant causal structure—that, as they put it, "there is some further causal mechanism, not represented in the graph, responsible for the statistical dependence" (1993, 31).

Does the common cause principle hold for the kinds of systems to which we wish to apply causal modeling techniques? One problem with the principle, which may initially seem more annoying than serious, but which I think is actually indicative of a deeper difficulty, has to do with the vagueness of the notion of a common cause. SGS assume that the systems with which they are dealing are pseudo-indeterministic which means roughly that they are deterministic, but look indeterministic because some variables have not been measured and no two of the measured variables depend upon the same unmeasured variable and none of the measured variables causally affect the unmeasured variables. Now, as Frank Arntzenius has recently shown (Arntzenius 1993), it follows from the assumption of determinism alone that there will always be a prior screener off for

**FIGURE 10.1**

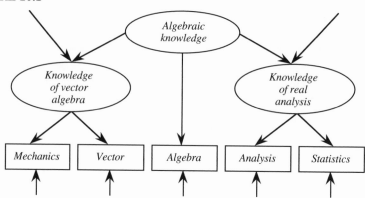

any correlation. However, this fact doesn't ensure this screener off will correspond to what we would intuitively regard as a common cause or causes—the screener off instead may be some unnatural compound, spatially discontinuous, highly gerrymandered set of factors. Consider a simple-minded example. Suppose that X and Y are statistically dependent and that this dependence is produced in the following way: X is caused by $X^*$ and only by $X^*$, Y is caused by and only by $Y^*$ and $X^*$ and $Y^*$ just happen to be correlated. X and Y are then independent conditional on the occurrence of the compound event $Z^*$ consisting of the co-occurrence of $X^*$ and $Y^*$. Does this compound event $Z^*$ count as a common cause? Intuitively $Z^*$ is exactly the sort of occurrence we mean to rule out when we talk of a common cause, but this judgment seems to rely on some unexplained idea about what counts as a natural or non-gruesome choice of variables or way of dividing up the world for purposes of applying the common cause principle.

This particular example may seem completely artificial, but consider now one of SGS's own examples, involving a data set consisting of the grades of a group of students on five mathematical subjects, including mechanics, vectors, algebra, analysis and statistics (1993, 199). Applying their vanishing tetrad test for latent variables, SGS suggest the latent variable structure in figure 10.1. As far as this example goes, we know nothing about postulated latent variables other than that they act as common causes explaining the correlations among the measured variables. Of course it may very well turn out that there is some single, substantive factor with an ungerrymandered representation corresponding to each of these unmeasured variables. But it also seems to me to be equally possible—perhaps even more likely—that one or more of these unmeasured causes is really more like the compound event $Z^*$ in my previous example. What SGS call "knowledge of real analysis," for example, may in fact involve a number of independent causal factors like $X^*$ and $Y^*$ in the previous example, that just happen to coexist within some highly complex correlated structure. Information associated with this structure may indeed screen off the variables "analysis" and "statistics" from one another, but it isn't obvious that the structure will correspond to anything that we intuitively regard as a common cause—or at least it looks as though an argument that it does so correspond is going to have to appeal to something more than the completely general connection between statistical structure and probabilities postulated in the working assumption. As this example illustrates, without some prior information about what can count as a common cause, it isn't clear what we are buying into—other than an underlying assumption of determinism—when we accept the common cause principle.

This example also draws attention to another point which although obvious when stated in the abstract is nevertheless worth making explicit. The common cause principle, in the form stated above, tells us that if X and Y are correlated and there is no direct causal connection from X to Y, then a set of common causes for X and Y must exist and X and Y will be independent conditional on this set. The principle does *not* say that whenever X and Y are correlated and not directly causally connected, then any variable Z such that X and Y are independent conditional on Z (or such that Z is in a set for which X and Y are independent conditional on this set) thereby qualifies as a common cause of X and Y. To adopt the latter view is in effect to hold that there is nothing more to a variable's being a common cause than its standing in the appropriate statistical relation to X and Y—i.e., to adopt a purely statistical definition of what it is for a variable to be a common cause. It is the latter view that leads us to the conclusion that the compound event $Z^*$ or "knowledge of real analysis" automatically qualify as bonafide common causes. Even if one adopts the common cause principle in the form given in Glymour's essay, the examples in this section seem to me to provide for doubts regarding this latter idea.[2]

This conclusion is reinforced by a closely related observation—one to which SGS themselves draw attention. The observation is that if one is given two variables X and Y which are statistically dependent it is always possible to construct new variables by taking linear or Boolean combinations of X and Y which will be statistically independent and vice-versa. For example, if X and Y are correlated, then we may standardize them to form new variables, X' and Y' with unit variance and zero mean. Then the variables X' − Y' and X' + Y' will be uncorrelated. Conversely, if X, Y are uncorrelated (or independent), the variables X + Y, X − Y will be correlated (dependent) and similarly X, X + Y will be correlated (dependent).

These facts also seem to me to fit uncomfortably with the idea that the common cause principle is a completely general principle of causal inference requiring no substantive or domain specific information for its application. If I can make statistical dependencies appear or disappear at will just by choosing appropriate combinations of variables, it seems hard to believe that all such dependencies must reflect causal connections or that all screeners off count as common causes in any substantial or intuitive sense of cause. In the second example above, I suppose one might regard the variables {X, Y} as common causes of the correlated variables X + Y, X − Y—for the latter variables are certainly independent conditional on the former, but the relationship between X and X + Y is not, or at least need not be, what one would intuitively regard as a causal relationship. It again looks as though a substantive non-trivial version of the common cause principle will require, not that all statistical dependencies not involving direct causal connections must be explained by common causes but at best only that some will demand this sort of explanation. Additional information will be necessary to guide our judgments about the sorts of correlations for which we should expect common cause explanations and to distinguish sub-stantive common causes from bogus candidates like $Z^*$. This at least raises the possibility that this additional information will be domain specific and hence that to employ the common cause principle in a way that produces sensible results we must combine it with domain specific background knowledge.[3]

A closely related point is that, as several writers have observed, unless some restrictions are imposed on what is to count as an exogenous variable, it is in-consistent to demand, as SGS do, that all exogenous variables be statistically independent.[4] Given a set of statistically independent exogenous variables we have seen that it will always be possible to define other variables from this set that must be dependent. If the demand for uncorrelatedness of exogenous variables is to be a substantive empirical thesis, and not just a sort of convention (when constructing a causal graph choose or define your variables in such a way that variables of zero indegree turn out to be independent), one again apparently needs to appeal to the idea that some choices of variables are natural or possess a substantive causal interpretation, while others do not. Here again, it looks very much as though domain specific causal background knowledge will be required for this purpose—using the common cause principle successfully in inference will require using it in combination with such background knowledge, and not as a completely autonomous principle of inference.

Of course, even if it is true that additional information is required for the in-formative application of the common cause principle, it does not necessarily fol-low that all of the domain specific background information demanded by more conventional approaches is required. Because philosophers haven't made much progress in understanding what is involved in saying that a choice of predicates or properties is natural, I think it is simply unclear what additional information is needed to apply the common cause principle. Perhaps such information, although domain specific, will turn out to be far weaker than that demanded by traditional approaches to the underdetermination problem. If this is the case, SGS will have made real progress in showing us how, provided certain general principles are satisfied, we can dispense with the need for a lot of domain specific background information, even if we can't dispense with all such information.[5]

I remarked above that the systems to which SGS apply their inference procedures are assumed to be pseudo-deterministic. It is interesting to note—again the observation is due to Arntzenius (1990) and is noted as well in Glymour's essay (p. 208)—that the common cause principle does not hold in general for indeterministic systems. Nor does the principle fail in indeterministic contexts only under very artificial or restricted conditions. Instead, as Arntzenius shows, it fails for some very simple Markov systems. Since it seems to be a live possibility that in the case of some social and behavioral systems, there simply may not exist deterministic relationships that can be formulated between variables at the level of analysis that characterizes theories in the social and behavior sciences, this restriction on the scope of the principle is far from trivial.

I turn now to another issue regarding the application of the common cause principle which is raised by an example of Wesley Salmon's (cf. Salmon 1984). This example, which is discussed by SGS in their book (1993, 61ff.), although not in Glymour's essay, involves a collision (C) between the cue ball and two other billiard balls which sends the first ball into the left-hand pocket (call this event A) and the second ball into the right-hand pocket (B). Because of the conservation of linear momentum, information about whether C and A have occurred will yield a better prediction about whether B has occurred than information about whether C alone has occurred. That is, even though C is a common cause of A and B, and A and B are not directly causally connected, nonetheless A and B are not independent conditional on C.

Does this example involve a violation of the Markov condition? SGS claim that it does not. They claim that if the prior event of the collision were more precisely specified—if we were to specify "the exact momentum of the cue ball on striking the two target balls" (call this C'), then A and B would be independent conditional on C'. Assuming that the billiard ball system conforms to the laws of classical mechanics, this response is of course correct. The Markov condition fails for the system consisting just of the variables A, B, C because we fail to identify or measure the correct variable C' and measure instead the imperfectly correlated proxy variable C or, alternatively because we have collapsed many variables which differ in causally relevant ways—variables specifying different possible values for the momentum of the cue ball into the simple dichotomous variable—collision or no collision—in a way that discards relevant information. If instead we measure C' the Markov condition is restored.

The point to which I want to draw attention—and I stress that this is not anything which, as I read them, SGS would deny—is that there is every reason to believe that the sort of failure illustrated by Salmon's example is extremely common in the social scientific contexts in which causal modeling techniques are used. In such contexts we are frequently unable to accurately measure a set of variables for which the Markov condition would be satisfied. Indeed, as noted above, there may not *be* such a set of variables at the relatively coarse-grained level of analysis at which we are working or at least no such set that we are able to conceptualize or measure. In other words, typical variables measured in social science data sets may be like C in Salmon's example rather than the more discriminating C'. For example, we may wish to ascertain the causal influence of education on earnings, but we may have no measure for education other than a proxy variable like number of years of schooling. Since institutions differ widely in educational quality and educational quality is probably relevant to earnings, the result of using the number of years of schooling variable is to collapse a number of variables that are relevantly causally different into a single category, just as in Salmon's example, with a likely consequent failure of the Markov condition for data sets involving this variable.

As I have expressed it, this line of assumption has so far neglected a complicating consideration: Glymour's working assumption does *not* say that the connection between the causal relationships that hold among the variables we observe and their independence relations must conform to the Markov condition but rather that, as he puts it, "the distribution of observed variables . . . is the marginal of a distribution" satisfying this condition. Does this complication avoid the difficulty posed by Salmon's example? While I certainly don't want to deny that there will be cases in which a set of measured variables stands to some larger set in exactly the relationship described in the working assumption, it seems to me doubtful that this will usually be the right way to think about typical cases involving Salmon's difficulty. The problem with trying to infer causal connections from the statistical relations among A, B, and C in Salmon's example is not just that a common cause C' has been left out. C' isn't an *extra* causal factor in addition to C, that can be invoked to explain the residual correlation between A and B that remains once C is taken into account and which can be just added on to the graph representing A, B, and C. Rather the correct causal theory is one that *replaces* C with C'. Our problem is not just that we have omitted to measure some causally relevant variables, but rather that among the variables we do measure, some are somehow the wrong variables, at least from the point of view of achieving any deep causal understanding. In other words, contrary to the working assumption, the distribution of A, B, and C is *not* the marginal of the distribution associated with the underlying correct graph, since that graph should not include the variable C. I suspect that a similar point holds in many of the social scientific contexts in which causal modeling techniques are used.

Suppose that in Salmon's example, we measure C and fail to measure C' and that, as a result, a residual correlation remains between A and B when we control for C. Consider first a methodological framework, like the one ascribed to Tufte above, that insists that we are not automatically entitled to postulate causal relationships whenever there are statistical dependencies, but only when there is, in addition, some independent causal rationale grounded in domain specific background knowledge for this postulation. In this case, the additional correlation, remaining after we control for C, will not necessarily lead us astray—for we may have no independent justification for giving that correlation a causal interpretation. This corresponds to the fact that within a conventional causal modeling framework, the usual expectation is that the result of measuring imperfectly correlated proxies for causally connected variables rather than those variables themselves will be that estimated causal relationships are attenuated, but not necessarily that mistaken new causal claims are introduced. By contrast, I worry that within Glymour's framework, such residual correlations may lead to the postulation of additional causal structure—e.g., we may be led to postulate some additional common cause, distinct from C, to explain the residual correlation or failing that, that the output of the SGS inference techniques will be, in the sense described above, relatively uninformative—it will be compatible with a very large number of distinct causal graphs.

My suspicion is that what is true in this example may be true more generally. The sorts of problems concerning the use of proxies, aggregation of relevantly different variables, and so on that we have been discussing are ubiquitous in the social sciences and can leave us with lots of residual statistical dependencies even when we control for apparently relevant variables. Relatively atheoretical causal modeling procedures, guided by the idea that statistical dependencies always indicate causal connections, will either tend to postulate causal relationships where none exist in such environments if they are constrained to produce a single hypothesis as output or, if they are like the SGS procedures in outputting all alternatives among which it is impossible to distinguish, they may be uninformative even if they would yield reliable causal knowledge in more ideal environments in which all the right variables are correctly measured, and the other conditions set out in the working assumption are satisfied. More conventional approaches which refuse to postulate causal connections whenever there are such statistical dependencies may perform better in such environments, provided of course—and I concede that this is a large *if*—background knowledge of the right sort is available.

Another class of cases in which the Markov condition may fail, or at least in which its application may be uncertain, involves mixtures of causally heterogeneous subpopulations. There is an extremely rich and interesting discussion of such cases in both SGS and Glymour's essay from which I have learned a great deal, and I again rely very heavily on this discussion in what follows. As both SGS and Glymour note, the Markov condition may be violated for populations of units, each of which has the same directed graph, but which are represented by different functions, and it may also be violated by populations of units which do not share the same directed graph. SGS also note that attributes which are independent in each of several different subpopulations will typically become dependent when the subpopulations are "mixed" in a combined population. In such a case the independence relations which follow from the causal graphs characterizing the subpopulations will not obtain in the combined population, although it may be possible to restore the Markov condition by introducing a further variable representing the cause of membership in these subpopulations—a point to which I will return below.

If I have understood them correctly, SGS argue that there is an interesting asymmetry with respect to the results of mixing heterogenous populations. While mixing subpopulations in which variables are correlated can produce an outcome in which they are uncorrelated in the whole population, this will happen only in rather special cases—a set of cases which have measure zero under some natural measure. (Such cases involve a violation of what SGS call "faithfulness"— a well-known example is the original source for "Simpson's paradox" [SGS 1993, 64 ff.].) By contrast, when different subpopulations are mixed, the "usual" outcome is that variables uncorrelated in the subpopulations will be correlated in the whole population—this will be the outcome unless rather special conditions obtain. Mixing causally heterogenous systems thus has a bias toward making dependencies appear rather than disappear—a point to which I will return below.

As Glymour notes, in some cases in which the Markov condition fails because of heterogeneity, it may be possible to restore the condition by introducing a classifier variable that takes different values for the different classes of units, but he also remarks that whether this variable "is regarded as a cause or as a fiction will depend on context and, no doubt on the regarder" (p. 207). This remark, to which I will return below, appears to acknowledge a point made earlier in my discussion of the common cause principle: that it looks as though something more is involved in ascertaining whether a variable is legitimately regarded as a common cause (or in deciding whether one can legitimately regard a population of units as causally homogenous or as described by a single causal graph) than is captured by the statistical dependence and independence relations in which it stands to other variables.

To see what is at issue here (or at least what I think may be at issue—I confess to considerable confusion on this score) consider another of SGS's examples (1993, 60). Contemporary neurophysiologists perform statistical studies on groups of patients with brain damage to ascertain whether there are correlations in their scores on tests designed to measure different kinds of cognitive deficits. Their hope is to discover some common cause—some common kind of brain damage or neurophysiological dysfunction—that underlies the correlated cognitive deficits. Suppose that the population sampled consists of a mixture of two subpopulations with different kinds of brain damage and that the result of this mixing is to produce a correlation between measured cognitive skills, even though these skills are distributed independently in each subpopulation. Then of course one can always introduce a variable V taking the values 0 or 1 depending on whether a unit comes from the first or second subpopulation. It may then be possible to invoke this variable as a common cause for the correlation in question and treat the combined population as representable by a single causal graph in which this variable is present as a common cause. Although it is not easy to say what underlies this judgment, I think that virtually everyone will agree that this restoration of the Markov condition is too easy and that, at least as described, V is somehow something less than a full-blown cause. SGS appear to agree for they remark that when such a correlation is present in the combined population the common cause need not be a cause in any sense that would interest neurophysiologists. They say that the common cause "need not be any functional capacity—damaged or otherwise—that causes both skills." Instead, in the sort of case I have envisioned, they suggest that the common cause will be "only a variable representing membership in a subpopulation" (1993, 60).[6]

This suggestion seems to acknowledge that something more is involved in a variable's being a real cause than is captured by the statistical independence and dependence relations in which it stands to other variables. By Glymour's own account, this something more makes some variables legitimate candidates for common causes while others instead are in Glymour's words "fictions" or mere devices for classification or for representing facts about group membership. Relatedly, it again looks as though if the Markov condition is to have substantive content, it requires some sort of independent restrictions on which variables can count as causes.

Let me conclude this section by underscoring an important observation made by Glymour at several points in his essay—I do this in part because I'm afraid that the observation goes by so quickly that its full significance may be missed. It is a plausible conjecture that the difficulties for the working assumption discussed in this section having to do with causally heterogenous populations and with measuring the wrong variables are very common in social science data sets. We have seen that the effect of such heterogeneity and mismeasurement will typically be to produce lots of correlations that do not reflect causal structure. Moreover, when one looks at typical, relatively atheoretical causal modeling exercises in many areas of social science, one very often finds causal graphs that are complete or nearly so. That is, almost everything ends up being causally connected to everything else or at least edges are only omitted when there is some commonsense rationale like temporal order for doing so. The Blau and Duncan (1967), and Rodgers and Maranto (1989) examples in Glymour's essay provide illustrations of this. Obviously many researchers take such graphs seriously but my intuitive reaction (and I find, also, the reaction of a number of social scientists to whom I have talked about the matter) to such graphs is that this near completeness itself signals that something has gone badly wrong and that the resulting causal structure shouldn't be taken seriously. Glymour's discussion gives us a deep and valuable insight into why this intuitive reaction is probably correct. Of course it is possible that in the domain under investigation (nearly) everything really is connected to everything else, but it will usually be a more plausible hypothesis that the near completeness of these graphs in part reflects the fact that populations from which they come are heterogenous and the wrong variables are being measured. And, as we have seen, these are among the conditions under which one can't successfully extract reliable causal conclusions from the statistics. This is just one of several points at which Glymour's discussion enables us to better understand the limitations of causal modeling techniques.

## II

I turn now to some brief remarks about the faithfulness condition which, along with the Markov condition, also plays an important role in solving the undetermination problem for SGS. Intuitively, a distribution P is faithful to a graph G if all of the conditional independence relations in P follow just from the Markov condition alone as applied to G—i.e., if all the independence relations follow directly from the graphical structure of G and not from special parameter values associated with the edges in G. Thus, to use SGS's own example (1993, 95) consider the following two models which are assumed to be represented by linear equations. (See figure 10.2.) In (2.2) with the error terms uncorrelated (as the working assumption requires), A and C will also be uncorrelated. But A and C will also be uncorrelated in model (2.1) if it should happen that the linear coefficients associated with each edge are related as ab = –c. Model (2.2) is faithful, while (2.1) is not—the zero correlation between A and C in (2.1) doesn't follow

**FIGURE 10.2**

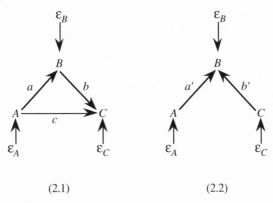

(2.1)                                            (2.2)

from the graphical structure alone, but rather is an "accidental" consequence of the particular values that the parameters a, b, and c happen to assume.

What is the justification for assuming faithfulness? SGS's basic argument seems to be measure-theoretic: they show that the set of cases in which faithfulness is violated has measure zero under some natural measure (cf. *SGS* 1993, Theorem 3.2, 68–69). Informally, the idea is that for a structure like (2.1), most choices of parameter values will not produce a correlation of zero between A and C; this will happen only if a, b, and c take on very special values. By contrast, when an independence relation follows just from the graphical structure as in (2.2), it will hold for all sets of parameter values.

Faithfulness is an interesting and ingenious idea with remarkably powerful consequences. Nonetheless, I have some misgivings (or at least questions) about the rationale for its adoption and the way it is deployed in SGS's model selection procedures. With regard to the first issue, the crucial question is exactly how the measure-theoretic result described above justifies the adoption of faithfulness. One possible line of argument is a Bayesian one—in view of the measure theoretic result we should assign zero prior probability to models violating faithfulness. But as David Freedman reminds us in his essay, there are well-known objections to such an argument and, in any event, the argument is not one that SGS themselves make. A more interesting and promising line of argument, alluded to in passing by Glymour in his essay, instead focuses on parameter instability. Suppose, as seems in fact to be the case, that the parameter values in most structural equation models in the social and behavioral sciences are not like physical constants but are instead relatively unstable—they are likely to change over time, from population to population, with changes in background conditions and so on.[7] In the case of a model like (2.1), the smallest change in any of the parameters a, b, and c (unless this is exactly and fortuitously balanced by a corresponding change in the other parameters) will upset the relationship $ab = -c$, producing a correlation between A and C. Thus (one might argue) cases in which a model implies conditional independence relation because of special parameter values should be rare and unstable—such structures are unlikely to persist over time or across space. It is this fact which justifies focusing on faithful distributions.

Let us explore the implications of this line of argument in a bit more detail.[8] Consider a simple recursive model represented by the following equations

$$(2.3) \quad \begin{array}{l} x_2 = ax_1 + u_1 \\ x_3 = bx_2 + u_2 \end{array}$$

with $u_1$ and $u_2$ uncorrelated, and $u_1$ uncorrelated with $x_1$ and $u_2$ uncorrelated with $x_1$ and $x_2$. This is an acyclic model which (let us assume) satisfies SGS's working assumption with respect to the observed distribution. But of course one can think of this model as a particular instance of much more general model—e.g.,

$$(2.4) \quad \begin{array}{l} x_2 = ax_1 + cx_3 + u_1 \\ x_3 = bx_2 + u_2 \end{array}$$

in which the coefficient c just happens to have the value of zero. If we apply the rationale for assuming faithfulness described above and assume that parameter values are generally unstable, shouldn't we then expect that the usual case is one in which all of the parameter values in system (2.4), including c, are not equal to zero? That is, won't the usual upshot of parameter instability be to replace systems like (2.3) with systems like (2.4) in which $c \neq 0$? However, systems like (2.4) will be cyclic and will violate the working assumption. More generally, don't these considerations show that the working assumption itself restricts our attention to a set of models which have measure zero in the set of all possible models? (See Freedman's essay for a similar point). If it is justifiable to ignore distributions that are unfaithful on the grounds that they are measure zero, what justifies us in focusing on just acyclic, or recursive models when they also have measure zero in the set of all possible models?

We can ask a similar question about why zero should have a special status in connection with the SGS assumptions about statistical independence and the Markov condition. As cited above, the working assumption requires that all exogenous variables be independent. In general SGS assume that statistical dependencies (between, for example, exogenous variables) don't just occur accidentally, but always have causal explanations of some sort. If two variables are not causally connected in some way (directly or via a common cause), SGS assume that they will be independent. But if SGS's measure theoretic argument for ignoring faithful distributions is a good one, why can't we argue, in a parallel way, that statistical independence itself is a special case, involving a special choice of parameter values? In the linear case if we are given two causally unconnected variables, it is of course logically possible that the correlation between them can take any value between 1 and –1. Why can't we then argue that Cor (A, B) = 0 is the measure zero case, and that the usual case will be one in which Cor (A, B) ≠ 0? Similarly, given a system of linear equations why can't we argue that the usual case is one in which the error terms will be correlated across equations and with the explanatory variables in the equations in which they figure, and that the sort of assumptions about uncorrelatedness that SGS make and which are embodied in the Markov condition correspond to a rare or unusual case?[9] And why not argue on similar grounds, that the common cause principle corresponds to a rare or unlikely case—given two variables that are causally unconnected, it would be an unusual coincidence if they turned out to be exactly uncorrelated. More generally, isn't there at least a tension between the rationale SGS offer for faithfulness and the Markov condition? That is, if the justification for ignoring distributions which are not faithful is that these are measure zero, why should we not also assume that the distributions with which we are dealing violate the Markov condition on the grounds that satisfaction of the Markov condition corresponds to a measure zero case?

Finally, we can also ask similar questions about the status of the assumption that the units in the population to which the SGS techniques are applied are causally homogeneous. If the rationale for faithfulness has to do with coefficient instability in the manner suggested above, shouldn't we regard it as unlikely that the same constant coefficients will characterize all units in the population? Won't the "usual" situation rather be one in which different coefficients or even different functional relationships characterize the behavior of different individuals in the population, with all of the attendant difficulties this sort of mixing of causally heterogeneous systems creates for the Markov condition and for causal modeling methods generally?

My intent in raising these worries is not to suggest that they are necessarily fatal for faithfulness (or the Markov condition). It may very well be that there are good reasons why we should not think of the absence of a directed edge from A to B as a special case of a more general model in which there is such an edge, but the coefficient along it has been set equal to zero. Similarly, there may be good reasons why we should not think of zero as just one special value, among many, that the correlation between two causally unconnected variables can take and as therefore corresponding to a measure zero case. Perhaps there are strong independent reasons for adopting the Markov condition that completely override the measure zero argument considered above. My point is merely that SGS need to set out these reasons in more detail if they are to fully motivate the use they make of faithfulness and the Markov condition.

## III

I turn now to a difficult and closely related issue that I have so far avoided: What conception of causation is assumed in SGS's work? In his essay, Glymour is noncommittal about this issue. He says that "while I will rely on some intuitions about relevant features of the notion of causation, I avoid throughout any jejune attempt at definition." In their book, SGS are, if anything, even less explicit about what interpretation of causality is intended. It is easy to sympathize with what I take to be the motivations that underlie this refusal. The philosophical literature is full of attempts to provide reductive definitions of causal notions—attempts which have not only been unsuccessful on their own terms but which have often served simply to distract attention from more tractable and intellectually worthwhile epistemological and methodological issues. Nonetheless there is something rather troubling about SGS's reluctance to be more explicit about what they mean by causation. After all, SGS's central claim is that they have discovered reliable new procedures for inferring causal conclusions from statistical data, given very general assumptions. How can one possibly assess this claim or even understand it unless one knows what SGS mean by a causal relationship and whether the "causal relationships" discovered by their inference procedures coincide with causal relationships in some more ordinary sense? (An account that addresses these questions need not take the form of a reductive definition—it might instead proceed by exhibiting the relationship between the intended interpretation of causation and other causal notions and by contrasting the intended interpretation with other, rival interpretations. Glymour's own informal remarks about the relationship between causation and manipulation are an example of such a strategy.)

Glymour contends, in his commentary, that it is "misconceived" to raise issues of this sort because, among other things, "whatever you take 'causal' to mean, if you assume that the [Causal Markov and Faithfulness] conditions apply, the theorems we prove follow necessarily." However, one can agree that SGS's theorems are genuine theorems and still think that there are legitimate questions that arise in connection with SGS's program about how one should understand causal claims. For one thing, the Causal Markov condition and the working assumption both connect causal claims (or graphical structure when interpreted causally) and probabilities. Thus the answer we give to the crucial question of how widely and on what circumstances these assumptions hold is surely going to depend in part on what we mean by 'cause'. That is, it seems likely that different accounts of causation will yield different answers to the question of the conditions under which the Causal Markov condition and working assumption hold. To explore this issue we need to formulate these alternative accounts, and try to understand the relationships among them.

Moreover, several substantial traditions in econometrics, described below, do just what Glymour claims it is misconceived to do—they regard the project of providing an interpretation of causality as an essential part of an overall treatment of causal inference. This is true both of the program of causality testing associated with Clive Granger and of the tradition associated with investigators like Haavelmo and Frisch, that connects causal relationships (or at least the kinds of causal relationships that are scientifically interesting) with "structural," "autonomous," or "invariant" relationships. (Readers unfamiliar with this last tradition should know that it persists despite Glymour's suggestion that "proving" that a relationship is stable "would require a refutation of Hume.") By providing an interpretation of causation both traditions give one a sense of the sort of relationships that an investigator ought to be aiming at or trying to discover when he conducts a causal investigation—the point or goal of the inquiry, so to speak. One can then ask how successful the proposed inference techniques are in achieving this goal. It seems to me to be perfectly appropriate to raise analogous questions in connection with SGS's program and to ask what the relationship is between their program and these other traditions.

To set the stage for my subsequent discussion, let me distinguish rather quickly and roughly between two different notions of or ways of thinking about causation. I will then try to deepen and sharpen the contrast as my discussion proceeds.

1) To motivate the first way of thinking about causation, consider that whatever else it represents, one can always think of a graph satisfying the conditions in the working assumption as a device for perspicuously and economically representing independence and dependence relations. According to this first conception, this is all there is or at least all there is that is clear and scientifically useful to the notion of causation—causation is simply a device for representing facts about independence and dependence relations. I will call this the informational conception of causation, since I take it that the guiding idea is that causal relationships have to do with the informational relevance of various variables to others. A specific version of this general idea is that the relationship between graphical structure and facts about independence relations spelled out in the Markov and faithfulness conditions serves as a sort of implicit definition of the

notion of causality—any system of relationships that can be associated with a directed acyclic graph and which satisfies the working assumption qualifies as causal in this sense.[10] Obviously, one might generate other specific versions of the informational notion by making other assumptions about the relationship between causal structure and independence relations. To anticipate my discussion below I think that, roughly speaking, the notion of probabilistic causation defended by the philosopher Patrick Suppes (1970), the notion of causation one finds in Judea Pearl's book (1989), and the notion of Granger causation within econometrics are all informational notions of causation, although they of course differ importantly in detail in various ways.

2) A second way of thinking about causation instead attempts to connect with the notion of manipulation and control. It is difficult to state this connection in a way that is plausible and precise, but Glymour himself gives as clear and succinct a statement of the basic idea as any that I know in his informal remarks about the interpretation to be placed on the directed edges in a causal graph. He writes:

> The idea of the graphical representation is roughly that a directed edge X →Y in a graph G indicates that there is a possible (I don't know in exactly what sense of possible) intervention that would change the value of X, and possibly hold constant some decendents of X, and that through and only through the change in X the intervention would also change the value of (or the probability of some value of) Y and would do so even if all other variables that are not descendants of X were to remain unchanged. (p. 206)

To anticipate my discussion below, I associate the control concept with the econometricians' notion of a structural model; that is, with the idea that causal relationships should be invariant or autonomous relationships. I might also add, for what it is worth, that my own view is that only the second notion deserves the name "causation." However, this claim will strike many as tendentious, and in an effort to avoid becoming embroiled in a purely terminological dispute, I will use the word "cause" in connection with both notions and devote my energy to exploring the relationship between them.

In the informal remarks cited immediately above, it seems that the interpretation of causation intended by Glymour is the control notion. But other portions of both Glymour's essay and SGS—I am thinking here both of the connections between causation and probability that are assumed in the Markov and faithfulness conditions and many of the particular examples discussed—instead often left me wondering whether the intended interpretation of causation wasn't instead something more like the informational notion. This raises a crucial question:

What exactly is the relationship between these two ways of thinking about causation? Are the two notions identical or at least extensionally equivalent or are there cases in which the first notion applies but not the second? In particular, does the fact that we can give a causal interpretation in the informational sense to the edges in a graph ensure that the control interpretation also applies? On the face of it, the informational notion seems far broader than the control notion. Indeed, a standard criticism of many causal modeling exercises is that they conflate the informational and control notions even though these are fundamentally distinct notions of causation—that such models are really just representations of independence or dependence relations in a population which are illegitimately taken to provide information about what would happen if we were to intervene to manipulate certain variables in that population.[11]

As an illustration of what looks like a purely informational notion of causation, consider Judea Pearl's recent book, *Probabilistic Reasoning in Intelligent Systems* (1988). Although Pearl's work differs in many respects from SGS's, there are nonetheless important parallels and as SGS acknowledge, Pearl's work has been an important stimulus to their own thinking. In Pearl's work we also find the use of directed graphs to represent facts about statistical independence and dependence relations and the use of the Markov condition to establish a connection between graphical structure and independence relations. Similarly, as SGS note, their notion of faithfulness is suggested by Pearl's notion of a perfect map (SGS 1993, 35–36). Moreover, like SGS, Pearl also gives a causal interpretation to the graphical structures he associates with independence relations. However, he makes it fairly clear that he thinks of causality (or at least his "causal networks") as nothing more than a device for perspicuously representing in a computationally tractable form facts about statistical relationships. He writes:

> It is this computational role of identifying what information is or is not relevant in any given situation that we attribute to the mental construct of causation. Causality modularizes (our knowledge) as it is cast from experience. By displaying the irrelevances in a domain, causal schemata minimize the number of relationships that need to be considered while a model is constructed and in effect legitimizes future inferences. (Pearl 1988, 52)

Elsewhere, he writes that "causation is not a property of nature but a mental construct devised for the efficient organization of knowledge" (p. 382) and adds, in a passage quoted by Freedman, that "formally speaking, probabilistic analysis is indeed sensitive only to covariation so it can never distinguish genuine causal dependencies from spurious correlation, i.e., variations coordinated by some common unknown causal mechanism" (p. 396). Pearl also seems to suggest that the association of causation with control is either misplaced or that at bottom claims that one variable can be used to control another are really just claims about "covariation observed on a larger set of variables" (p. 396).[12]

Now of course the fact that the relationships represented by Pearl's causal networks are interpreted by him as causal in the informational sense doesn't show that the relationships discovered by SGS's inferential procedures are causal in the informational sense only, and not the control sense. Still, Pearl's treatment seems to show that it is at least possible to think of the graphical structures associated with statistical relationships via the working assumption as nothing more than devices for representing those statistical relationships. This in turn raises the question of what the grounds are for believing that the SGS procedures find causal relations in some stronger sense. There is also the closely related question, suggested by the passage quoted from Pearl above, of whether the two conceptions are really sharply distinct or whether instead facts about controllability (insofar as this is a clear notion) are reducible to facts about covariation.

Reasons for thinking that, contrary to Pearl's apparent suggestion, there are important differences between the informational and control conceptions of causation emerge rather clearly when one looks at the econometrics literature on structural equations. This difference is worth spelling out in sure detail, since I want to return to it later in my discussion. An influential tradition in econometrics, reaching back to writers like Haavelmo (1944) and Frisch (1938) and reflected in contemporary writers like Lucas (1976), claims that equations that represent causal relationships should be stable or invariant or as it is sometimes said, "autonomous" under some class of hypothetical changes or interventions. Obviously invariance in this sense will always be relative to a class of changes or interventions—causal relations in the social sciences are not fundamental laws of nature, invariant under all possible interventions, but rather will typically be invariant under certain interventions but not others. To illustrate the basic idea, consider again a system of simultaneous equations like (1.1). If these equations are structural, then the idea is that the relationship expressed by each equation—its functional form, its coefficients, and so on—should remain unaltered under some relevant classes of changes, including at a minimum, changes in the values of the independent variables in the equation and changes in the value of the error term. Moreover, as Duncan (1975) argues, in the context of systems of linear equations, it often will be plausible to impose a distinct, additional invariance or autonomy requirement; that it should be possible to intervene in such a way as to separately change each of the coefficients in each equation, and each of the coefficients in each of the equations should be invariant under interventions that produce changes in any of the other coefficients. To illustrate, consider the following elementary example, drawn from Duncan (1975). Suppose that the true structural model (the model that captures the causal relationships) for some system is

$$(3.1) \quad \begin{aligned} X_3 &= b_{31} X_1 + b_{34} X_4 + U \\ X_4 &= b_{42} X_2 + b_{43} X_3 + V \end{aligned}$$

Then the following model (3.2)—which is just the reduced form equations for (3.1)—will be "observationally equivalent" to (3.1), in the sense that it will have exactly the same solutions and imply the same facts about statistical relationships among the measured variables.[13]

$$(3.2) \quad \begin{aligned} X_3 &= a_{31} X_1 + a_{32} X_2 + U' \\ X_4 &= a_{41} X_1 + a_{42} X_2 + V' \end{aligned}$$

$$where \quad a_{31} = \frac{b_{31}}{\Delta}, a_{32} = \frac{b_{34} \, b_{42}}{\Delta}, a_{41} = \frac{b_{43} \, b_{31}}{\Delta}, a_{42} = \frac{b_{42}}{\Delta}, U' = \frac{U + b_{34} V}{\Delta},$$

$$V' = \frac{b_{43} U + V}{\Delta}, \Delta = 1 - b_{34} b_{43}$$

Despite their observational equivalence, there is, as Duncan notes, an important difference between (3.1) and (3.2). Each of the coefficients in (3.2) is a "mixture" of several of the coefficients in (3.1). Since (3.1) is by assumption a structural model, each of the coefficients in (3.1) should be invariant or autonomous in the sense that each can change without any of the other coefficients in (3.1) changing. But since the coefficients in (3.2) are mixtures of the coefficients in (3.1), if one of the a's in (3.2) changes, this would presumably mean that one or more of the b's in (3.1) had changed, which would mean in turn that a number of the other coefficients in (3.2) would change as well, since each b is involved in the expression for several a's. In short, if (3.1) is structural, (3.2) will not be structural—the coefficients in (3.2) will not be stable or invariant under changes in the other coefficients. In this sense the b's are, as Duncan puts it "more autonomous" (1975, 153) than the a's.

There is an obvious connection between these ideas about structure and invariance and the control conception of causation. If the relationship between C and E will remain invariant under some significant class of changes (including, crucially, changes in C itself and changes elsewhere in the system of which C and E are a part) then we may be able to avail ourselves of the stability of this relationship to produce changes in E by producing changes in C. If, on the contrary, the relationship between C and E is not invariant, so the relevant changes in C itself or in background circumstances will simply disrupt the previously existing relationship between C and E, then we will not be able to make use of this relationship to bring about changes in E by manipulating C.

To illustrate this idea in the context of a system of simultaneous equations, suppose, as before, that the true system of structural relations is represented by (3.1). Then if, for example, I wish to alter the relationship between $X_1$ and $X_3$, I can intervene to change $b_{31}$ and because (3.1) is structural the result of this will not be to produce any changes elsewhere in the system (3.1) which may undercut the result I am trying to achieve. By contrast, the coefficients in (3.2) are, as it were, entangled with each other in a way that makes the relationships described in (3.2) unsuitable for purposes of manipulation and control. Suppose one wishes to alter the relationship between $X_1$ and $X_4$ in (3.2) by altering the value of the coefficient $a_{41}$ (or suppose some natural change occurs that produces this result). As I have noted, to accomplish this result, one or more of the b's in (3.1) would have to be different. However, not knowing the true structural model (3.1), one will not know exactly how the coefficients in (3.2) will have to differ in order to produce this change in $a_{41}$. Thus, while we may expect that if $a_{41}$ were to change, some of the coefficients in (3.2) would probably change as well, it will be impossible to say (in the absence of knowledge of the true structural model) which coefficients will change and exactly how they will change.

Another way of putting the contrast is this: There is hypothetical manipulation associated with (3.1) which is unproblematic and well defined. If, say, I intervene to change the value of $b_{31}$, it is clear from the independence of the coefficients in (3.1) how this change will ramify through the system and thus what the result of this change will be. By contrast, the hypothetical experiment or manipulation associated with changing $a_{41}$ to, say, $a'_{41}$ will be indeterminate and not well defined. The result of this experiment will depend upon how this change in $a_{41}$ has been produced, and without knowing this (i.e., without knowing how the true structural coefficients have changed so as to produce this change in $a_{41}$), one can't say what the result of this experiment would be. Clearly this makes (3.2) in comparison with (3.1) an unsuitable instrument for manipulation and control. This in turn makes it plausible to suggest, on a control conception of causation, that (3.1) captures genuine causal relationships in a way that (3.2) does not.

Yet another way of getting at essentially the same intuition appeals to the idea that causal relationships should be local and modular rather than holistic. If the equations in (3.1) are structural, then each relationship between a dependent variable and an independent variable—e.g., the relationship between $X_3$ and $X_1$ in the first equation (3.1)—should describe a distinct relationship or mechanism. If this is so, then it should make sense to think in terms of doing something that affects just this relationship or mechanism and none of the other relationships specified in (3.1). Interfering with the relationship between $X_3$ and $X_1$ by altering $b_{31}$ alone should be an allowable hypothetical experiment, even if it is not one that, as a practical matter, we are able to carry out. Moreover, since there is a distinct mechanism connecting $X_3$ and $X_1$ the result of this experiment should depend just on what happens to this mechanism and not on the relationships that hold elsewhere in the system. In this sense, individual causal relationships should be modular or context-independent, rather than dependent in an indeterminate and holistic way on the other relationships and mechanisms holding elsewhere in the system. The system (3.1) exhibits this sort of modularity, while the observationally equivalent system (3.2) does not—the relationships described in (3.2) are entangled in a non-local, non-modular way.

The notion of causation associated with invariance and autonomy in the above systems seems different from and stronger than what we have been calling the informational notion. As we have seen, (3.1) and (3.2) imply exactly the same facts about statistical dependence and independence relations, which is the basic resource out of which informational causal claims are constructed. Where (3.1) and (3.2) differ is in terms of what they imply about what would happen under various hypothetical interventions, and this is not a difference that will be fully caught by actual statistical relationships.

We can also bring this out by contrasting the above ideas about structure and invariance with a well-known conception of causation due to Clive Granger, which, as remarked above, embodies an informational notion of causation. Granger's basic idea is that causes should carry information about their effects. Put roughly and informally, $Y_t$ causes $X_{t+1}$ if, as Granger puts it, "we are better able to predict $X_{t+1}$ using all available information than if the information apart from $Y_t$ had been used" (1969, 376).

The general attitude within the econometrics literature seems to be that the control notion of causation associated with invariance and Granger causation are very different. For one thing, distinct but observationally equivalent structural models may yield exactly the same claims about which variables Granger-cause others but different claims about which relationships are structural or invariant under various possible manipulations. Kevin Hoover, in his recent survey of neoclassical economics (1988), remarks that "clearly, Granger causality and causality as it is normally analyzed (he takes this to be the control conception) are not closely related concepts" (p. 174) and gives a number of examples to illustrate this claim. Granger himself distinguishes his account of causation from the controllability or manipulability conception and explicitly rejects the latter. He writes:

> The equivalence of causation and controllability is not generally accepted, the latter being perhaps a deeper relationship. If a causal link were found and was not previously used for control, the action of attempting to control with it may destroy the causal link. (1990, 46)

By contrast, the heart of the control conception is the denial of this last claim. According to the control conception, if the relationship between C and E is such that any intervention to alter C (including even an ideal intervention that affects only C and no other variables that are not effects of C) has the result of destroying the relationship between C and E, then that relationship is not causal.

The contrast between informational and control conceptions of causation is also apparent in the philosophical literature on probabilistic theories of causation. Theories like Patrick Suppes' (1970) which seek to define causation in terms of facts about conditional probabilities clearly represent informational conceptions of causation—roughly speaking, for such theories a causal relation is just a statistical dependence relation that doesn't disappear when one conditionalizes on other sets of factors (where these additional factors are not themselves picked out by causal considerations). A contrasting formulation of the probabilistic theory of causation in the spirit of the control conception would involve the idea that a cause must not only raise the probability of its effect but that this relationship must be invariant when one intervenes to introduce C into a situation from which it had previously been absent (i.e., to change the situation from $\bar{C}$ to C). That is, the idea would be that it must be the case that the relationship $P(E/C) > P(E)$ continues to hold as we change from $\bar{C}$ to C or, more generally, as we change the frequency of C. Clearly it is only in this last case that one can use C to manipulate or control E. (For a formulation of a probabilistic theory of causation which also emphasizes the idea of invariance, but which differs in detail from the one given above, see Arntzenius [1990].) From the point of view of the control conception, the problem with a Suppes-style definition of cause in terms of conditional probability relations is that the mere fact the C raises the probability of E (and satisfies Suppes' other conditions for non-spuriousness) does not seem to guarantee that if we intervene to manipulate the value of C this will change the value of E.

Essentially the same contrast is also drawn in David Freedman's essay in this volume. Freedman distinguishes between two procedures—in the first, the investigator simply observes or selects units with X = x and observes the average value of Y for these units. In the second procedure, the investigator actually intervenes and sets the values of the units to X = x and then observes the average value of Y. Freedman's point is that in general the results of these two procedures needn't coincide—there is no general guarantee that when one carries out the first procedure and observes the value of Y, this will tell one what would happen to Y if one were to carry out the second procedure. The difference between these procedures corresponds roughly to the difference between what we have been calling the informational and control conception of causation.

What is the relevance of all of this to Glymour's discussion? The basic issue I want to raise can be put quite simply. Is there some reason to think that the inference procedures described by SGS can be used to discover relationships that are structural or invariant or autonomous—that is, relationships that correspond to the control conception of causation and not just to a conception of causation which is informational in the Suppes-Pearl-Granger sense?[14] If the informational and control notions are distinct in the way suggested above, then it looks as though the working assumption might be satisfied with respect to some population when "cause" is interpreted in the informational sense but not when it is interpreted in the control sense. Of course one may agree that if the working assumption is satisfied for some population under an interpretation according to which "causal relations" in the statement of that assumption means or implies "causal in the control sense" or "causal in the sense of structural or invariant," then SGS's techniques will discover causal relationships in just that sense. My question is whether SGS believe that the working assumption is ever or very frequently satisfied when so interpreted for the sorts of systems to which they apply their techniques, or whether they think it is only likely to be satisfied when causation is interpreted in the informational sense, so that their techniques will typically yield causal conclusions only in this latter sense. Or is their view perhaps instead that the informational and control conceptions coincide at least extensionally and that if the working assumption holds when interpreted according to the former conception it will also hold according to the latter? If so, what is the argument for this claim? More generally, when they apply their techniques to a body of statistical data and output a set of graphs which they then claim represent causal relationships, do they wish to remain completely non-committal about the sense in which those relationships are causal and if not, what particular interpretation of 'causation' do they have in mind?

In addition, we can also ask a number of questions about the connection between SGS's program and the more traditional econometric programs that focus on using statistical relationships as an aid to finding structural or invariant relationships. Suppose that the working assumption is statisfied (under some relevant interpretation of causation). Does it then follow that among the outputs of the SGS inference procedures will be a model that qualifies as the correct structural (or invariant or autonomous) model for the domain under investigation? If so, what is the relevant interpretation of causation that must be satisfied for this to be the case? Do SGS think of the goal or object of causal inquiry in the way that Haavelmo does—that is, as the identification of invariant relationships? Or do SGS mean instead to reject this whole way of thinking about causal relationships and the goal of causal inquiry—either on the grounds that the idea that causal relationships must be invariant relationships is mistaken or confused or on the grounds that it is unreasonable to expect causal inference procedures to identify such relationships?

A closely related issue is this: As we have observed, the notion of invariance is always relative to a class of interventions—a relationship may be invariant under certain interventions and not under others. One of the things one would most like to know about putative causal relationships in the social sciences is their scope of invariance—the hypothetical experiments they will support. For reasons that will be explained in more detail below, it seems undeniable that there will be a number of quite general circumstances having to do, for example, with the presence of various kinds of temporal instabilities, or with the presence of switches or non-reversible relations, in which the relationships discovered by the SGS procedures will fail to be invariant in important respects. So presumably if the SGS inference procedures discover relationships which we have good reason to believe are invariant, these will at best be relationships that are invariant under some limited class of interventions. So then a further question is whether we can say something in general to characterize this class.[15]

Related questions about the relationship between the control and informational conceptions of causation, and about whether the inference procedures in SGS really pick out structural relationships, arise naturally when one looks at the details of SGS's discussion. We have already noted, for example, that when a statistical dependency between two variables arises because we have mixed two causally heterogenous populations in which those variables are independent, it may be possible to save the Markov condition by introducing an additional variable representing group membership. Intuitively such a variable will often be a mere classificational device, rather than corresponding to anything we may wish to regard as a cause.

While I don't pretend to have a clear account of what this contrast between cause and classifier consists in, it is at least tempting to think that it has some connection with the contrast between the two notions of causation distinguished above. A classifier variable C which tells us merely that a unit is in group 1 rather than group 2 can certainly convey information about the value of some other attribute A that screens off information carried by other variables about A. However, knowing the value of the classifier variable will typically not be helpful if we are interested in manipulation and control. For example, in studies of American voting behavior, it is common to include a classificational dummy variable, which takes the values "south" and "non-south"—people in the south just behave differently from their northern counterparts even when one controls for other relevant variables like socioeconomic status, party identification, and so on. But while knowing that Arkansas has the classification "south" may convey information about its likely voting patterns, it tells us nothing about how to manipulate or change those patterns. There is no well-defined hypothetical experiment that consists in moving Arkansas from the south to non-south classification. Similarly, a variable V that classifies brain-damaged patients in a mixed population into one of two groups may serve as a common cause of a correlation between test results in that population, but V may be only an informational or Granger-cause—knowing V may allow one to predict a patient's score, but V may not inform us about any feature of the patients such that if we could change or manipulate it, this would affect their scores. Glymour acknowledges, both in his paper and commentary, that his techniques will not distinguish between causal and classificatory relationships but also claims that they should not, on the grounds that both are in some broad sense causal. In doing so he seems to adopt an interpretation of causal that, for the reasons described above, is distinct from and in some tension with, the manipulationist conception to which he also seems to be sympathetic.

One's suspicion that the class of relationships that is picked out by SGS's inference procedures often corresponds to the informational conception of causality and that this is a larger class than the control conception is also strengthened when one looks at many of the particular examples discussed in SGS. In a number of these examples the inference procedures seem to pick out as causal, relationships that signal classificational or conceptual dependence or predictive relevance but which don't seem to coincide with the control conception. Paul Humphreys, in his essay for this volume, draws attention to an example involving the AFQT Test in which the PC algorithm generates a graph in which a variable representing the weighted average of the scores on various tests is adjacent to variables representing the scores themselves. Here it is plausible to interpret the edges in the graph as relations of informational relevance or dependence, but rather less plausible to regard them as either direct causes or as indicating the presence of a common cause in the control sense of cause.[16] Similarly consider the following graph, described in Glymour's essay and SGS (1993, 139) and represented in figure 10.3, which is the output of the PC algorithm. Here there is a directed edge from YCIG (whether a woman smokes cigarettes at a young age) to AGE (age of women at which first child is born) and ED (education at time of marriage). As David Freedman notes in his essay, it is plausible that YCIG is correlated with, or serves as a proxy for, variables measuring the socio-economic background and also perhaps variables measuring personality traits which in turn causally influence (in the control sense) decisions regarding school leaving or age at which first child is born. It is thus plausible that YCIG carries information relevant to ED and AGE. Moreover, statistical investigation reveals that YCIG remains correlated with and hence continues to carry information about AGE and ED even when other variables measured in the model are controlled for. It is thus intuitively plausible to interpret the edge from YCIG to ED and AGE as representing causal relationships in the informational sense. It is considerably less

**FIGURE 10.3**

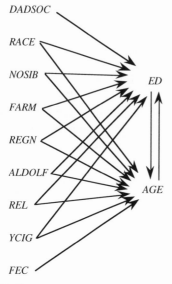
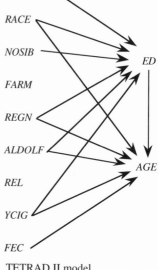

Rindfuss, et al. theoretical model;                    TETRAD II model
AGE –> ED coefficient not statistically significant

plausible that in a controlled experiment manipulating the age at which someone smokes and holding all else constant will affect the age at which she leaves school, i.e., that the directed edge for YCIG to ED represents a causal relationship in the control sense. It is perhaps slightly more credible that smoking produces physiological changes that influence AGE, but again it seems to me to be an enormous leap to interpret the correlational data from which this graph is derived as licensing this conclusion. This might of course be true—perhaps early smoking of cigarettes somehow produces physiological or psychological changes that alter the age at which women become sexually active or at which they conceive—but it is at least not obvious that the correlational data from which this graph is derived licenses this further conclusion.[17]

Of course one obvious response to this example and many of the other examples described below in which the SGS procedures apparently fail to output invariant relationships is that they involve violations of the working assumption when "cause" in that assumption is interpreted in either the control or informational sense—the examples involve omitted common causes, mismeasured variables, heterogeneous populations and so on. If so, it is hardly a surprise or a fair criticism of the SGS inference procedures that they fail to pick out structural relationships if the conditions required for those procedures to function reliably are not met. I accept this response but would add an additional observation. To the extent that the SGS project is intended to have a practical pay off, it is surely of some interest to observe that when it is applied to actual social scientific data, it often seems to produce claims about causation in the informational but not in the control sense—that is, it produces something that is more like a representation of the dependence and independence relations present in a population than anything that tells us about the results of hypothetical experiments.

I can further flesh out these worries about the notion of causation that is assumed in SGS by turning to a portion of their discussion which focuses most explicitly on the connection between causation and manipulation—their manipulation theorem. SGS say in their book that the manipulation theorem is a "consequence" of the Markov condition (1993, 80). A careless reading of this remark—not, I hasten to add, a reading that is in any way licensed by SGS's own surrounding discussion which is quite clear—is that it follows just from the fact that the graph for some not yet manipulated system satisfies the Markov condition that one can use the graph and the Markov condition to predict the results of various hypothetical manipulations. If this were correct, the theorem would establish the connection between the informational and the control conception of causation that we have been seeking. But, as I read the theorem—and here again I'm subject to correction by Glymour—it does not say anything like this. Instead the theorem requires as *input* the assumption that some relevant set of relationships represented by a causal graph are invariant or structural. It is only given this input, and provided various other conditions are met, that the theorem allows one to make use of the factorizability properties that follow from the Markov condition and the structure of the graph to predict the result of the manipulation of various variables.

The conditions that must be met for the application of the manipulation theorem are stated or represented in various ways (they are, at least in part, built into the definitions of key terms like "ideal intervention" or "manipulation" or into the requirement for the application of the theorem that there be a single combined graph that represents both the manipulated and unmanipulated systems) but basically what they do is to establish various connections between the causal graph representing a system prior to manipulation and the graph that represents the system once a manipulation occurs. For example, when a manipulation or ideal intervention occurs with respect to a variable X, this is taken to break the edges which are into X, but to leave other edges in the graph representing the original system undisturbed, and to leave unaltered the distribution of variables which are not effects of X. To use SGS's own example (1993, 76 ff.), suppose that in the U.S. population, prior to manipulation, the causal structure in figure 10.4 is correct. Suppose then that a ban on smoking is introduced and that in SGS's words,

**FIGURE 10.4**

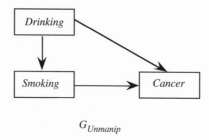

$G_{Unmanip}$

**FIGURE 10.5**

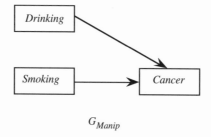

$G_{Manip}$

it is "completely effective, stopping everyone from smoking, but [does] not affect the value of drinking in the population." The graph for the manipulated system is represented in figure 10.5. In this case the edge from drinking into smoking is broken, but other edges (e.g., the edge from drinking to cancer, and the edge from smoking to cancer) are left unaltered, as is the distribution of variables like drinking that are not effects of smoking. Now the basic point I want to make—and I assume that SGS would agree—is that these assumptions or rules about which edges would be broken and which would not be under various manipulations are invariance assumptions in just the sense described earlier in this essay—they describe what will remain stable or unchanged when we intervene in a system to make various changes. These assumptions don't seem to follow just from the claim that the Markov assumption holds for the unmanipulated graph, but are rather built into the characterization of an ideal manipulation and the other assumptions that are made about the relationship between the unmanipulated and manipulated graph in the conditions for the application of the manipulation theorem. Indeed, as nearly as I can see, when we assume that there is some possible "ideal intervention" which will produce results respecting the assumptions described above connecting manipulated and unmanipulated graphs, we are in effect assuming that those graphical representations behave like true structural models in the sense described above—that is, like model (3.1) rather than model

(3.2) on p. 289 of this essay. Thus, when we assume that manipulating smoking produces just the changes in graphical structure described above, we are in effect assuming that the original graph $G_{unmanip}$ correctly describes a set of local, autonomous, disentangled causal mechanisms, and that the results of this manipulation follow just as one would expect from the causal structure of the graph. That is, we are assuming that by banning smoking we can interfere just with the relationship between drinking and smoking without at the same time necessarily disrupting the relationship between drinking and cancer or smoking and cancer, etc. In effect we are assuming that the edges in the graph behave as though they represent causal relations in the control and not just the informational sense.

This raises again the question of where these suppositions about invariance and structure come from and what justifies them. Is the idea that, given the working assumption, the SGS inference procedures themselves under appropriate conditions can or will generate graphs for which these invariance assumptions automatically hold? That is, is the idea that, assuming causal sufficiency, if the algorithms deliver a pattern representing a class of directed graphs and all of the portions of these graphs which are relevant to predicting the results of some manipulation share a common graphical structure, then we can assume that the relationships represented in the shared graph are structural and thus that the conditions required for the application of the manipulation theorem will hold? Or is the idea instead that, even given the working assumption, the inference procedures don't by themselves necessarily deliver the required invariances, and that they must instead be imported from some other source—from background knowledge of some kind? Some of SGS's language suggests this second view (see especially SGS 1993, 80–81 and 213), but obviously if the invariance assumptions connecting manipulated and unmanipulated graphs must be imported from some other source, then it seems that the SGS inference procedures do not by themselves deliver causal knowledge in the structural, control sense of causation. If the claim is instead that the inference procedures themselves yield information about invariances, then again one would like to know how exactly the argument for this claim goes.

When one considers specific examples of the application of the PC and other algorithms, it certainly looks as though they sometimes yield causal graphs that don't conform to the conditions required for the application of the manipulation theorem. Of course this again may simply be a result of the fact that in these cases, some of the conditions associated with the working assumption are violated, but it may also again reflect the fact that the sense of 'causal' in which those algorithms discover causal relationships is not the structural, control sense. Thus, in one example discussed in SGS (1993, 147 ff.), when the algorithm is applied to data asking school boys to judge whether or not they are members of the leading crowd (A) and their attitudes toward this crowd (B) at an earlier time t and then similar questions (C, D) are asked at a later time t + Δ, the algorithm produces a directed edge from A to C. But this edge surely does not represent a relationship that is stable under a very large class of manipulations. Suppose, for example, subjects are induced to join the leading crowd by being promised large financial incentives for joining at t, but these incentives are then withdrawn at t + Δ—perhaps a fine is even imposed for belonging. Or perhaps a manipulation of A is imposed which forces people (including people who don't want to be in the leading crowd) to join it or to believe that they are in it—perhaps they are required to be members in some club that is sufficient for being in the leading crowd at t, but then allowed to withdraw if they wish at t + Δ. It seems doubtful that under these manipulations of A, membership at t or judgment that one is a member at t will be invariantly related to membership or judgment of membership at t + Δ, at least for all units in the population. What may very well be true is that if A and C are produced by processes ordinarily operative in an unmanipulated population—people are allowed to voluntarily form relationships, to sort themselves into leading and non-leading crowds, to form judgments about who is in what group and so on, then A will be useful for predicting C, even when B and D are taken account of. But if this fact about predictability is all that the A → C edge represents, the presence of the edge in the graph for the unmanipulated population isn't going to tell us much about the sorts of manipulations of A under which the directed edge from A to C will be stable.[18] In this case, as in many other cases discussed in SGS, the effect of the manipulation of a variable A on some other variable B is highly sensitive to the way in which the manipulation of A is produced, which is just another way of saying that the model fails to be structural under many interventions and that there is no well-defined hypothetical experiment associated with the manipulation of A per se.

## IV

Let me conclude by drawing attention to three additional points about invariance. The first is that, quite apart from the specific examples discussed above, there are many general kinds of circumstances in which it is plausible to think that the relationships discovered by SGS's inference procedures may fail to find invariant relationships. A very simple kind of case is one in which those relationships are temporally unstable—obviously there is nothing in the SGS procedures that guarantees that if there now is an edge from A to B, this relationship will continue in the future. Nor—to mention a possibility I will come back to below—is there anything that guarantees that if we find such a relationship in one population, it will be present in other populations, i.e., as nearly as I can see there is nothing about the procedures that makes it likely that the relationships they find will be invariant across populations.

A more interesting kind of case in which invariance fails is the following: a structure corresponding to what SGS call a "switch" is present in the population under investigation. When the switch is off certain relationships obtain among the other variables; when the switch is on, the relationships among the other variables is different. The values of the variables when sampled all come from conditions under which the switch is off, but one result of intervening is to turn the switch on, changing the relation between other variables. One might think of the leading crowd example as a case in point, in which the effect of forcing subjects to join the leading crowd is to turn on a switch, which is off when the subjects are merely passively observed and allowed to choose whether or not to be members. Obviously such switch structures may be common in connection with many social phenomena and nothing about the SGS inference procedures allow us to tell when the relationship we find will fail to be invariant because such switches are present.

Another kind of case in which there will be a failure of invariance that is unlikely to be detected by the SGS inference procedures, or for that matter by any causal modeling exercise that doesn't rely heavily on a lot of reliable background theory, involves relationships that fail to be reversible in the following sense: X and Y are two correlated variables, with the $X_1$ value of X usually followed by the $Y_1$ value of Y, the $X_2$ value of X by the $Y_2$ value of Y and so on. If we observe units of the population which previously had the values $X_1$, $Y_1$ for these variables and now have values $X_2$, $Y_2$, it may be tempting to think that if we were able to intervene to change the $X_2$ values back to $X_1$, this would change the $Y_2$ values back to $Y_1$—obviously this is the pattern of dependence the observed values of X and Y suggest. But equally obviously this expectation may be mistaken. It may be that history matters—that returning X to the value $X_1$ after it has been $X_2$ produces quite different effects from assigning a unit a value of $X_1$ when it has not yet been $X_2$—i.e., that even if the relationship between X and Y is invariant under a change from $X_1$ to $X_2$, it doesn't remain invariant under a change back from $X_2$ to $X_1$. Here again, there are probably many social phenomena exhibiting this sort of structure. Consider the political behavior (Y) of a group of people who are first allowed a low level of civil liberties and democratic participation (X) at time t and whose liberties and participation are then allowed to increase to some higher level at t + ε and then, at a later time t + 2ε reduced again to their original level at t. It seems unlikely that the relationship between the levels of X and Y will remain exactly the same at t + 2ε as it was at t. Non-reversible relations of this sort will represent a systematic kind of failure of invariance that atheoretical causal modeling techniques will not be good at detecting.[19]

My second observation is that there is also a systematic, subject-matter specific reason, illustrated by many of the examples in SGS, why one ought to expect that invariant relationships will fail to exist between many of the variables that figure in standard social science causal modeling exercises. It is a plausible conjecture—I take this to be the central claim of methodological individualism or of rational choice approaches to social science—that for a great many social phenomena, many of the truly causally relevant variables—the place where at least a substantial part of the causal action resides—are the sorts of variables mentioned in microeconomics, game theory, and theories of individual decision making. These variables have to do with facts about people's preferences and beliefs and expectations, and about the incentives and various constraints on production possibilities that they face, and the institutional frameworks in which they act. In many cases, it will only be when we devise models that include these variables that we will find relationships that are structural or invariant under the interventions that most interest us.

Many purported social scientific causal models, especially in areas like sociology, either omit such variables entirely or at least introduce them in a form which is incomplete, theoretically muddled, or otherwise unperspicuous. Thus, in the previous example, while attitude toward the leading crowd is certainly relevant to membership and hence presumably to judgment of membership, many other relevant facts about the preferences and beliefs of the subjects have been omitted from the model e.g., their beliefs and preferences regarding the costs and benefits of being or not being a member, beliefs and preferences regarding other available options and so on. Our main reason for thinking that the directed edge from ( judgment of) membership at earlier and later times does not reflect a structural, causal connection is precisely that there is no reason to think that this edge will remain invariant under changes affecting these omitted variables—e.g., no reason to think that the edge from A to C will remain unaffected if we alter subjects' beliefs about the costs and benefits of membership. But this is just to say that a structural theory about membership in the leading crowd—a theory that would allow one to predict how membership would be affected by various hypothetical manipulations—will need to include these omitted variables. Similarly, in another example discussed in SGS (p. 147, the data are from Reiss, Banwart, and Foreman), the PC and SGS procedures produce the following pattern (i.e., representation of a class of graphs):

$$E - V - C$$

where E is attitude toward extra-marital intercourse, V is virginity, and C is use of a contraceptive clinic among a population of female undergraduates. SGS suggest that "[o]ne sensible interpretation is that attitude affects sexual behavior which causes clinic use" (1993, 147). But surely if anything "causes" use of a contraceptive clinic, it is not virginity or non-virginity per se, but rather, at least in large measure, a subject's beliefs about whether she will have intercourse in the future, her desire to have or not have children, and other related belief/desire variables. Virginity is at best loosely correlated with, or is a highly imperfect proxy for, expectations about or desire for future sexual behavior and it is presumably this fact that produces the undirected edge from V to C. Relatedly, if we were to try to use the hypothesized edge from V to C for manipulation, we would find that it is not invariant for many possible manipulations of V.

Similarly, in the case of the Blau and Duncan study, *The American Occupational Structure* (1967), discussed in Glymour's essay, even if we put aside worries about unit heterogeneity it seems extremely implausible that the directed edges from father's education to respondent's education or from the latter to respondent's occupation are invariant under many possible changes in (beliefs about) the costs and benefits of education, or under economic and technological changes that affect the attitudes of employers toward the educational qualifications of their employees. Such changes not only can occur, but have occurred on a very extensive scale in the United States over the past few decades. Thus, for example, the comparative rate of return on a college or post college, as opposed to a high school education, has risen enormously over the past twenty-five years, presumably making it much more attractive for those with the opportunity and resources to do so to attend college. If, instead, economic and technological changes had acted so as to reduce the wage differential between high school and college graduates, presumably children of the college educated would be less likely to get college educations themselves. Similarly, changes in the costs of or access to higher education or quality primary and secondary education can and presumably have affected the edge from FE to ED. Thus it is really facts about the respondent's beliefs and preferences and the resources to which she has access, rather than variables like father's education or father's occupation, which causally affect the choice of educational level—but the former variables do not figure in Blau and Duncan's model.[20]

Now there is perhaps nothing in principle about causal modeling techniques that requires the omission of rational choice variables having to do with belief and desire. Nonetheless, it is not surprising that many causal models omit such variables—they are often difficult to measure and their interactions are often highly complex and nonlinear and difficult to trace without sophisticated theories. But in the absence of good stories about the beliefs and choices of individual actors, the factors affecting them and how they in turn affect social outcomes, it will very often be difficult to predict with much confidence the effect of various possible interventions—a moral that I think is illustrated by the above examples.

More generally, if one begins with the "wrong" variables, there is no reason to think SGS's techniques (or any other causal modeling technique) will successfully discover structural relationships from correlational data. There may just not be any relationships among these variables that are causal in the control sense, although as long as there are correlations among these variables, techniques like SGS's will deliver relationships that are causal in the informational sense.

Finally, consider by way of conclusion an issue alluded to briefly above. Under the control conception of causation, part of what interests us when we worry about whether the relationship between A and B in a particular population is causal is whether it will remain invariant under some set of interventions in that population and if so, what the scope of invariance of the relationship is. But it also seems to me that the claim that a causal relationship holds in some particular population ought to have some implications, even if only very weak ones, for what we should expect to see in other populations and circumstances. Of course causal relationships in the social sciences are rarely, if ever, formulated as laws of nature; it is perfectly possible for a system of causal relationships to hold in one population and for a quite different set of relationships among the same variables to hold in some other similar population. Still, it seems to me that, if the relationships in the original population are genuinely causal, there must be some limit on the extent to which this can happen. Even if we do not expect to find the same pattern of causal relationships in all other populations in which the same variables are exhibited, there should be a qualitatively similar pattern in some other populations in the sense that we can find many of the same qualitative relationships (the same graphical structure) among variables in other populations even if quantitative details—the exact values of coefficients in the associated system of equations and so on—are different. Relatedly, one ought to find that genuine causes sometimes exhibit the same qualitative mechanisms or modus operandi across populations. To take an extreme possibility, if someone claims that A causes B in population P, but there are no other circumstances or populations outside of P, even those which apparently closely resemble P in which we observe or are able to experimentally produce an association between A and B at all, it seems to me that one has considerable grounds for skepticism that the original association in P is causal.

I take it that this idea that causal relationships should exhibit some degree of invariance or generalizability *across* populations, as well as invariance *within* populations is also naturally suggested by the idea that causal relationships are relationships that support manipulations—if wiggling C is really a way of changing E, then one expects that C will be capable of producing this effect in some contexts other than the original population P in which the association between C and E is estimated.

The relevance of all of this to causal modeling is that it seems to me that it is one of the functions of background theory in disciplines like the social sciences and epidemiology to supply information about the behavior of putative causes outside the population of interest and that it is a somewhat problematic feature of the SGS inference procedures that, while they certainly allow us to incorporate such information if it is available, they do not in any way require it. That is, as nearly as I can see, there is nothing about those inference procedures or the conception of causality adopted by SGS that makes the behavior of the relationship between C and E outside the population under investigation relevant to the question of whether C causes E within that population.[21]

To illustrate the role of such background information in causal inference consider the question of whether smoking causes lung cancer. In a well-known review article, Cornfield et al. (1959) describe evidence which, as of 1957, was regarded by the authors as strongly supporting the hypothesis that smoking causes lung cancer. They particularly emphasize the point that, although quantitative details vary, there is a qualitative association between smoking and lung cancer that is stable across many different populations and changes in background circumstances. For example, they note that some association appears between smoking and lung cancer in every well-designed study on sufficiently large and representative populations with which they are familiar. There is evidence of a higher frequency of lung cancer among smokers than among non-smokers, when potentially confounding variables are controlled for, among both men and women, among people of different genetic backgrounds, among people with quite different diets, and among people living in quite different environments and under quite different socioeconomic conditions (Cornfield et al. 1959, 181). The precise level or quantitative details of the association do vary—for example, among smokers the incidence of lung cancer is higher among those in lower socioeconomic groups—but the fact is that there is some association or other which is stable or robust across a wide variety of different groups and background circumstances. A similar stable association is also found among laboratory animals. In controlled experiments animals exposed to tobacco smoke and other tobacco products have a higher incidence of various kinds of cancer than unexposed controls. Moreover, as they also note, there is a distinctive mechanism or modus operandi that is stably associated with this putative cause and which behaves in just the way that one would expect if smoking were causing lung cancer. Thus, for example, heavy smokers and inhalers in all populations show a higher incidence of lung cancer than light smokers and non-inhalers, groups that have smoked the longest show the highest incidence of lung cancer and the incidence of lung cancer is lower among those who have smoked and stopped than among relevantly similar groups who are still smoking.[22]

Epidemiologists frequently appeal to qualitative information of this and related sorts in constructing arguments that smoking causes cancer. Their arguments have often been statistically unsophisticated and in fact come in for heavy criticism in chapter nine of SGS. Some of this criticism is fully justified, but it also seems to me to be very hard to deny that the above sorts of facts about the stability of the relationship between smoking and lung cancer across populations are highly relevant to whether an association between smoking and lung cancer in some particular population is interpretable as a causal connection. Indeed, even by the time of Cornfield's study, the evidence supporting the claim that smoking causes lung cancer looks to me to have been far stronger than the evidence for the causal relationships postulated in many of the social scientific causal models discussed in this essay, despite the greater statistical sophistication of the latter. What this suggests is the importance of qualitative non-statistical arguments of the sort described above in establishing causal conclusions.

It seems to me that it is an important methodological defect in many causal modeling exercises that they make no attempt to discover or incorporate such across-population information about stable relations or, more generally, to incorporate qualitative, non-statistical considerations of the sort that played such an important role in the discovery that smoking causes lung cancer. Thus, for example, in the Rodgers and Maranto (1989) study or the Rindfuss and Bumpass study, no attempt is made, at least as reported by SGS, to find out whether an even vaguely similar pattern of relationships holds in other populations and circumstances—we know nothing about the behavior of these variables that is like what we think we know about the relation between smoking and lung cancer.

These concerns about the behavior of causal relationships across populations are closely related to two other themes in my discussion. A common feature of many causal models is that, apart from some claims about how they are allegedly measured, we know almost nothing about the various putative causes that figure in them other than the statistical relationships in which they stand to other variables in the model. Although we give the vertices in various graphs names like "ability" or "graduate program quality" or "extroversion" or "algebraic knowledge" or "political exclusion" this is not accompanied by any well-founded theory—however vague and informal—about other features of these variables. That is, we know nothing about the mechanisms or intervening links by which these variables produce the effects represented in the model, other effects these variables might have that are not represented in the model, or whether and in what circumstances we should expect these variables to produce similar effects in other populations—indeed even whether the measures we use will pick out the same variables in other populations. All of this information about the behavior of these variables outside the population of interest plays no role in the investigation.

This omission is of course particularly conspicuous in the case of some of the arbitrary constructions introduced in our discussion of the Markov condition—the compound event $Z^*$ which was introduced as a common cause of the correlated variables X and Y or the variable whose values represented membership in group 1 or group 2. Surely one of the main reasons we think of these variables as something less than real causes is that we lack the kind of information about them that the epidemiologists have in the case of smoking—we have no idea what else these variables do, what other effects they might have, indeed how even to determine whether they are present or absent in other circumstances. Instead, the content of these variables is completely exhausted by the statistical relationships in which they stand to other variables in the particular population under investigation, and something more than this—some story about features or behavior that generalizes to other populations or circumstances—seems to be required before we can say that we have identified causes.

These considerations also bear on what Glymour calls Fisher's condition. As I understand it, this condition denies that statistical dependencies can obtain just as a matter of chance or coincidence—i.e., that A and B can be statistically dependent, but that it is not the case that A causes B or that B causes A or that A and B have some common cause or causes. Let us grant for the sake of argument that, as we obtain more and more extensive evidence of a pervasive association that holds across different places, times, populations, and background circumstances, the pressure to find some causal explanation for the association grows. What I think is less clear is that Fisher's principle is a reliable principle of inference in connection with associations that ( for all we know) hold in just one particular population of the sort with which causal modelers deal.

Consider a standard complaint against relatively atheoretical causal modeling—that it involves "data-mining." Such complaints are rarely accompanied by clear explanations of what "data-mining" consists in or why it is wrong. Sometimes this worry is, in effect, a worry about statistical inference—about whether the particular sample of data the investigator is working with is really representative of the population from which it is drawn or about whether a specification search has been conducted in a way that deprives standard statistical tests of their usual interpretation. But I think in many other cases, the worry is at least in part different from (although it may be connected to) this. In effect, the worry is that Fisher's condition is violated—that although X and Y are associated in the particular population under investigation (and not just in the sample) and this association is not due to some common cause or causes (at least in any non-informational sense of cause), it is nonetheless a mistake to infer there is a direct causal connection between X and Y. For this worry to make sense one must think that the claim that X causes Y has more content than just that a certain pattern of statistical association exists between X, Y and other variables in the population. At least in part, this additional content will have to do with whether the relationship between X and Y is invariant under some class of changes or interventions and this in turn will be connected with the question of whether it will persist in at least some other populations or circumstances. Put differently, the worry about data mining is that one's model "capitalizes on chance," exploiting a non-variant relationship which holds by chance or coincidence in this particular population, but doesn't generalize to other populations. Accidental associations in this sense are certainly logically possible (surely there is nothing about the control sense of causality, at least, that automatically rules out this possibility) and the frequency with which they occur and hence whether this sort of worry deserves to be taken seriously are, at bottom, empirical matters. Artificial variables like $Z^*$ or the group membership variable V are particularly vulnerable to this sort of worry for it is completely unclear what, if anything, they imply about other populations and circumstances. To the extent that these worries associated with data mining have a real basis, the sorts of considerations to which I draw attention above—knowing how the cause acts outside the population of interest—will be of obvious relevance to causal inference.

## NOTES

I would like to thank Frank Arntzenius, Nancy Cartwright, David Freedman, Clark Glymour, David Grether, Paul Humphreys, David Papineau, Richard Scheines, and Peter Spirtes for very helpful comments and conversations regarding earlier drafts. I should also add that this essay was originally written for oral presentation. Rather than trying to completely rewrite it for the written proceedings of this conference, it seemed preferable to me, for a variety of reasons, to retain the somewhat loose and informal character of the original. Also, a number of the questions and issues raised below were addressed by SGS during the conference proceedings. Rather than rewriting my remarks to fully take account of their comments, I have in most cases stuck to my original version in the interests of giving SGS a stationary target at which to aim their reactions.

This paper is a commentary on a paper presented by Clark Glymour at the Notre Dame conference in 1993. The version of Glymour's paper that appears in this volume is a considerably altered version of the paper presented at the conference and I did not have access to this version in writing up my commentary. As a result my paper refers at various points to material from his earlier version that I believe is not included in the present version. This is particularly true in connection with my remarks in section 3 about the interpretation of causation assumed in Glymour's paper. In the earlier version Glymour often seemed agnostic about this issue; the present version is more clearly sympathetic to what I call in my paper a manipulationist conception of causation. A very similar position is taken by Richard Scheines in his paper for this volume, also written after the conference. As a result some of my queries about this issue have been answered in Glymour's and Scheines's papers.

1. It is true that these general background assumptions are often supplemented, in the treatment of particular examples, by subject matter specific considerations—in particular, as Paul Humphreys notes in his contribution to this volume, SGS often appeal to temporal order and to other kinds of commonsense causal knowledge to rule out alterna-

tive models. But it is nonetheless clear that one of the primary motivations that SGS offer for their program is that it reduces the need for the extensive reliance on subject matter specific causal background assumptions that characterizes more conventional approaches. SGS regard this as desirable because they believe that, particularly in the social and behavioral scientific contexts in which causal modeling techniques are used, background information is often unreliable and too weak to solve the underdetermination problem. Although I think that SGS are sometimes unduly pessimistic on this score, I do not doubt that their skepticism is often warranted. The key question, to be explored below, is how reliable and effective the SGS procedures are when we lack such background knowledge.

2. The distinction between the two ideas will of course be blurred if, when one refers to a putative common cause, one means nothing more than "whatever variables are responsible for such and such a correlation." Perhaps this is all that "knowledge for real analysis," etc. means in SGS's example. My reasons for regarding this sort of move as methodologically undesirable are described in more detail in section 4.

3. For a somewhat similar conception of the status of the common cause principle, see Sober (1988). Like Sober, I don't mean to deny that it will often be heuristically useful and reasonable, when one discovers a correlation, to search for a direct causal connection or common cause which produces it. My claim is rather that the fruitful use of the common cause principle (or, as I would prefer, common cause heuristic) will require background knowledge—about which correlations are worth trying to explain causally, about which sorts of variables are plausible candidates for causes, and so on. Moreover, as I see it, this heuristic is highly defeasible—there isn't some exceptionless or near exceptionless general rule of methodology that somehow guarantees that all correlations must have a causal explanation. Instead, what the heuristic suggests is that, given that there is such a correlation, it will often be worthwhile to look for a causal explanation. But we may sometimes fail to find such an explanation and reasonably decide instead that the correlation is purely accidental.

4. See, for example, Arntzenius (1990). A closely related observation concerning the status of the error term in regression models occurs in Pratt and Schlaifer (1984).

5. What implications do these observations about the choice of variables have for the SGS inference procedures? Glymour and Spirtes point out in an interesting unpublished paper that, if we consider all transformations of the same general form employed in the above example—i.e., transformation of the variables A' and C' of form XA' + ZC' and YA' + WC', the effect of such transformations will be to replace correlated variables with uncorrelated variables only if a very special condition is met: only if $XY + WZ + P_{A'C'}(ZY + XW) = 0$. This condition of course will not hold for most choices of X, Y, W, and Z. If I have understood them correctly they infer from this that the usual result of transforming variables is not to replace correlations with non-correlations, leading to the omission of true causal relationships but rather to produce correlations that are uninformative about true causal relationships—roughly because nearly everything ends up statistically dependent on everything else even when we control for other variables, and in such cases the SGS algorithms will produce as output all possible systems of causal relationships compatible with these dependency relationships which may be a very large class of possibilities indeed.

Assuming this claim about the typical effect of transforming variables is correct, it seems to me that the fundamental point made above still remains—the project of getting interesting, non-trivial causal conclusions out of information about statistical dependencies requires that one choose the right variables to begin with. Indeed, as Spirtes and Glymour themselves point out, what the above result seems to show is that for *most* choices of variables, statistical relationships between them will be uninformative. This seems to suggest that unrestricted versions of the common cause principle are at least not very helpful principles of causal inference—it is just too easy, by transforming variables, to produce endless statistical dependencies and most such dependencies will not be informative about the presence of common causes, at least in any very robust or substantial sense of cause.

6. For further relevant discussion here, see Caramazza (1986).

7. For a more detailed argument that this will be the case, and an exploration of some of the implications of this fact, see Woodward (1995).

8. SGS themselves note a number of these implications (or at least closely related implications), but perhaps do not accord them quite the weight I would accord them.

9. To put the point in another way, the formulation adopted above in which it was said that some independence relations follow just from graphical structure while others do not is a bit misleading, at least in the usual context in which one assumes the existence of unobserved error terms. It is the combination of graphical structure and assumptions about the distribution of the error terms that entail or fail to entail independence relations. The question then is why, as the working assumption seems to require, we should give a special status to cases in which the independence relations follow from the graphical structure and uncorrelatedness assumptions, but ignore models in which the observed statistical relationships follow from the graphical structure and assumptions about correlations among the error terms.

10. Consider in this connection a result which was originally established by Pearl and to which both SGS (1993) and Paul Humphreys (this volume) draw attention. This is that if we are given a positive distribution and some ordering of the variables (which may be simply an ordering by time) there will be one and only one directed acyclic graph for which the distribution satisfies the Markov and minimality conditions. Here it certainly looks as though the directed graph associated with this distribution need be nothing more than a device for representing independence and dependence relations in the context of an ordering relation of some kind, which intuitively may have nothing to do with causality. On the face of it, there seems to be no reason to believe that the relationships represented by such a graph must be structural or autonomous or invariant in the sense described below.

11. SGS's claim that it is possible to infer causal conclusions from purely statistical information and the very general conditions embodied in the working assumption is perhaps somewhat less surprising if the intended interpretation of causation is only the informational conception and this turns out not to imply the control conception. But I hasten to add that it is still of course a difficult and worthwhile task to figure out what can be learned about causal structure in the informational sense from facts about indepen-

dence and dependence relations, especially without the assumption of causal sufficiency, and to produce algorithms for the discovery of such structures.

12. Pearl writes

> But what is the operational meaning of a genuine causal influence? How do humans distinguish it from spurious correlation? Our instinct is to invoke the notion of *control;* e.g., we can cause the ice to melt by lighting a fire but we cannot cause fire by melting the ice. Yet the element of control is often missing from causal schemas. For example, we say that the rain caused the grass to become wet despite our being unable to control the rain. The only way to tell that wet grass does not cause rain is to find some other means of getting the grass wet, clearly distinct from rain, and to verify that when the other means is activated, the ground surely gets wet while the rain refuses to fall.
>
> Thus, the perception of voluntary control is in itself merely a by-product of covariation observed on a larger set of variables [Simon 1980], including, for example, the mechanism of turning one's sprinkler on. In other words, whether X causes Y or Y causes X is not something that can be determined by running experiments on the pair (X, Y), in isolation from the rest of the world. The test for causal directionality must involve at least one additional variable, say Z, to test if by activating Z we can create variations in Y and none in X, or alternatively, if variations in Z are accompanied by variations in X while Y remains unaltered.

I agree that we can test a relationship between X and Y for causal direction by embedding it in a larger structure in the way Pearl describes. However, it doesn't follow that all there is to X's causing Y in the control sense is that X, Y and various other variables like Z stand in certain statistical relationships. Among other things, this second view seems to have the consequence, as Pearl himself recognizes (1988, 397), that the causal direction between X and Y is ill-defined if the additional variable Z standing in the appropriate statistical relationships to X and Y fails to exist. It seems to me, on the contrary, that even in a world in which only X and Y exist, it can be true that X causes Y rather than that Y causes X. Finding an additional variable Z that causes X but not Y is a way of testing the claim that the causal order runs from Y to X rather than from X to Y; it doesn't bring that order into existence.

13. In his oral comments on this paper at the Notre Dame Conference, Glymour noted that if the system of equations (3.2) is to reproduce the observed covariance matrix, the error terms in (3.2) must be correlated in a very precise way. If I recall his argument correctly, he took this to show that (3.2) violated the working assumption (in particular the Markov condition) and, more generally, that (3.2) corresponded to a measure zero case. I also take it that (3.1) violates the working assumption since it contains a cycle. My intent in introducing (3.1) and (3.2) was simply to illustrate how the idea that causal relationships are invariant relations gives rise to an underdetermination problem. However, I would also want to resist the suggestion that we are justified in ignoring models like (3.1) and (3.2) because they violate the working assumption. Indeed, without wishing to belabor the argument given in section 2 above, a natural reaction to (3.2) is that

*any* particular assumption about the covariance structure among the errors in a system of linear equations (including the assumption that those errors are uncorrelated) corresponds to a measure zero special case and that no such assumption should be privileged over any other on abstract methodological grounds. That is, why ignore (as contrary to the working assumptions) models in which the observed independence relations follow only if one assumes that the error terms have some special non-zero correlation, but not ignore models in which these independence relations follow from assumptions about the uncorrelatedness of the error terms (together with assumptions about graphical structure)?

14. I realize that the notion of causation associated with a directed graph in the SGS framework does not coincide exactly with Suppes' conception or with Granger's—my query has to do rather with whether the SGS's conception is merely informational in a sense that contrasts with the structural conception.

15. One rather simple-minded way of thinking about the difference between the informational and control notions of causation is as follows: The mere fact that a set of dependence or independence relations holds in a population appears to carry with it no modal or subjunctive commitments—it apparently says nothing about what would happen if we were to intervene in the population in various ways. By contrast, a structural model does carry with it exactly such commitments—different structural models make claims about what would happen to various independence and dependence relations that we observe in a population if we were to carry out various hypothetical manipulations. *Prima facie* information of the former sort doesn't seem to fix the truth values of claims of the latter sort. Do SGS think that once one adds additional general conditions like Markov and faithfulness to information about dependence relations, this does fix the truth values of at least some of these modal claims? If so, which modal claims?

16. As described in David Freedman's contribution to this volume, there are a number of other problems with SGS's treatment of this example—see Freedman's essay, especially pp. 151 ff.

17. As David Freedman points out in his essay, there are a number of additional problems and complications (having to do, among other things, with mistakes in data entry), with SGS's treatment of this example that go unrecognized in the above remarks. (Among other things, Freedman notes that with the mistaken data the coefficient between YCIG and ED is positive; with the correct data one recovers the expected result that early smoking is a good predictor of early school leaving. My remarks above assume that SGS procedures yield this expected result.) Rather than trying to rewrite this section of my discussion to recognize all of the complications to which Freedman draws attention, I thought it best to keep my discussion in something like the form originally delivered at the Notre Dame conference. I do think that the basic points made above are simply reinforced by Freedman's more detailed observations. For example, Freedman reports that the SGS model produces strange relations among the regressors—"the model says that race and region cause region of residence" (Freedman, p. 131). Quite apart from the technical problems, to which Freedman draws attention, created by having a dichotomous variable like region occurring on the l.h.s. of Freedman's equation (7) (p. 131), one surely wants to say that any causal connection from race to region is causal in the informational sense

only. Here we see additional evidence that the SGS techniques, when applied to real data, seem to deliver results about informational dependencies, not reliable predictions about the results of hypothetical experiments.

18. One of course can imagine a variety of responses that SGS might give to these observations. Perhaps one should view the proposed interventions as manipulations not just of variable A (belief about membership at time t) but also as manipulations of variable B (attitude toward the leading crowd at t) and perhaps contrary to what I just suggested, the interventions leave both the A $\rightarrow$ C and the B $\rightarrow$ C edges intact, the apparent non-invariance of the A $\rightarrow$ C edge being accounted for by the fact that changes in B are producing changes in C independently of A. However, it is unclear to me how the correlational data from which the graph is derived could ever be used to establish that this is the right analysis, in terms of what happens to structure, and that my initial account in which the A $\rightarrow$ C edge is broken is mistaken. Or perhaps we should instead think of this as one of those cases in which we have collapsed causally different variables into some single causally heterogenous aggregate variable—perhaps what is really causally relevant is not membership per se or judgment of membership, and instead we should distinguish among different ways in which (judgment of) membership can be brought about—perhaps (judgment of) coerced membership is not the same variable as (judgment of) freely chosen membership. Or perhaps what has gone wrong in this example is that other relevant causal structure has been omitted in a way that involves violation of the working assumption—perhaps, as I suggest in a bit more detail immediately below, a more adequate specification of the model will need to talk more explicitly about the perceived costs and benefits of membership in the leading crowd.

In the last two cases, failures of invariance can perhaps be attributed to violations of the working assumption, i.e., it perhaps can be argued that if the working assumption were satisfied, there would be no failure of invariance. But, however the example is to be analyzed, my point is that, as a matter of practical methodology, it is dubious that the application of the SGS inference procedures to the data provided in the example yields information about invariant relationships. Although I lack the space for further discussion I believe that a similar conclusion is plausible regarding many other examples treated in SGS.

19. These observations about non-reversible relationships are not original with me, but rather come from Stanley Lieberson's much more detailed discussion (1985).

20. For similar observations about the Rindfuss et al. model, see Freedman's essay (p. 130 ff.). As Freedman notes, to assume that there is a causal arrow running from, e.g., respondent's father's occupation to respondent's education which does not involve variables having to do with father's or respondent's beliefs, attitudes, and choices is in effect to assume that the effect of father's occupation on respondent's education will be the same whether or not fathers are allowed to freely choose their occupations or are forcibly assigned to them. Similarly, the causal model connecting virginity and contraceptive use discussed above seems to imply that the relationship between virginity and contraceptive use will be the same whether or not one becomes a non-virgin through one's free choice or in some other way (e.g., as a result of rape). The implausibility of this last claim is simply a reflection of the fact that the really causally relevant variables here are belief/

attitude variables with which virginity is imperfectly uncorrelated. The more general problem is that many of the relationships of most interest in the social sciences are the result, at least in part, of people's choices (and beliefs and preferences). The assumption that some relationship between social variables (e.g., between family background and education) which results in part from some particular pattern of beliefs and preferences would be the same if people were assigned values of those variables independently of their beliefs and preferences will be often quite implausible.

21. I believe that this is one of several points at which SGS's project would benefit from a more explicit discussion of what the causal claims discovered by their inference procedures mean and what sort of commitments they carry with them. On their conception of causality does the truth of the claim that C and E are directly causally connected in population P or under background conditions B carry with it any implications about their relationship outside of P or B?

22. In correspondence, David Freedman has drawn my attention to a survey of more recent studies of the relationship between smoking and lung cancer (International Agency for Research on Cancer 1986) which casts doubt on some of the claims described above. For example, the IARC survey notes that, while some studies find a positive association between inhalation and risk of lung cancer among light smokers, other studies show a negative relationship between inhalation and lung cancer among heavy smokers—one suggestion is that high inhalation leads to more smoke being deposited in peripheral parts of the lung rather than in the bronchi, which is the tissue affected by cancer of the lung (p. 179). Similarly, the evidence from animal studies is rather mixed, although there is consistent evidence that exposure to tobacco products has a mutagenic effect on animal tissue (p. 196). Nonetheless, it seems to me that the general point made in the text above is supported by this more recent survey. The general conclusion for which I was arguing was the importance of evidence about stable qualitative patterns and mechanisms across different populations and background conditions in showing that a relationship in a particular population is causal. The claim that a relationship in a particular population is causal is often greatly strengthened by showing that the relationship behaves in a way that is qualitatively similar to what is known of its behavior in other populations, even though quantitative details vary. The IARC study provides considerable evidence for such qualitative stability and its role in causal inference. For example, the authors describe, in the form of "factors" on which the "observed incidence of lung cancer may depend," the following qualitative patterns in the relationship between smoking and lung cancer:

(i) *The daily dose of tobacco*

It is consistently found that, among otherwise similar cigarette smokers, there is a direct relationship between the daily dose and the excess risk of lung cancer in both men and women. The observed relationship is, in many studies, one of approximate linear proportionality.

(ii) *The duration of regular smoking*

Because damage to the lung accumulates with continuous smoking, the incidence of lung cancer depends strongly on the duration of smoking. So,

(1) those who start to smoke in adolescence and continue to smoke are at the greatest risk of developing lung cancer in adult life;

(2) there is a delay of several decades between the widespread adoption of cigarette smoking by young adults and emergence of the full effects on national lung cancer rates;

(3) even among people who have been smoking for many years, those who have not already developed lung cancer (or some other disease) can, by ceasing to smoke, avoid most of their subsequent lifelong risk of tobacco-induced lung cancer.

(iii) *The form in which tobacco is smoked (cigarettes, pipes and cigars)*

It is generally found that, among otherwise similar smokers, those who have used only cigarettes have lung cancer risks much higher than those who have used only pipes and/or cigars, although the latter materials do cause some appreciable risk. (p. 243)

The IARC monograph also notes that the use of cigarettes which deliver less tar or which involve filters is associated with a decreased risk of lung cancer, just as one might expect.

As the IARC study emphasizes, all of these qualitative features of the relationship between smoking and lung cancer persist across populations. This contrasts with a significant amount of quantitative variability across populations in, for example, relative risk ratios (cf. p. 205).

# REPRESENTATIONS AND MISREPRESENTATIONS

## Reply to Humphreys and Woodward

## *Clark Glymour*

Skepticism is trivial: without some assumptions limiting the possibilities, any interesting inferences—whether about causal relations or about the existence of other minds—are underdetermined by data, and no reliable, informative inference procedures are possible. Dogmatism is easy too; make the background assumptions strong enough and reliable (under the background assumptions) inference is trivial. Real work in methodology investigates the trade-off between the strength of background assumptions and the existence of reliable, informative, and feasible inference procedures. In that spirit, *Causation, Prediction, and Search* and the subsequent work reviewed in my essay are marked by the following attitude towards understanding causal inference and prediction:

1. Rather than "analyzing" the "concept" of causation, give axiomatic characterizations of assumptions implicit in large segments of practice.
2. Investigate mathematically the conditions necessary, sufficient or, if possible, necessary and sufficient for inference that, asymptotically, gives as much correct information as possible in view of the axioms.
3. With regard to the possibilities or impossibilities of discovery or prediction procedures, find algorithms where they exist and characterize their reliability and complexity, and where no algorithms exist to solve a problem, or solve it feasibly, prove as much.
4. Where discovery procedures are found, implement them in computer programs and investigate their small-sample behavior with well-designed simulation studies.
5. Apply the results of these inquiries to a range of problems presented by scientific practitioners; where independent confirmation is available, test the reliability of discovery procedures.

6. Investigate conditions under which the axioms can fail.
7. Generalize and iterate.

I thank Paul Humphreys and James Woodward for the thoughtfulness, effort, and good will of their closely related comments. Humphreys has since published a review of this work that has none of these virtues, but I will address it elsewhere. Many of their objections are motivated by their dissatisfaction with the fact that the work I have described does not pursue a strategy prominent in philosophy since Plato:

1. Propose analyses (in the present case, of "cause") by Boolean combinations of other predicates.
2. Find counterexamples.
3. Iterate.

I do not believe that analyses of the Platonic kind are likely to prove interesting or useful, but I may be wrong in that and my approach does not preclude such philosophical theories. Woodward and Humphreys also correctly point out some inherent limitations to algorithms that infer causal structure or predict the effects of interventions from observational data; many of these limitations are shared by experimental methods as well, and most of them are noted explicitly in *Causation, Prediction, and Search*. Some of their objections demand reliabilities that are demonstrably impossible. Not much enlightenment is gained, I think, by fixating on the fact that algorithms do not do what it is evidently impossible to do. I will address in more detail some of their questions and objections.

Humphreys concludes his essay with four questions:

1. In what sense are DAGs (directed acyclic graphs) representations of causal relationships, rather than of conditional probabilistic dependency relations?
2. Is there a consistent non-epistemic interpretation of the d-separation condition?
3. (a) How much causal knowledge is actually needed to apply the algorithms?
   (b) In particular, will aggregation problems require background causal knowledge to select the correct level of causal analysis?
4. Can these methods separate causal graphs from classificatory graphs?

Here are my answers:

*1. In what sense are DAGs representations of causal relationships, rather than of conditional probabilistic dependency relations?*

One sense in which the DAGs represent causal relationships is that they represent features of probability distributions that would obtain were ideal interventions to be made. The theory of prediction presented in chapter 7 of *CPS* develops the mathematical basis of that connection, which was anticipated in the work of Robins, an epidemiologist, and has been since expanded in papers by Pearl and by Heckerman and Geiger. Humphreys discusses a theorem of Pearl's which asserts that for every strictly positive distribution there exists a unique directed graph satisfying the minimality and Markov conditions. This theorem does not imply, and it is not true, that there always exists a graph satisfying the faithfulness condition, which is assumed in the proofs of correctness of our algorithms.

### 2. Is there a consistent non-epistemic interpretation of the d-separation condition?

D-separation is a three-place relation among disjoint sets X, Y, Z of vertices in a directed graph. There is nothing "epistemic" about it. Pearl, Geiger, and Verma showed that if G is a directed acyclic graph, X and Y are d-separated given Z if and only if in every distribution that satisfies the Markov condition for G, X is independent of Y conditional on Z. The initial value of the result, which does not hold for cyclic-directed graphs, is that d-separation is easy to compute, so given a directed acyclic graph it is possible to compute the conditional independence constraints the graph implies under the Markov condition. Lauritzen et al. (1993) proved an equivalent property that is very differently described, which makes it a straightforward graphical realization of a condition for the independence of X, Y given Z, namely that the joint density can be factored as a product of the joint density of X, Z and the joint density of Z, Y.

If your interpretation of probabilities is subjective or epistemic, then there is no "consistent non-epistemic interpretation" of this theorem about independence and d-separation, nor of the Causal Markov condition either, nor of any relation that involves probability distributions, as d-separation does. If your interpretation of probability is frequentist, then of course there is a non-epistemic interpretation of d-separation.

### 3. (a) How much causal knowledge is actually needed to apply the algorithms?

I interpret Humphrey's question to mean "what conditions logically guarantee that the output of the algorithms be correct?" That question is answered in our book: Given population conditional independence relations, the Causal Markov and Faithfulness conditions suffice; conditions slightly weaker than Faithfulness would do instead. Perhaps he means, as one passage in his essay suggests to me, that in the absence of knowledge that the Causal Markov and Faithfulness conditions obtain, what conditions ensure that the output of the algorithms is correct? Anything that provides independent evidence of the claims in the output. Thus applied to the AFQT data, our algorithms say that certain tests are not components of AFQT, and we established their correctness by finding out how the AFQT was constructed; in other contexts, such as the discussion in our book of the factors influencing Spartina growth in the Cape Fear estuary, experimental results may confirm the claims of the algorithms. Or, finally, the intent of the question may be: how can it be determined from the data that the assumptions are false, if they are? It is often straightforward to determine that faithfulness has failed, either by finding deterministic relations among measured variables or by finding that the models produced assuming faithfulness entail conditional independence relations or other constraints not found in the data. Failures of the Markov condition—especially through feedback—are more difficult to detect, but there are specific patterns of independencies that, if faithfully produced, can only be generated by cyclic structures that violate the Markov condition. These facts generated the work by Spirtes and Richardson that I briefly describe in my essay, work that beautifully illustrates the spirit I described at the outset of this comment.

*3. (b) In particular, will aggregation problems require background causal knowledge to select the correct level of causal analysis?*

One interpretation of the general question is about what happens when search algorithms are applied to a population that is a mixture of two or more subpopulations having different causal structures or different probability relations or both. This is a good question that is only partially answered in *Causation, Prediction, and Search*, and since then my colleagues and I have done further work on this question. The answer depends on the characteristics of the subpopulations:

(i) if the subpopulations share the same causal structure but differ in probability distributions, and the Markov and Faithfulness conditions are satisfied within each subpopulation, the FCI algorithm described in chapter 6 of *Causation, Prediction, and Search* gives correct output, including indication that there is an unrepresented variable, which in this case represents any feature that indexes the various subpopulations;

(ii) if in addition to the conditions in (i), the parameters are all independent of each other and the other random variables, the FCI or PC algorithms will find more informative output without an unrepresented or latent variable indexing the subpopulations;

(iii) the same conclusions as in (i) hold if no two subpopulations have two variables linked respectively in opposite directions, as in A → B and A ← B.

(iv) when condition (iii) fails, the proper representation of the combined population is with a cyclic graph and a latent variable. Thomas Richardson has found a correct algorithm adequate for both cyclic and acyclic structures in the linear case. He is investigating algorithms for cyclic structures with latent variables.

Another interpretation of the question asks what the units of statistical analysis should be when causal relations are between individuals rather than properties. I don't know the answer.

### 4. Can these methods separate causal graphs from classificatory graphs?

In one sense the answer is no, and they should not. Most cases of the distinction between "classificatory" graphs and "causal" graphs seem to turn on whether the dependency of one variable to another is an act of will or intended human practice rather than an asocial or unintended relation. Acts of will and intended human practice are as causal as acts of nature. AFQT scores, for instance, are computed by adding together certain component scores and not others, and that is a causal fact about how any AFQT score is obtained, and our procedures discover it.

Some of Woodward's concerns are the same as Humphrey's and the responses above serve both. Here I offer some remarks that have a particular bearing on Woodward's comments. Woodward devotes a lengthy first section of his essay to when and why we should believe the Causal Markov condition to apply. Now, that's an interesting problem because it is easy to imagine circumstances in which the Causal Markov condition fails, and the condition doesn't seem to follow from any deeper or more general or astochastic principles about causality. Some of the ways the Markov condition can fail invite lots of work in the very spirit I have described, and I urge Woodward to join in. The Markov condition fails for cases of feedback, as I noted in my essay, and exciting research on feedback systems and the connections among conditional independence, cyclic graphs, and the limiting distributions of time series is under way. The Markov condition can fail because of certain sorts of determinism, but no one has investigated search algorithms that are robust under deterministic cases. As Wermuth and Cooper have emphasized, the Markov condition can fail—and arguably sometimes does when missing data occur in experimental trials—because the property by which a sample is selected is influenced by the variables under study, but Cooper and Spirtes have gone on to investigate what can be determined about causation under what assumptions when sample selection bias may be present.

Another section of Woodward's comments is devoted to what could be meant by "cause," with whether I and my colleagues have an "informational" or a "control" notion of cause, and with how we could possibly know from probabilities that causal structure is stable enough to permit prediction. These comments seem to me ill-conceived for two reasons. First, the Causal Markov and Faithfulness conditions state assumptions about causal systems. Whatever you take "causal" to mean, if you assume the conditions apply, the theorems we prove, including the possibilities of discovery algorithms given population conditional independence relations, follow necessarily. It would be gratuitous for us to have insisted on some analysis of causation as a precondition for applying our theorems. Second, the whole of chapter 7 of *Causation, Prediction, and Search* is devoted to an elucidation of the differences between conditioning and intervening, and to the development of theorems that relate conditional probabilities satisfying a Causal Markov condition for a directed graph to probabilities that result from suitably idealized interventions. Woodward also correctly points out that a general recipe for knowing in advance when interventions are ideal is not provided, nor is there a proof that the output from our procedures, supposing them to be true descriptions of the processes that generated the data, are stable. Such a demonstration would require a refutation of Hume. But it is certainly not a criticism of an algorithm that it does not do the impossible.

There is one minor error of reading that Woodward makes repeatedly, and that I wish not to let pass only so that assiduous readers of this volume will not be misled: contrary to what he says, we do not always assume that the systems under study are "pseudo-indeterministic." We produce theorems that apply to any systems that satisfy the various axioms sets we discuss; pseudo-indeterminism with independent errors is one way to satisfy the Markov condition.

# IV

PUTTING
PROBLEMS
ABOUT
CAUSAL
INFERENCE
IN
PERSPECTIVE

# THE ROLE OF CONSTRUCT VALIDITY IN CAUSAL GENERALIZATION

## The Concept of Total Causal Inference Error

### *Larry V. Hedges*

Freedman's criticisms of structural equation modeling are important in putting mathematical methods in the proper perspective. His work in this volume (and his prior work, especially the excellent 1987 article in the *Journal of Educational Statistics*) clarifies one set of assumptions that lead to valid causal inferences from structural equation models and does so at a mathematical level that is accessible to practicing social scientists. This work serves as a useful palliative to those who feel that statistical methods are somehow magical. Freedman has helped to show that statistical methods can be used to evaluate the consequences of substituting assumptions for data, but if the assumptions themselves are not verified there is little reason to expect the results of such an exercise to be valid.

Glymour et al. have shown that a different set of assumptions than those stated by Freedman can lead to discovery of structure and valid causal inferences from structural equation models. Much of the debate between them can be conceived as a debate about which sets of assumptions are most plausible. The issue of which assumptive structures may be considered as the starting point for inference is obviously an important one for social science. However, explicitly analytic (deductive) activity such as research in logic or mathematical statistics is unlikely to resolve this issue.

In many ways the debate is similar to that among the various schools of thought concerning causal inference from quasi-experiments under selection models (see Wainer 1986; 1989). There are two prominent approaches to the problem. One involves modeling of the selection mechanism (see, e.g., Heckman and Robb 1986) and involves making assumptions about the form of the function relating the chance of selection to the outcome variable. The other involves modeling of the distributions of the outcome variable given selection and making assumptions about the distribution of the outcome among those not selected (see, e.g., Glynn, Laird, and Rubin 1986). Both approaches (and others as well) involve making assumptions which cannot be verified from the data collected and hence the choice between approaches cannot be resolved by mathematics alone. Mathematical approaches can be useful in transforming the problem so as to change the nature of assumptions required for valid inferences (see Wainer 1989), and this may lead to practical progress.

While the study of assumptive structures that can lead to valid causal inferences or to discovering valid causal models for a given set of data is an important one, it is not the only important problem in the study of causality in the social sciences. Another important theoretical tension is between causal inference and valid causal generalization. While randomized experiments are widely admired for their ability to control biases, serious questions have been raised about the generalizability of experimental evidence and hence of the ultimate adequacy of experimental data for informing public policy (see Heckman and Robb 1985; Heckman and Hotz 1989; LaLonde 1986; LaLonde and Maynard 1987).

In this chapter, I consider the problem of causal inference (from both true randomized experiments and from quasi-experiments) in a somewhat larger context that includes both valid causal inference in the narrow sense considered by Freedman and Glymour et al., and the problem of generalizing from a necessarily limited study to the broader context in which the research information will be applied.

## Local Causal Inference, Causal Generalization, and Global Causal Inference

I do not believe we try to gather causal knowledge to collect isolated facts as ends in themselves. We collect this knowledge because we wish to relate it to other things we know or because we wish to apply it. The most typical applications of social knowledge are for the purpose Morris Janowitz has characterized as "enlightenment" rather than "engineering," although applications of social knowledge for engineering purposes do occur, particularly in certain areas of applied psychology (e.g., human factors research), economics (e.g., monetary theory), sociology (e.g., design of social interventions), and political science (e.g., deterrence theory). Either purpose requires some ability to generalize beyond the particularities of the study from which the empirical evidence was derived. Contemporary examples of the enlightenment function of social knowledge can be found in connection with the use of evaluative data about social programs as a means of informing policy decisions. A great deal of recent attention in economics has been devoted to the development of estimates of causal effects of social policies such as manpower training or compensatory education programs.

## Local Causal Inference

Freedman has considered the question of whether typical applications of structural equation modeling techniques in the social sciences are adequate to establish a causal relation between the variables, *as measured*, in the settings and populations of people, *as examined*. This is a reasonable and important question that would be subsumed under the category of internal validity by Campbell and Stanley (1966) or Cook and Campbell (1979). I refer to this as the problem of "local causal inference."

## Causal Generalization

To be useful, causal relations demonstrated within research studies must be valid in settings other than those explicitly studied. This poses additional requirements that can be framed in terms of construct validity and generalizability (or external validity). Failure of causal relations to generalize across contexts or to larger constructs can invalidate useful causal inference as easily as can inaccurate causal attribution within a research study (see Cook and Campbell 1979). Assuring causal generalization can be just as daunting a problem as the internal validity problem in causal inference.

Note that it is quite possible for purportedly causal relations to be invalid, but to generalize across contexts and measurement operationalizations. That is, a particular research design might systematically underestimate the relation between two variables, yet provide relatively consistent results across contexts and measures (all underestimates of the true relation). Thus it is important to distinguish consistency of causal estimates from bias.

## Global Causal Inference

Thus useful causal inference must be both locally (internally) valid and generalizable (externally valid) across a relevant domain of contexts. Note that the universe of generalization need not be particularly broad—but it is typically broader than the narrow context of any single research study. For example, the relevant context might be as narrow as third-grade school students in Chicago public schools or as broad as all elementary school students in the United States, depending on the application of the research finding.

The validity of a global causal inference depends on both freedom of local causal inferences from bias and generalizability of those inferences across contexts and conditions. Thus causal inferences that are to be useful for policy or scientific purposes must be both locally valid and generalizable.

In this essay, I discuss a model for global causal inferences that incorporates both the idea of internal validity of causal inferences and their generalizability across contexts. It takes into account both systematic sources of error (bias) and statistical uncertainty (variance) in its description of the total uncertainty of global causal inferences. The model is similar in spirit to that underlying Kish's idea of total survey error (see Kish 1965).

## Local Causal Inference

Local causal inference (internal validity) is a familiar concept that has been discussed at length by Campbell and Stanley (1966), Cook and Campbell (1979), and Rubin (1974). In its simplest form, the local causal inference problem is one of estimating the effect of a treatment on an outcome. Typically this involves estimating the average ( for an identifiable population of persons) of the treatment effects for that group of people on a specified outcome variable. Here the treatment effect for any particular person might be estimated as the difference between that person's outcome value given that she received the treatment and that person's outcome value given that she did not receive the treatment. Of course this definition involves a counterfactual (the same person cannot both receive and not receive the treatment), but under reasonable assumptions a randomized experiment can produce unbiased estimates of the average treatment effect defined in this way (Rubin 1974; 1978).

### A Statistical Model of Local Causal Inference

One might denote the (average) treatment effect parameter (the true treatment effect) as

$$\delta(\mathbf{x,y,z})$$

where $\mathbf{x}$ represents a vector of characteristics that define the subject population, $\mathbf{y}$ represents a vector of characteristics that describe the context, and $\mathbf{z}$ is a vector of characteristics that describe the outcome measure.[1] The notation $\delta(\mathbf{x,y,z})$ is meant to imply that the treatment effect $\delta$ is a function of $\mathbf{x}$, $\mathbf{y}$, and $\mathbf{z}$. One might denote the sample *estimate* of the treatment effect by

$$d(\mathbf{w,x,y,z}),$$

where $\mathbf{w}$ is a vector that describes the characteristics of the method used to obtain the estimate (e.g., the type of study design). The average treatment effect estimate obtained by method **w** in context **y** on outcome **z** for subject population **x** could be expressed as

$$E_X\{d(\mathbf{w,x,y,z}) | \mathbf{w,y,z}\},$$

where $E\{\cdot | \cdot\}$ is the conditional expectation operator.

Any given estimate $d(\mathbf{w,x,y,z})$ will differ from the true treatment effect $\delta(\mathbf{x,y,z})$ by an amount that could be decomposed into two components. The first is an *error of estimation*, defined as the difference between the observed and average treatment effect:

$$e(\mathbf{w,x,y,z}) = d(\mathbf{w,x,y,z}) - E_X\{d(\mathbf{w,x,y,z}) | \mathbf{w,y,z}\}.$$

The second component is the *local causal inference error*, defined between the average treatment effect estimate and the average true treatment effect for subject population **x** in context **y** on outcome **z**:

$$h(\mathbf{w,x,y,z}) = E_X\{d(\mathbf{w,x,y,z}) | \mathbf{w,y,z}\} - \delta(\mathbf{x,y,z}).$$

The bias of a method **w** for obtaining causal inferences could be defined as the average value of the local causal inference error, where the average is taken over the subject population **x**. In more formal terms the bias could be expressed as

$$\eta(\mathbf{w,x,y,z}) = E_X\{h(\mathbf{w,x,y,z}) | \mathbf{w,y,z}\}.$$

A method **w** of obtaining causal inferences is unbiased for subject population **x**, in context **y**, using outcome **z** if the average bias or conditional expectation of $h(\mathbf{w,x,y,z})$ is zero.

Note that the contention between Freedman and Glymour et al. concerns the conditions under which $\eta(\mathbf{w,x,y,z}) = 0$. I will argue later that while the magnitude of $\eta(\mathbf{w,x,y,z})$ is an important consideration in total causal generalization, other components are also important in determining the validity of global causal inferences. Given that other components are almost certainly *not* zero, the question of whether $\eta(\mathbf{w,x,y,z}) = 0$ is not of overriding importance.

## Causal Generalization

Causal generalization has received less attention in the methodological literature than has local causal inference. Campbell (1957) and Campbell and Stanley (1966) used the term 'external validity' to refer to the generalizability of findings across populations of persons, contexts, and measured variables. Cronbach

(1982) provided a conceptual model of causal generalization based on its psychometric counterpart. Cook and Campbell (1979) contributed to this literature by noting that the generalization across measured variables could be construed as a construct validity problem. Most recently Becker ( forthcoming) has formalized many of the concepts discussed by Cronbach in a statistical model of causal generalization. In this section, I have adapted some aspects of previous work on causal generalization, choosing a consistent terminology.

## Breadth of Variables

We can distinguish between variables *as measured* and the variables *as conceptualized* in theory. This is the distinction we might make between unobservable theoretical constructs and observable operationalizations of those constructs. I intend something deeper here than just the conventional idea that variables as measured are certainly measured with error. In classical measurement theory the variables, as measured, are divided into a "true score" (reliable or systematic) component and a "measurement error" (unsystematic measurement error) component. I intend that the systematic, reliable component of the variable, as measured, is partly irrelevant to the construct it supposedly measures. To put it another way, the true score component is only a partially valid measure of the corresponding construct.

This conception is at least consistent with Donald Campbell's multiple operationism (e.g., Campbell and Fiske 1959). I believe it is also consistent with conventional thinking about how research informs us about the world beyond the particular experimental or nonexperimental data collected. Social theory is not about the relationship of *particular* SES measures, *particular* measures of educational attainment, and a *particular* measure of income. The research findings are clearly intended to suggest what would happen if any of a class of similar measures were used.

In any application, the generality of social knowledge is critical to its usefulness because we tend to think in terms of broad social variables, rather than narrow ones. A useful example is the field of education. The concepts are accessible even to nonspecialists and there is a long tradition of theory-driven systematic measurement in education. In fact, psychometrics measurement theory was developed at the cusp of education and psychology and was often called educational and psychological measurement before diffusion of knowledge led to the more general term "social measurement." For example, when we think about the goal of improving American education, the objective is most often described broadly: to improve critical thinking, written communication, and mathematics skills. If I were to suggest the use of a single narrowly defined measure (e.g., a particular achievement test battery) as the sole indicator of educational accomplishment, I would expect immediate protest. Partly, this is because of multiple goals for social institutions like schools in pluralistic societies. But it is also because any particular constituency agreeing on a particular goal would see any such potential measure as *flawed*, *incomplete*, and *partially irrelevant* to that goal.

One source of evidence that particular narrowly defined measures fail to capture relevant variables of interest is provided by discussions that arise about "corruption of indicators." By this we mean that once there are incentives to institutions to perform well on a narrow indicator, those institutions typically find ways to perform better on that indicator. That is, schools find ways to make themselves look better on the measures used to evaluate their performance. This typically does not mean teaching the *particular items* on the test, but rather teaching the *kinds of skills and applications* on the test. If the narrowly defined measure really reflected the variable of interest, this would not be "corruption," but a successful control strategy leading to appropriate alignment of curriculum with desired educational goals.

It is perhaps interesting that the National Assessment of Educational Progress (NAEP), perhaps the most extensive educational measurement enterprise ever undertaken, has made massive investments in the development of professional consensus (not always shared beyond the development committees) on descriptions of multidimensional content domains in subject matters such as mathematics. Since the number of test items is so large—the development of highly incomplete item sampling designs and analytic techniques to go with them has also been a substantial enterprise—virtually every measurement specialist in the country has been involved in one way or another. It is perhaps revealing that the most contentious aspect of the NAEP is the attempt to summarize the data on subject matter (e.g., math, reading, or writing) performance on a single scale and then attach some meaning to numbers on that scale. People find it easier to agree on what to include as relevant test questions than to agree on a *particular* measure derived from them. (They say "Yes, this is *part* of what I mean by mathematics achievement," but cannot easily agree on a single scale.)

## Context

I have argued that for social knowledge to be useful it must involve variables that are relatively broad in their conception. The same kind of argument can be made about the importance of application across a (specifiable) range of contexts. Sociologists and social psychologists have long studied the effects of social context as a primary substantive interest, while other social scientists have regarded it as a source of either nuisance or moderator variables. In fact a whole genre of statistical models (so-called hierarchical linear models) has come into fashion to study the joint effects of context and individual variables and to characterize the generalizability of findings across contexts (see, e.g., Bryk and Raudenbush 1992). The reason for the attention is that, to be useful, social knowledge must be sufficiently generalizable across some specifiable range of social contexts.

## Persons

The most obvious aspect of generality, and the one to which much of statistical inference in social science is addressed, is that useful social propositions must generalize across specifiable groups of persons to be useful. Because this form of generalization is so fundamental and because so much attention has been devoted to it, I will have little to say about it except that the formal aspects of generalization across people are very similar to those of generalizing across contexts and measures, but are usually treated differently in social science.

## Variable Breadth Determines Generalizability

Although it does not have to be formulated in this way, the problem of breadth of variable or, if you prefer, construct definition, can be characterized in terms of universes of operations (hypothetical "variables as measured"). One form of multiple operationism would define the construct by specifying the range of corresponding operations. That is, the construct is defined by specifying the universe of operations that are admissible as measures of it.

To say that the universe of corresponding operations defines the construct is not to say *exactly how* it is defined and there are different possibilities. One approach is to precisely define the construct from the universe of operations by saying that the construct is some sort of optimal linear combination of all the corresponding operations, the "common factor" or first principal component. This is the idea implicit in multiple indicator models involving simultaneous factor analysis of several indicator systems to estimate relations between latent variables (see, e.g., Joreskog 1971). This approach is appealing because it takes seriously the idea that the construct is what is common among the operations and uses a mathematical model to extract an estimate of this common variable. Relations between common variables corresponding to different constructs can then be estimated directly, net of any components due to irrelevancies in operations.

Another approach to defining constructs from operations is much more common in social science however. This approach treats any single operation as a representative of the construct that is partially irrelevant, but "exchangeable" in some sense with any other. That is, we regard any operation sampled at random as an equally good representative of the construct, albeit one containing randomly irrelevant components. (There is another version of this approach that recognizes structure in the universe of operations, so that some operations are "better representatives of the construct" than others, but this notion introduces a complication that does not illuminate any crucial details so I will not consider it further.) This second approach is implicit in studies that use single indicators of constructs, as did the studies Freedman considered.

How are the effects on relations between variables of the irrelevancies in the operations handled in the second approach? Typically they would be handled by appeal to the principle of fortuitous combination. Namely, that certain combinations of irrelevancies (in both variables) tend to increase strength of relations among operations, others tend to decrease them, so the average strength of relation between operations is not far from the strength of relation between constructs (except perhaps for the attenuating effects of genuine measurement error). In effect we say that the observed relation between any pair of operations is, on the average (across pairs of such operations), a good estimate of the relation between constructs. I do not offer this explanation as a marvel of scientific ingenuity, but as an attempt to give a modestly plausible account of the logic of our practice.

This account is consistent with analytic practice in integrative research reviews in the social sciences. Reviews determine what is known about relations between constructs by summarizing the relationships between operations derived in different studies. Reviews are generally regarded as more credible when operations, contexts, and types of persons vary across studies. In the last two decades, there has been a minor revolution in methods for reviews (see Cooper 1984; Rosenthal 1984; Hedges and Olkin 1985; Hunter and Schmidt 1990; Cooper and Hedges 1994) which has led to more clearly defined standards for methods in research reviews. These changes in standards have, if anything, increased the emphasis on the average relation between operations (the average effect size) as the appropriate index of relation between constructs.

The clearest statement of this principle is made in the literature on methods for reviewing test validities, often called "validity generalization" (see Schmidt and Hunter 1977 or Hunter and Schmidt 1990). Here the individual studies produce findings that are relations (correlation coefficients) between a psychological test and a criterion. Different studies may use different tests that measure the same construct and the criterion typically consists of different measures of the same construct such as "employee productivity."

Statistical methods for reviewing test validities were developed precisely because there was a widespread belief in the situational specificity of test validities. That is, test validities were believed to vary substantially across contexts so that generalizations about the magnitude of test validities were potentially suspect. The average of the validity coefficients (perhaps corrected for artifacts of study design such as measurement error) is explicitly specified as the best indicator of the relation between the underlying constructs. Between-study variation in the correlation parameters (the between-study variance component) is viewed as a measure of the generalizability of test validities across contexts.

## A Conceptual Framework for Causal Generalization

I argue that a framework for useful causal inference must describe generalization of relationships between *constructs* (not variables as measured) across contexts and persons. We have just seen that the most common approach to relations between constructs is to generalize across relations between operations. Thus the framework for useful causal inference can be seen as a framework for causal generalization: across operations, contexts, and persons. In an important recent paper, Becker ( forthcoming) has drawn on Cronbach's (1982) work on causal generalization to demonstrate how the psychometric theory of generalizability (Cronbach et al. 1972) can be used to describe the causal generalization problem.

The essence of her formulation is that we may think of any hypothetical study in terms of three characteristics: the set of operations chosen, the context, and the persons represented. In principle, each of these characteristics may affect the estimate of the relation between constructs obtained in the study. We define a causal relation of interest by defining the universes (or populations) of the three characteristics. This point is crucial:

> The *definition* of the relation between constructs is the average of the relation obtained between operations in the relevant operation universe over the contexts in the relevant context universe for the persons in the relevant person universe.

Adequacy of causal generalization must be measured against this definition. Note that this definition easily handles notions like conditioning on particular subclasses of persons, contexts, or operations by simply restricting the appropriate universe.

Of course it is somewhat limiting to talk about a single value of the relation between constructs. It is probably more useful to talk about the distribution of relations across the universe of operations, contexts, and persons. Becker proposes trying to estimate components of variance in indices of relations (e.g., correlation coefficients) attributable to variation in each of the universes and their interactions. These components of variance can then be used to estimate the uncertainty of the relation between constructs produced by any particular research design.

Becker has demonstrated in detail how such generalizability analyses could be conducted in the context of studies that estimate path models, where the correlation matrices used to construct path coefficients are modeled by a random effects model, with variance and covariance components for differences across contexts, persons, and operations of each construct (see Becker 1992; Becker and Schram 1994). Her analyses show that components of variance associated with differences in operations and contexts are not always negligible. The implication is that, *even if the relation among variables as measured is causal, that relation may be highly uncertain as a reflection of the relation among constructs.*

## A Statistical Model of Causal Generalization

One can express the model for causal generalization in terms of the effect parameters discussed in connection with local causal inference. There we described each individual study in terms of a subject population, $x$, a context, $y$, and a measure, $z$, used in the study. Questions of generalizability necessarily involve a range of $x$, $y$, and $z$ values, over which generalization is desired. Let $X$, $Y$, and $Z$ be the universes (hyperpopulations) of subjects, contexts, and measures over which generalization is desired. Use the term 'universe value' of the causal effect to mean the average of the causal effect over $X$, $Y$, and $Z$, and use the term 'generalizability' of the causal effect to mean the variance of the causal effect over $X$, $Y$, and $Z$.

To make these ideas unambiguous requires us to define not only the set of possible $x$, $y$, and $z$ values contained in $X$, $Y$, and $Z$, respectively, but also to indicate their relative frequency. That is, we must put a frequency (probability) distribution on $X$, $Y$, and $Z$. Thus the universe value $\delta$ and the generalizability $\sigma_\delta^2$ (which are both functions of $X$, $Y$, and $Z$) can be defined by

$$\delta = E_{XYZ}\{\delta(x,y,z)\}$$

and

$$\sigma_\delta^2 = VAR_{XYZ}\{\delta(x,y,z)\},$$

where $E\{\cdot\}$ and $VAR\{\cdot\}$ are the expectation and variance operators, respectively. The *generalization error* of generalizing causal effect $\delta(x,y,z)$ to the universe conditions $X$, $Y$, $Z$ could be defined as

$$g(x,y,z) = \delta(x,y,z) - \delta.$$

## Global Causal Inference

A conceptual model for global causal inference starts with the definition of the global causal effect $\delta$ discussed in the last section: the global causal effect is the expected value (average) of the perfectly locally valid (unbiased) estimates of the causal relation between operations, averaged over the relevant range of contexts, operations, and subjects. Given this definition of the global causal effect, one can define the *error* of any particular estimate of the global causal effect as the difference between that estimate and the global causal effect. The conceptual model we have discussed so far implies that the error of any estimate of the global causal effect can be decomposed into three parts: an estimation error, a local causal inference error, and an error of causal generalization.

Of course any particular estimate is a random quantity subject to chance fluctuations due to sampling. For this reason it is useful to define the *accuracy* or typical error of an estimator (a process for getting estimates) of the global causal effect in terms of the expected (average) size of the errors, for example by introducing a decision theoretic "loss function" of some kind (see, e.g., Ferguson 1967). A conventional expression for the accuracy (actually the *in*accuracy) is the squared error loss or "mean squared error," which is the average value of the squared errors. The square root of the means squared error is roughly the typical difference between an estimate and the true global causal effect. The mean squared error can be decomposed into two components, one due to bias and the other due to variance. Specifically:

Inaccuracy = Bias² + Variance.

Estimation error will contribute to the variance term. In most cases limited internal validity will contribute some variance, but primarily bias. In most cases, limited generalizability across contexts will contribute primarily variance, but also some to the bias. Which of the two components contributes most profoundly to overall accuracy depends on the details of the research methodology and the substantive problem.

## A Statistical Model for Global Causal Inference

The components we have already developed can be combined to give a statistical model of global causal inference. The *total causal inference error* in generalizing from an estimate based on a particular estimate $d(\mathbf{w},\mathbf{x},\mathbf{y},\mathbf{z})$ using method $\mathbf{w}$ on subject population $\mathbf{x}$ in context $\mathbf{y}$ with outcome measure $\mathbf{z}$ could be decomposed into three components: an error of estimation, $e(\mathbf{w},\mathbf{x},\mathbf{y},\mathbf{z})$, a local causal inference error, $h(\mathbf{w},\mathbf{x},\mathbf{y},\mathbf{z})$, and a generalization error, $g(\mathbf{x},\mathbf{y},\mathbf{z})$. Dropping the qualifying arguments $\mathbf{w}$, $\mathbf{x}$, $\mathbf{y}$, and $\mathbf{z}$, the total causal inference error for estimated effect d, can be expressed as:

$$d - \delta = e + h + g.$$

The relative size of the components e, h, and g will vary for different designs and universes to which inferences are desired. The component whose properties have been studied most conventionally is the error of estimation—the statistical theory of estimation is devoted to the formal study of such errors. Consequently, a great deal is known about the behavior of errors of estimation and how to minimize them. For many study designs, the bias and variance of estimation errors for all conventional estimates of the causal effect are well understood.

Local causal inference errors are the topic of this volume and of much other literature on causal inference. Causal generalization errors have received less attention, in part because the problem has not usually been formulated analytically. It has usually been treated relatively informally, and often not very well. The next section outlines some of what is known.

## How Large Are Causal Generalization Errors?

There is nothing, in principle, that would prevent the use of conventional sampling theory as a method for studying and quantifying errors of causal generalization. This approach has been taken by Becker and other meta-analysts. Becker's analytic methods require, in principle, designed experiments in which elements of the universes of constructs, contexts, and persons are sampled at random and then assigned at random to studies which yield values of relations between operations. The problem of planning the generalizability study to estimate variance (and covariance) components is therefore just an exercise in planning a random effects experiment. The particular choices of the operations, contexts, and people can be thought of as levels of factors that can be taken as random effects. One might make practical compromises in design to afford greater practicality, but in principle the design and inference is straightforward.

It is all straightforward *provided* that the levels (elements from universes across which we generalize) are sampled at random and assigned at random to studies. This requirement is almost never met, with the possible exception of a very few surveys (e.g., some aspects of NAEP, where instruments are assigned by design to a probability sample of people in a probability sample of contexts). The sampling of operations and contexts in any given study is usually entirely unknown, and is certainly not random probability based. Thus the causal generalization problem is essentially always one of drawing inferences about a population distribution (e.g., the average and variance) from nonrandom samples drawn by unknown sampling mechanisms (see also Cronbach 1982; Cook 1993).

If the mechanism for sampling of operations and contexts (let alone persons) were known, classical (frequentist) statistical arguments could be used to justify causal generalization in a completely straightforward way. Without the assurance provided by knowledge of the sampling mechanism, generalizations "as if random sampling had been used" are perilous. Indeed one of the major difficulties encountered in meta-analysis is precisely accounting for the effects of *possible* sampling mechanisms (such as systematic selection in the publication system) that might bias the results of naive generalizations from collections of studies (see, e.g., Begg and Berlin 1988; or Hedges 1992). The problem clearly plagues other attempts to make causal generalizations: biases created by selection apply just as profoundly to single studies as to groups of studies (see Hedges 1984).

An interesting and important paper by Heckman (1979) suggests a link between the problems of local causal inference and those of causal generalization. Heckman construes the problem of causal generalization from nonrandom samples as a problem of omitted variables—misspecification of the causal model. This is generally the most difficult of the problems Freedman raised in assuring the validity of causal inferences from structural equation models. It is interesting that essentially the same conceptual problem (and its analytic manifestations) arise in both local causal inference and causal generalization.

### Justifying Causal Generalizations

How do we justify causal generalizations? One might be tempted to say that we assure ourselves that the samples (of persons, contexts, and measures) available to draw inferences from are *representative*, if not random. Unfortunately the idea of representative samples is not a technical term in statistics with a well-specified definition. Nor is it a term that has a reasonable approximate definition. Kruskal and Mosteller (1979a; 1979b; 1979c) have identified nine distinguishable meanings of the term 'representative sampling' in their extensive review of the concept. Considering these meanings, it seems appropriate to divide them into two categories, one that provides external criteria for deciding if a sample is representative and one that does not. The latter category of definitions for representative samples says little more than that a sample is representative if it is "good enough for a particular purpose" (p. 245), or "permits good estimation" (p. 245). These definitions are circular in the sense that they do not provide independent criteria for determining whether a given sample might be representative.

Meanings for the term, 'representative sample', included in the first category stress "absence of selective forces" (p. 245), "coverage of the population" (p. 245), that the sample is "a miniature of the population" (p. 245), or that the sample consists of "typical or ideal case(s)" (p. 245). These definitions suggest criteria for deciding whether a sample might be representative. One can imagine using elements of all of these meanings to evaluate the representativeness of a sample or to a construct a systematic, but representative, sample. All of them require knowledge of the structure of the population or universe to be represented.

I would suggest that a definition of 'representative' incorporating these elements should go something like this:

A representative sample must include cases from the full range of the universe with approximately the same density of observations as the universe in each *important region*. This may be accomplished by selecting prototypical or ideal cases from each *important region*, but in any event a range of cases should be selected in any important region so that *irrelevant attributes* vary in each *important region*.

I emphasize the terms 'important region' and 'irrelevant attributes' in this definition because they depend on the structural knowledge of the universe necessary to define representativeness. As Heckman has noted, the assumption that these are known *a priori* is analogous to the assumption that all the relevant causal variables can be known in advance in causal inference problems.

This analysis of criteria for causal generalization yields conclusions similar to those which Cook (1993) draws from the literature on construct validation. His principles of proximal similarity, of empirical interpolation, and of extrapolation, require reproduction in the sample of the most important regions of the universe. Cook's principles of heterogeneous irrelevancies and of discriminant validity correspond to avoiding selective forces when choosing cases within important regions.

## The Role of Weighting

It is likely that the sample of persons, contexts, and measures in studies actually carried out will *not* be a representative sample by any reasonable definition of that term. For example, researchers are likely to carry out studies using persons and contexts that are chosen at least partially for convenience. It would be fortuitous indeed if they proved representative of the universes of interest for later generalization. Moreover, different scientific or policy questions dictate different universes of generalization, which makes representative sampling a logical, as well as practical, impossibility.

However, the latter problem plagues any attempt to gather evidence for a variety of purposes which imply mutually contradictory universes. The problem is solved in multipurpose sample surveys by developing sets of sample weights that permit weighting evidence from a stratified sample in various ways to produce estimates relevant to different universes. Sample surveys usually have the advantage that the evidence gathered within a stratum is representative of the corresponding population stratum. However, weighting can still be used to produce estimates that are representative in the sense described above. Formal analysis would reveal that under most conditions such weighting would reduce bias (see Rubin 1973).

A weighting procedure could be used with one or more studies of a causal effect to estimate the magnitude of generalization bias associated with the study(s) by comparing the unweighted estimate with a synthetic estimate obtained from weighting to approximate representativeness (see Hedges and Waddington 1993). Of course the ability to use weighting depends on the variation across persons, contexts, and measures available in the data collected. Any single study is likely to be quite limited in the variation of contexts and measures represented. A collection of studies would be more likely to provide a sufficiently wide range of contexts and measures necessary to ensure the representativeness of the eventual synthetic causal estimate.

## Conclusions

Both local causal inference about relations between operations and causal generalization about relations between constructs make similar demands for *a priori* knowledge in order to make valid inferences. Yet in some research domains, the former is often perceived to be much more problematic than the latter. This certainly seems to be suggested by many of the previous essays in this volume.

It is revealing that some of the social sciences (e.g., experimental psychology) appear to be more concerned about local causal inference (internal validity), and others (e.g., much of sociology) more concerned with causal generalization. The former emphasize the use of randomized experiments to establish causal relations, but are less concerned about the problem of generalizability across measures and contexts. The latter emphasize the use of representative samples and are less concerned with the problem of inferring causal relations.

One possible explanation for these differences is that researchers in the relevant fields are, by and large, acting rationally. Investigators have limited resources to invest in research data. How should they allocate these resources to obtain the most accurate practical causal inferences? Limitations in both local causal inference and causal generalization contribute to the limitations of global causal inference, and both cost resources to overcome. In the language of Becker's generalizability theory, both contribute components of variance and bias which decrease the accuracy of estimates of global causal relations between constructs. Researchers can expend resources to increase internal validity or to increase generalizability. The problem is how to allocate resources between the two.

If the components of variance typically associated with variations across contexts are large relative to those associated with variations in internal validity, then overall accuracy may be minimized by expending resources to minimize variance associated with contextual variations. In other words, by concentrating on obtaining research data in more representative settings, researchers may obtain more accurate practical causal inferences even though they may pay a penalty in bias (or variance) in estimates of local causal effects.

If the components of variance typically associated with variance across contexts and measures is small, then the consequences of bias (or variation) associated with estimates of local causal effects may be more profound. In other words, researchers can increase overall accuracy of practical causal inferences by concentrating on reducing bias (and variability) in estimates of local causal effects.

Thus, if one believes that research communities are rational in their choice of methodology, the emphasis on local causal inference versus generalizability provides an indicator of the community's *a priori* beliefs about the relative magnitude of these sources of error. Nevertheless, there are at least some cases which appear to raise questions about the well-foundedness of research communities' *a priori* beliefs as instantiated in their choices of research method.

Academic medicine has moved steadily toward the position that internal validity should be the dominant criterion for the design of medical research. This movement is exemplified by the extreme emphasis on the randomized controlled trial as a means of evaluating the effectiveness of medical treatments (see, e.g., Lachlin, Tygstrup, and Juhl 1982). However, the growing effort to determine the global effectiveness of medical innovations seem to suggest that such randomized controlled trials may be poor guides to effectiveness (see, e.g., Silberman, Droitcour, and Scullin 1992).

## NOTES

1. To avoid an even more cumbersome notation than the one I have chosen, I use the symbol **x** both as a vector which describes fixed subject characteristics and as a random variable that describes the distribution of subject characteristics for a given subject population. There is a corresponding ambiguity in the formal meaning of the symbols **w**, **y**, and **z**. The precise meaning that obtains is usually clear from the context.

# WHAT IS A CAUSAL STRUCTURE?

## Nancy Cartwright

### I. Where Do Laws Come from?

This essay is about a spate of recent work using graph representations of causal structure to study methods for inferring causes from statistics, especially the work of Peter Spirtes, Clark Glymour, and Richard Scheines (1993). The ideas I argue for here make up one stage in my more general Aristotelian campaign to replace the ontology of natural law with the ontology of capacities. I begin with the question of the title: What is a causal structure? According to Spirtes, Glymour, and Scheines, a causal structure is

> an ordered pair $\langle V,E \rangle$ where V is a set of variables, and E is a set of or-
> dered pairs of V, where $\langle X,Y \rangle$ is in E if and only if X is a direct cause of
> Y relative to V. (1993, 45)

Alternatively, V can be a set of events. But we should not be misled into thinking we are talking about specific events occurring at particular times and places. The causal relations are supposed to be nomic relations between event-types rather than singular relations between tokens. I prefer to call this ordered pair a *complexus* of causal laws. The causal structure, I shall argue, is something different.

Where does this complex of causal laws come from and what assures its survival? Perhaps this seems a peculiar question. Surely the point of a law is that it just is true, and eternally so. No: these laws, like all laws, whether causal or associational, probabilistic or deterministic, are transitory and epiphenomenal. They arise from—and exist only relative to—a causal structure. So, what is a causal structure? Answer: a fixed (enough) arrangement of components with stable (enough) capacities that can give rise to the kind of regular behavior that we describe in our scientific laws. There are, of course, enduring facts about what kinds of capacities are associated with what kinds of characteristics. We may, if we choose, call our records of these facts "natural laws," but if we do, a natural law will look very different from the way we normally picture it, and certainly very different from the causal and probabilistic laws that appear in discussions of probabilistic causality.

## II. The Socioeconomic Machine

In figure 13.1 we have a typical graph of the kind Spirtes, Glymour, and Scheines employ to represent sets of causal laws. What we see represented in the graph are fixed connections between event-types. One event-type either causes another or it does not, and it does so with a definite and fixed strength of influence if Spirtes, Glymour, and Scheines's best theorems are to be applicable. We typically apply these methods to event-types like unemployment and inflation, income and parents' education level, divorce rates and mobility, or a reduction in tax on "green" petrol and the incidence of lung cancer. What could possibly guarantee the required kind of determinate relations between event-types of these kinds? You could think of reducing everything ultimately to physics in the hope that in fundamental physics unconditional laws of the usual kind expressed in generalizations will at last be found: proper causal laws and laws of association with modal operators and universal (or probabilistic) quantifiers in front. Apart from the implausibility of this downward reduction I think the hope is bound to be dashed. All the way down, laws are transitory and epiphenomenal, not eternal.

I ask "Where do sets of laws of the kinds represented in typical graphs come from?" I will tell you exactly how the particular laws pictured in figure 13.1 were generated. They come from the machine in figure 13.2. The machine gives rise to the causal relations and regularities depicted in the associated graph. The machine pictured in figure 13.2 has fixed components with known capacities arranged in a stable configuration. It is the fixed arrangement among the parts that ensures that the causal relations among the vertices of the consequent graph are determinate. (All the machines figured in this essay plus the associated causal graphs have been designed by Towfic Shomar.) My thesis is that we always need a machine like this to get laws—any laws, causal or otherwise. Sometimes God supplies the arrangements—as in the planetary systems—but very often we must supply them ourselves, in courtrooms and churches, institutions and factories. Where they do not exist, there is no sense in trying to pick out event-types and to ask about their nomological relations. That would be like asking: "Given that we drop six balls in a bucket, what is the probability that a second bucket ten feet away will rise by six inches?" That question makes no sense unless we have in mind some background machine to which the buckets are bolted.

It is interesting to note that this way of thinking about the two levels is similar to the way matters were originally viewed when the kinds of probability methods we are now studying were originally introduced into social inquiry.

**FIGURE 13.1**

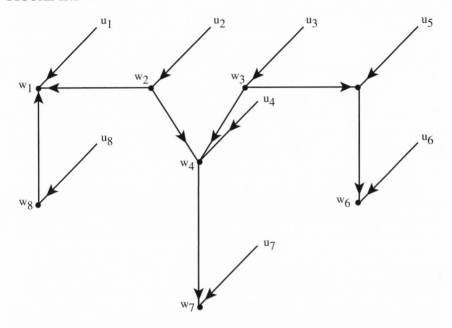

The founders of econometrics—Trygve Haavelmo and Ragnar Frisch—are good examples. Both explicitly believed in the socioeconomic machine. Frisch, for instance, proposed a massive *re*organization of the economy relying on the economic knowledge that he was acquiring with his new statistical techniques (cf. Andvig forthcoming). Or consider Haavelmo's remarks about the relation between pressure on the throttle and the acceleration of the car (Haavelmo 1944). This is a perfectly useful piece of information if you want to drive the car you have, but it is not what you need to know if you are expecting change. For that you need to understand how the fundamental mechanisms operate. My own hero Otto Neurath expressed the view clearly in his pre-war (that is, pre-*first* world war) criticisms of conventional economics. Standard economics, he insisted, made too much of the inductive method, taking the generalizations that hold in a free market economy to be generalizations that hold *simpliciter:* "Those who stay exclusively with the present will very soon only be able to understand the past" (quoted from Nemeth 1981, 51, trans. Nancy Cartwright). Just as the science of mechanics provides the builder of machines with information about machines that have never been constructed, so too the social sciences can supply the social engineer with information about economic orders that have never yet been realized. The idea is that we must learn about the basic capacities of the components; then we can arrange them to elicit the regularities we want to see.

**FIGURE 13.2**

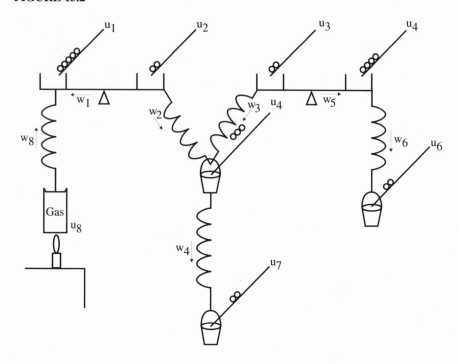

The causal laws we live under are a consequence—conscious or not—of the socioeconomic machine that we have constructed.

I have been developing my views about socioeconomic machines for some time, starting with *Nature's Capacities* (Cartwright 1989) and other papers on the role of Aristotelian natures in contemporary science (Cartwright 1993). Here I want to make a small advance in favor of the two-tiered ontology I propose. I will do this by defending a distinction between questions about what reliably causes what in a given context and questions about the persistence of this reliability. The first questions concern the causal laws that a causal structure will give rise to; the second are questions about the stability of the structure. Recall the machine in figure 13.2 that supports the graph in figure 13.1. There is a sense in which this is a "badly shielded" machine. Normally we want to construct our machine so that it takes inputs at very few points. But for this machine every single component is driven not only by the other components of the machine but also by influences from outside the system. We had to build it like that to replicate the graphs used by Spirtes, Glymour, and Scheines. In another sense, though, the machine is very well shielded, for nothing disturbs the arrangement of its parts.

My defense of this distinction in this essay will consist of three separate parts: (1) the first is a summary of a talk I gave with Kevin Hoover at the British Society for the Philosophy of Science (BSPS) in September 1991 (see also Cartwright 1994 and 1995); (2) the second is a local exercise to show the many ways in which this distinction is maintained in the Spirtes, Glymour, and Scheines work; (3) third is a very brief discussion of probabilities and chance set-ups.

## III. Stability

My discussion with Kevin Hoover focused on Mary Morgan's remark that the early econometricians believed that their statistical techniques provided a *substitute for the experimental method*. I defended their belief. As I see it, the early econometrics work provided a set of rules for causal inference (RCI):

RCI: Probabilities $\rightarrow$ Causal Laws

Why should we think that these rules are *good* rules? The econometricians themselves—especially Herbert Simon—give a variety of reasons; I gave a somewhat untidy account in my book on capacities. At the BSPS, I reported on a much simpler and neater argument that is a development of Hoover's own work following Simon and my earlier proof woven together. This was the thesis:

*Claim:* RCI will lead us to infer a causal law from a population probability only if that law would be established in an ideal controlled experiment in that population.

That is, when the conditions for inferring a causal law using RCI are satisfied, we can read what results would obtain in an experiment directly from the population probabilities. We don't need to perform the experiment—nature is running it for us.

That is stage one. The next stage is an objection that Hoover and I both raise to our arguments. The objection can be very simply explained without going into what the RCI really say. It is strictly analogous to a transparent objection to a stripped-down version of the argument. Consider a model with only two possible event types $C_t$, $E_{t+\Delta t}$. Does C at t cause E at t $+ \Delta t$? The conventional answer is that it does so if and only if $C_t$ is statistically relevant to $E_{t+\Delta t}$, i.e. (suppressing the time indices)

$RCI_{2\text{-event model}}$:     C causes E iff $P(E/C) > P(E/-C)$.

The argument in defense of this begins with the trivial expansion:

$$P(E) = P(E/C)P(C) + P(E/-C)P(-C).$$

Let us now adopt, as a reasonable test, the assumption that if C causes E then (*ceteris paribus*) raising the number of C's should raise the number of E's. In the two-variable model there is nothing else to go wrong, so we can drop the *ceteris paribus* clause. Looking at the expansion, we see that $P(E)$ will go up as $P(C)$ does if and only if $P(E/C) > P(E/-C)$. So our mini-rule of inference is vindicated.

But is it really? That depends on how we envisage proceeding. We could go to a population and collect instances of C's: if $P(E/C) > P(E/-C)$, this will be more effective than picking individuals at random. But this is not always what we have in mind. Perhaps we want to know whether a *change* in the level of C will be reflected in a change in the level of E. What happens then? Suppose we propose to effect a change in the level of C by manipulating a 'control' variable W (which for simplicity's sake is a 'switch' with values $+W$ or $-W$). Now we have three variables and must consider a probability over an enlarged event space: $P(\pm E \pm C \pm W)$. Once W is added, the joint probability of C and E becomes a conditional probability, conditional on the value of W. It seems that the $P(E)$ that we have been looking at then is really $P(E/-W)$. But what we want to learn about is $P(E/W)$, and the expansion

$$P(E/-W) = P(E/C-W)P(C/-W) + P(E/-C-W)P(-C-W)$$

will not help since from the probabilistic point of view, there is no reason to assume that

$$P(E/C-W) = P(E/C-W).$$

But probabilities are not all there is. In my view probabilities are a shadow of an underlying causal structure. The question of which are fundamental, causal structures or probabilities, matters because reasoning about causal structures can give us a purchase on invariance that probabilities cannot provide. Think again of our model (C, E, W) in terms of its causal structure. The hypothesis we are concerned with is whether C at t causes E at $t + \Delta t$ in a toy model with very limited possibilities: (i) W at some earlier time $t_0$ is the complete and sole cause of C at t; and (ii) C is the only candidate cause for E at t. In this case the invariance we need is a simple consequence of a kind of Markov assumption: a full set of intermediate causes screens off earlier causes from later effects.

Spirtes, Glymour, and Scheines have a theorem to the same effect, except that their theorem is geared to their graph representation of complex sets of causal laws. Kevin Hoover and I use the less general but more finely descriptive linear modeling representations. Spirtes, Glymour, and Scheines call theirs the "manipulation theorem." They say of it:

> The importance of this theorem is that if the causal structure and the direct effects of the manipulation . . . are known, the joint distribution [in the manipulated population] can be estimated from the unmanipulated population. (1993, 79)

Let us look just at the opening phrases of the Spirtes, Glymour, and Scheines theorem:

> Given directed acyclic graph $G_{comb}$ over vertex set $V \cup W$ [the expanded set that includes the "control" variable W, as well] and distribution $P(V \cup W)$ that satisfies the Markov condition for $G_{comb}$ . . . . (1993, 79)

So the Markov condition plays a central role in the Spirtes, Glymour, and Scheines argument as well. (In fact what Spirtes, Glymour, and Scheines call the Markov condition is far stronger than anything I would want to give that name to. But that is not immediately relevant.)

My main point can be brought out by thinking about a very informal version of the Markov condition: a full set of intermediate factors (in our two-variable case that will be C) screens off earlier factors (say W) from later effects (E). In our case this amounts to

MC:   $P(E/\pm C \pm W) = P(E/\pm C)$

What exactly are we presupposing when we adopt the screening-off condition? Two different things I think. The first is a metaphysical assumption that in the kinds of cases under consideration causes do not operate across time gaps; there is always something in between that "carries" the influence. Conditioning on these intermediate events in the causal process will render information about the earlier events irrelevant, so the probability of E given C should be the same regardless of the earlier state of W. But more is involved. Clearly the assumption that $P(E/\pm C)$ is the same given either $+W$ or $-W$ assumes that changes in W are not associated with changes in the underlying causal structure. This is required not just with respect to the qualitative causal relation—whether or not C causes E—but also with respect to the exact quantitative strengths of the influence, which are measured by the conditional probabilities (or functions of them). Now this assumption is built into the Spirtes, Glymour, and Scheines theorem in two ways: (1) The antecedent of the theorem restricts consideration to distributions $P(V \cup W)$ that satisfy the Markov condition, a restriction which, as we will see in the next section, has no general justification in cases where causal structures vary; (2) they assume in the proof that $G_{unman} = G_{man}$ are subgraphs of $G_{comb}$. That is, they assume that changes in the distribution of the externally manipulated 'switch variable' W are not associated with any changes in the underlying causal structure. That is the same assumption that Hoover and I pointed to as well. Where do we stand if we build this assumption in as Spirtes, Glymour and Scheines do? It looks as if the original thesis is correct: *if* changes in the variables that serve as control variables in nature's experiment are not associated with changes in the underlying causal structure *then* RCI (or the Spirtes, Glymour, and Scheines program) will allow us to infer causal laws from laws of probabilistic association.

That seems to be the situation with respect to the back and forth about causal inference between Hoover and me at the BSPS meeting. The reason for covering this again is to point out that central to the argument is the distinction between (1) the stable causal structure that gives rise to the laws (the socioeconomic machine) and (2) the laws themselves—both the causal laws and the associated laws of probabilistic association.

## IV. Causal Auxiliaries *vs* Causal Structures

There is an obvious strategy for trying to undo the distinction. That is to try to take the causal structure "up" from its foundational location and to put it in as an auxiliary in each of the emerging causal laws themselves, on the model of an auxiliary. Following Mill or Mackie (as Spirtes, Glymour, and Scheines do too) we can take the law connecting causal parents with their effects to have the form:

$$C_1 A_1 \vee C_2 A_2 \vee \dots \vee C_n A_n \text{ causes E,}$$

where the C's are the "salient" factors and the A's the necessary helping conditions that make up a complete cause. The idea is to include a description that picks out the causal structure that gives rise to the causal relation between $C_i$ and E in each of the disjuncts. That is, to assume for each $A_i$ in the above formula that:

$$A_i \supset CS_i,$$

where CS is a random variable whose values represent the range of possible structures. In the case of the graph represented in figure 13.1, this means that the values of CS must represent all different possible machines, if sense can be made of that.

The proposal is obviously impracticable. Just think of it narrowly from the point of view of the Spirtes, Glymour, and Scheines framework. Suppose we are studying a set of causally sufficient variables (so that every common cause of every member of the vertex set is in the vertex set). This is the case for which they get the most powerful theorems. If the description of the causal structure must be included at this level, the set would be radically insufficient and we would, artificially, no longer be entitled to the results of those theorems.

There is also a general philosophical issue: in Mackie's and Mill's accounts the relationship between the C's and A's on the one hand, and the E's on the other is a causal relation. That is not true of the relation between the causal structure itself and the set of laws (causal and probabilistic) that it gives rise to. Look again at our machine and its concomitant causal graph. The relations between vertices in the graph are causal; the relation between machine and graph is not. The machine causes neither the causal laws nor the effects singled out in the causal laws. The importance of this is not just metaphysical; there are real methodological implications. The two types of relation are very different and so too will be the methods for investigating them. In particular we must not think that questions about the stability of causal laws can be established in the same way that the laws themselves are established, for instance by including a variable that describes the causal machine in the vertex set and looking at the probabilistic relations over the bigger vertex set. This procedure has many defects, and what I want to do in this section is to look—again locally—at how these defects drain power away from the Spirtes, Glymour, and Scheines techniques. I will close with some very general considerations of why this way of looking at it must be wrong.

Almost all of the Spirtes, Glymour, and Scheines theorems presuppose that nature's graph—that is, a graph of the true causal relations among a causally sufficient set that includes all the variables under study—satisfy their Markov condition. But that assumption will not in general be true if the vertex set includes variables—like the causal structure variable—such that changing their values affects the causal relations between other variables. The problem is one that Spirtes, Glymour, and Scheines themselves discuss, although their aim is to establish causal relations not to establish stability, so they do not put it in exactly this context. Here is the difficulty: consider two different values (say $CS = 1$, $CS = 2$) for the causal structure variable. There are cases in which the true causal relations among a set of variables V (not including CS) will satisfy the Markov condition both when $CS = 1$ and when $CS = 2$; but when they are put together, the combined graph does not. This is an instance of the by-now familiar Simpson's paradox. (Spirtes, Glymour and Scheines call this "mixing." In the published version of *Causation, Prediction, and Search*, they agree with me in

**FIGURE 13.3**

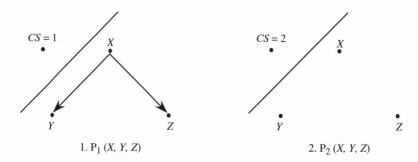

1. $P_1 (X, Y, Z)$
2. $P_2 (X, Y, Z)$

**FIGURE 13.4**

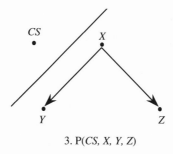

3. $P(CS, X, Y, Z)$

seeing mixing as a case of Simpson's paradox though earlier drafts treated the two as distinct). Take a simple case with two extremes (see figure 13.3).

When CS = 1, X causes both Y and Z deterministically. When CS = 2, X, Y, Z are all causally independent of each other. For Spirtes, Glymour, and Scheines a causal connection appears in the graph just in case it is in the true causal relations for *any* values of any of the variables. The graph over CS, X, Y, Z is shown in figure 13.4.

The probability P relates to $P_1$ and $P_2$ thus:

$$P(\pm X \pm Y \pm Z / CS = 1) = P_1(\pm X \pm Y \pm Z)$$
$$P(\pm X \pm Y \pm Z / CS = 2) = P_2(\pm X \pm Y \pm Z)$$

The following numerical example shows that $P_1$ and $P_2$ can satisfy the Markov condition relative to their graphs, but P does not do so relative to its graph. See figure 13.5.

**FIGURE 13.5**

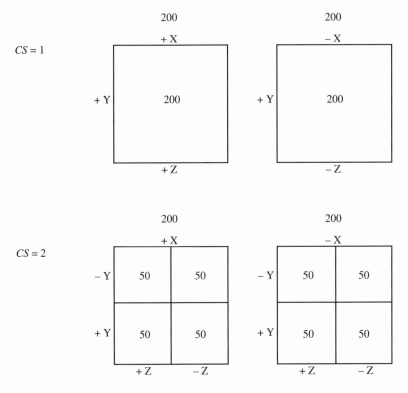

Consider $P(+Y + Z/+X)$; the others follow the same pattern.

$$P_1(+Y+ Z/+X) = 1 = 1 \cdot 1 = P_1(+Y/+X)P_1(+Z/+X)$$

$$P_2(+Y+ Z/+X) = \frac{50}{200} = \frac{1}{4} = \frac{100}{200} \cdot \frac{100}{200} = P_2(+Y/+X)P_2(+Z/+X),$$

but

$$P(+Y+Z/+X) = \frac{250}{400} = \frac{5}{8} \neq \frac{9}{16} = \frac{300}{400} \cdot \frac{300}{400} = P(+Y/+X)P(+Z/+X).$$

Simpson's paradox is the name given to the following fact (or its generalization to cases with more variables and more compartments in the partition):

There are probability distributions, P, such that a conditional dependence relation (positive, negative or zero) that holds between two variables (here Y, Z) may be changed to any other in both compartments of a partition along a third variable (here CS).

**FIGURE 13.6A**

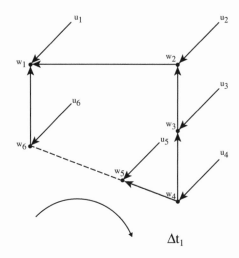

$\Delta t_1$

Defining $P_{\pm X}(CS \cdot Y \cdot Z) = P(CS \cdot Y \cdot Z / \pm X)$, we see that the example just cited exhibits Simpson's paradox for both the distribution $P_{+X}$ and for $P_{-X}$. A similar construction can be made for any of the conditional independences associated with the Markov condition. A necessary condition for Simpson's paradox to obtain is that the third variable (CS) be conditionally dependent on each of the other two (Y, Z). That is the case for each of the distributions $P_{+X}$, $P_{-X}$:

$$P_{+X}(Y/CS = 1) = 1 \neq 0.5 = P_X(Y/CS = 2),$$
$$P_{-X}(Y/CS = 1) = 0 \neq 0.5 = P_X(Y/CS = 2),$$

and similarly for $P_{\pm X}(Z/CS)$. The dependence between Y (or Z) and CS arises in $P_{+X}$ and $P_{-X}$ because, given either $+X$ or $-X$, information about whether we are in CS = 1 or CS = 2 (i.e., whether we are considering a population in which X causes Y [or Z] or not) will certainly affect the probability of finding Y (or Z).

The case I have considered here is an all-or-nothing affair: the causal law is either there and deterministic, or it is altogether *missing*. Figures 13.6 and 13.7 show another case where the causal graph gets entirely reversed, and not even because of changes in the structure of the machine but only because of changes in the relative values of the independent variables. But we do not need such dramatic changes. Any change in the numerical strength of the influence of X on Y and Z can equally generate a Simpson's paradox, and hence a failure of the Markov condition for the combined graph.[1]

Returning to the main argument, I wanted this foray into Simpson's paradox to provide a concrete example of the harmful methodological consequences

**FIGURE 13.6B**

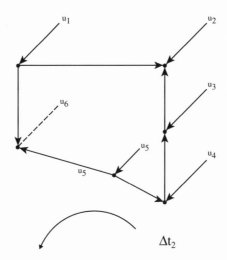

of a mistaken elevation of the description of the supporting causal structure into the set of causes and effects under study. This work by Spirtes, Glymour, and Scheines is very beautiful: they provide powerful methods for causal inference, and their methods are well suited to a number of real situations, and in those situations the methods work very well. But if we model the situations in the wrong way—with the causal structure in the vertex set—the theorems will in general no longer apply. In such circumstances no reason can be provided why these methods should be adopted.

## V. Concluding Remarks

I claim that *all* laws—not only causal laws, but laws of probabilistic and deterministic association as well—are derivative from causal structures. I have concentrated here on the causal laws that are depicted in Spirtes, Glymour, and Scheines–type graphs and have said virtually nothing about the probabilities that are generated in tandem. As a pointer to a larger discussion, I want to close with some brief remarks about what probabilities are, a discussion that will connect some familiar themes with my insistence that we require socioeconomic machines for the socioeconomic laws and socioeconomic probabilities of the kind that our standard methods treat. In *The Logic of Statistical Inference*, Ian Hacking (1965) urges that propensities and frequencies are obverse sides of the same phenomenon. Propensities give rise to frequencies; frequencies are the expression of propensities. This is a point of view that I want to endorse. A chance set-up may occur naturally or it may be artificially constructed, either deliber-

**FIGURE 13.7**

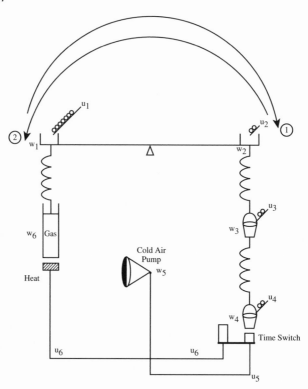

ately or by accident. In any case probabilities are generated by chance set-ups, and their characterization necessarily refers back to the chance set-up that gives rise to them. We can make sense of the probability of drawing two red balls in a row from an urn of a certain composition with replacement; but we cannot make sense of the probability of 6 percent inflation in the U.K. next year without an implicit reference to a specific social and institutional structure that will serve as the chance set-up that generates this probability. The originators of social statistics followed this pattern, and I think rightly so. When they talked about the "iron law of probability" that dictated a fixed number of suicides in Paris every year or a certain rising rate of crime, this was not conceived as an association laid down by natural law as they (though not I) conceived the association between force and acceleration, but rather as an association generated by particular social and economic structures and susceptible to change by change in these structures.

The same, I claim, is true of all our laws, whether we take them to be iron—the typical attitude towards the laws of physics—or of a more flexible material, as in biology, economics, or psychology. J. J. C. Smart (1963) has urged that biology, economics, psychology, and the like are not real sciences. That is because they do not have real laws. Their laws are *ceteris paribus* laws, and a *ceteris paribus* law is no law at all. The only real laws are, presumably, down there in fundamental physics. I put an entirely different interpretation on the phenomena Smart describes. If the topic is laws in the traditional empiricist sense of claims about nomological patterns of association or regularity, we have *ceteris paribus* laws all the way down: laws hold only relative to the chance set-up that generates them. What basic science should be aiming for, whether in physics or economics, is not to discover laws but to find out what stable capacities are associated with features and structures in their domains, and how these capacities behave in complex settings. What is fundamental about fundamental physics is that it studies the capacities of fundamental particles. These have the advantage, supposedly, of being in no way dependent on the capacities of parts or materials that make them up. That is not true of the capacities of a DNA chain or a person. But the laws expressing the regularities that arise when these particles are structured together into a nucleus or an atom are every bit as dependent on the stability of the structure as are the regularities of economics or psychology. A nucleus may be a naturally occurring chance set-up, but it is a set-up all the same. Otherwise the probabilistic laws of nuclear physics make no sense. I repeat the lesson about the dual nature of frequencies and propensities: Probabilities make sense only relative to the chance set-up that generates them, and that is equally true whether the chance set-up is a radioactive nucleus or a socioeconomic machine.

### NOTES

Thanks to Jordi Cat for very helpful discussions as well as to the participants of the British Society for the Philosophy of Science Conferences in 1991 and 1993 and the participants of the Causality in Crisis Conference. This research was supported by the LSE Research Initiative Fund and aided by the LSE Centre for the Philosophy of the Natural and Social Sciences.

   1. As an aside, let me make a remark about the Spirtes, Glymour, and Scheines scheme. They say they are interested in qualitative relations—whether or not X causes Y—rather than in the quantitative strength of the causal relation, and that they do not like results that depend on particular values for those strengths. Nevertheless the plausibility of their Markov condition presupposes that they are studying sets of causal relations that are stable not only qualitatively, but also across changes in the quantitative strengths of influence as well.

# THE BIG BROAD ISSUES IN SOCIETY AND SOCIAL HISTORY

## Application of a Probabilistic Perspective

### *Stanley Lieberson*

When an important event occurs, we want to understand it in the sense that we want to know what causes it. This sentence involves questions that I will not try to define in an elaborate way. By "important" I simply mean an event that the larger society (or some subset) defines as such. It can also be "important" if it contributes to social science because of the question's theoretical relevance, such as linking up to empirical issues that have not been resolved. By "understand" I have in mind gaining a greater sense of the event in a variety of possible ways such as what the experience is like for those involved, the attitudes of society towards it, institutional and societal predecessors and/or correlates, the strategies and tactics used by the individuals and groups involved, the historical background, and the like. One specific kind of understanding involves "causality," which I shall view as determining the condition(s) which affect the likelihood of occurrence of the observed event.

My concern here is not about elaborate definitions of these terms. I will not focus primarily on the difficulties that a discipline such as sociology encounters when applying routine quantitative procedures to problems that generally involve a relatively large number of cases. Typical examples of this are studies of the factors influencing internal migration, unemployment, marriage, divorce, intergenerational occupational mobility, infant mortality, prejudices and other attitudes. Answering these questions is difficult enough; there are often serious questions about the logic of social research even under these relatively ideal circumstances (as other essays in this volume make clear). I will not delve into these matters other than to mention in passing that we often try to pursue experimental analogies based on nonexperimental data. Since it is difficult enough to do first-class experiments, just visualize the added difficulties when working with nonexperimental data. This is the typical situation in such primarily nonexperimental social sciences as sociology, economics, political science, and social anthropology, and it leads to a wide variety of difficulties. For example, there are extremely sticky difficulties in creating suitable controls, taking into account unmeasured selectivity, finding suitable comparisons that are themselves unaffected by the independent variable, or dealing with asymmetrical causal processes ( for a more elaborate discussion of these issues, see, for example, Lieberson 1985; Singer and Marini 1987; Freedman 1991, and the literature cited therein).

I will start here with a different set of issues: namely, important kinds of events that, for one reason or another, cannot be studied through the analysis of large numbers of cases. As we shall see, there are special weaknesses in the methods that are customarily used for analysis in such instances. In considering these weaknesses, I will develop a probabilistic perspective that is relevant for analyses of social processes—regardless of data size. It is a perspective which has implications that I believe are not fully appreciated even when researchers work with large data sets. Some of the implications will be illustrated with counterfactual examples drawn from sports. From these examples, I then draw three major conclusions which—if valid—are applicable to all kinds of social science problems, regardless of the type of data. These involve a perspective on how social processes are best conceptualized given the limitations of our methodological tools (what I shall call a "multilayered probabilistic approach"), a discussion of the role of chain models in understanding events, and—returning to our initial question—how we might best approach the task of causally explaining specific events.

### Small N's When There Are Just Not Enough Cases for a Statistical Analysis

There are many questions about society that do not readily lend themselves to straightforward quantitative studies. Often only a small number of cases exist in the entire universe. (Remember that "small" and "large" are not fixed, because the minimum necessary size will vary by the number of variables, the complexity of the models, the strength of the relationships, and the like.)[1] In some problems, even if the total number of cases might be adequate, sometimes there are only a few that we know enough about to permit sufficient description and measurement. In other problems, the amount of effort necessary to research even one case is massive and it is impractical to expect researchers to go beyond one or two cases (as might well be the problem when intensive field work over the course of several years is necessary as in some sociological and anthropological studies). Or years in the library may be necessary to research an enormous number of historical documents to adequately study one case, and a sufficient number of cases for a statistical analysis may be a biological impossibility under present life expectancy for a single researcher. In any case, for one reason or another, the usual large-N statistical analyses are not easy for such questions as, for example: the consequences a specific social policy has for society; the reason a given piece of legislation is passed; the causes of a given war; the causes of revolutions; the decline of empires; the reason two nations differ on a specific social policy; the origins of capitalism; the political consequences of presidential assassinations in the United States; the causes of current black-white differences in income; the reason why there is an exceptional level of violence in the United States; the rising tide of nationalism in Europe; the economic success of several new nations in South East Asia; or the absence of economic progress in most African nations.

Incidentally, the social organization of research is not a trivial consideration here. What keeps a team of researchers, say a group of anthropologists, from each studying a different community with the goal of pooling the results at the end and doing a statistical analysis? Perhaps antipathy towards statistical analysis is itself a sufficient reason. In any case the cross-cultural files are constructed only as each individual researcher does his or her study.

In any case, we are considering problems about society which do not lend themselves to the customary statistical analyses. Under such circumstances, there are typically three approaches taken. One is to rely on a descriptive case study of one or two such events, with the notion that such a deep analysis will permit one to ferret out the causes of the events. (Although such work is sometimes labeled as descriptive, involving no explicit appeals to the notion of causality, I think the distinction may be less meaningful in practice because the "describer" in such a work makes a judgment about what circumstances leading up to an event are relevant or important enough to be emphasized and discussed. Likewise, analyses of the consequences of an event implicitly involve causal notions.) Incidentally, a case study of this sort can be quantitative and include complicated statistical models. A second "solution" to the problem is what I would call an artificial forced variance approach. The third approach involves what *appears* to be an application of the logic of causal inference described by John Stuart Mill in his *System of Logic* (1872).

All three face serious problems. The single case study is a powerful instrument for deeply delving into a given situation and can be extraordinarily illuminating and suggestive. But it is usually very difficult to truly sort out and isolate the causal factor(s). Most often, there are almost infinite possibilities for what is accounting for the observed relationship. Moreover, it is difficult to have much confidence about the generalizability of the results. Is this the only cause of the dependent variable? Is it the most common cause? Is it a relatively rare cause? All of these questions are difficult to answer with the case study. Incidentally, a study based on a large N, with all sorts of variables and elaborate statistical models, can still be viewed as a small-N study if it deals with only one unit but means to generalize beyond that. As my colleague John Campbell observed in a discussion of these issues, if there is a one-country analysis of taxes over time, then it is a study of one case (even if there are elaborate data sets for a long period of time) if it is meant to generalize about tax policies in other nations.[2]

The use of forced variance is not an uncommon approach because it allows the researcher to use standard quantitative procedures when facing a situation with either a small number of cases in the universe or a small number of available cases. By "forced variance" I mean a situation where the research problem is modified so as to increase the number of measurable instances. In point of fact, it normally involves creating a different problem (even when the variables are the same) because it pursues analysis at a different level than that at which the original problem was formulated. An example is described in Lieberson (1985, 110–14). The United States has an exceptionally high homicide rate when compared to nations with comparable levels of urbanization and industrialization. Why is this? It is suggested that racial differences in socioeconomic position cause criminal violence.[3] Based on data gathered for 125 large metropolitan areas in the United States, it is found that indeed there is a significant correlation between the level of violent crime in the metropolitan areas and the level of socioeconomic differences between the races. The study, you will notice, has shifted from an issue about differences *between* nations to an issue about differences *within* a nation. The variables are the same ones we would use if there were cross-national data: some indicator of violent crime and some indicator of racial gaps in socioeconomic circumstances. But we have dropped to a different level of analysis—a lower one which could hold but not be applicable to what occurs cross-nationally. The nation effect, per se, is dropped. I will not go into details about the assumptions that this step involves since they are detailed in Lieberson (1985, chapter 5). The point here is that one cannot blithely drum up some added variance by dropping down a level of analysis. It is not interchangeable with the evidence that would occur at the appropriate plane.

The third method, which involves applying the method of agreement and the method of difference proposed by Mill (1872), has been used in the area of comparative and historical research when the number of cases are limited (see the illustrations provided in Lieberson 1991). The method is appealing since it allows one to logically eliminate possible causes if they do not correctly account for all of the small number of cases under examination. It is also appealing because there are no shades of gray in evaluating evidence: either the theory holds or it fails. In other words, all of the evidence is assumed to be strong evidence—indeed, exceptionally strong—of a magnitude rarely encountered in social science. As a matter of fact, the method never allows for multiple causes of a given event.[4] Multiple causes are indistinguishable from non-causes. If more than one cause operates in a set of situations, then whether the method of agreement or the method of difference is applied, the causes will be rejected because they will be indistinguishable from variables which have no influence on the outcome.

Unfortunately, as Mill himself observed in a section of his book that seems to be ignored by those who insist on using these procedures, the method of difference and the method of agreement are not applicable to the kinds of problems encountered in the social sciences (see Mill 1872, Book VI, chapter VII, 469–78). If the operation of multiple causes is indistinguishable from non-causes, for example, then the method cannot work in the typical social science circumstance in which nonexperimental data involve more than one causal factor affecting the dependent variable. If all variables present in two or more cases are the same except for one, it is almost certain—as Mill argues—that there has to be a complex set of reasons underlying the fact that two societies, say, are identical on all counts but this one variable (a circumstance which these methods cannot address). I will not review the criticisms I have raised elsewhere (Lieberson 1991) or the excellent and related comments in an earlier paper by Nichols (1986). But there is one issue that should be noted in passing here. A central weakness of attempts to apply Mill's methods—in spite of his objections—is the inability of the method of difference or the method of agreement to yield probabilistic causal models. As a consequence, the methods are obliged to use deterministic assumptions about social processes even when they are patently inappropriate (Lieberson 1991, 309–12).[5] Also, causal inferences based on small data sets have difficulty guarding against conclusions reflecting random combinations of variables. Typically, a small number of cases are used to evaluate the influence of several possible causal variables on a given explanandum. The information on each variable is usually presented in a dichotomous form (e.g., present-absent in each cell, or high-low). If we visualize the score for each dichotomy being randomly assigned to each cell in such a table, there is a good chance that some "meaningful" conclusion will be obtained under this methodology such that only one variable is found to account for the outcome. This is particularly the case when variables are combined if they do not seem to hold individually. The structure of the analysis allows for meaningless conclusions to be generated.

In short, to my knowledge the three major procedures for dealing with small-N situations are deeply limited, at least as presently practiced. The assumptions used to infer causality in small-N situations are often unrealistic, but are driven by the fact that small-N situations are not suited to a probabilistic approach. It is a case of the perspective being driven by the data. Indeed, the all-or-nothing quality of causal ascriptions on Mill's methods virtually precludes an ability to identify causal factors at work in large or small data sets dealing with social events.

## A Probabilistic Perspective

There are a variety of reasons for taking a probabilistic approach even if we think in deterministic terms. Bear in mind that a deterministic perspective refers to a view that a given factor, when present, will lead to a specific outcome. By contrast, a probabilistic perspective is more modest in its causal claim, positing that a given factor, when present, will increase the likelihood of a specified outcome. When we say, "If $X_1$, then $Y$," we are making a deterministic statement. When we say, "The presence of $X_1$ increases the likelihood or frequency of $Y$," we are making a probabilistic statement. Obviously, if given the choice, deterministic statements are more appealing. They are cleaner, simpler, and more easily disproved than probabilistic ones. One negative case ($Y$'s absence in the presence of $X_1$) would quickly eliminate a deterministic statement (Lieberson 1991, 309).

The subjects studied in the social sciences often involve substantial measurement errors that are not easily avoided and are often of a non-random nature. Complex multivariate patterns are often present in naturally occurring data and these are likely to make a deterministic approach unproductive. In addition, we are usually unable to measure all of the variables that we believe are actual (or possible) influences on the problem at hand. Finally, there is the role of what we might call "chance" in affecting outcomes.[6] (See Lieberson 1991, 309–10 for a more elaborate discussion of how these factors work against using a deterministic model.) Here "chance" or "random" can be taken as referring to idiosyncratic factors at work that we cannot deal with, either because we do not know them or because there are so many that we could not manage them all without, in effect, trying to explain the history of the universe. In a certain practical sense, it is almost irrelevant whether they could all be taken into account given sufficient energy and effort—although we will see below that there is reason to suspect that they cannot be.

However, there are two additional reasons for comparing the probabilistic and deterministic approaches. First, explanations of particular events, even where the research driving an explanation is itself probabilistic and based on very large N's, are often formulated in an inappropriate deterministic manner. Second, a probabilistic approach suggests a different way of conceiving social processes and of theorizing about them that is appropriate for both large- and small-N work, for both what passes as quantitative and qualitative research.

## A Sports Example

The path leading to a championship in a competitive sport provides a convenient opportunity to consider the role of "chance" factors and is a relevant model for thinking about social events more generally. From a deterministic perspective, the champion team is the best team. If we could visualize a counterfactual condition such that we roll time back to the beginning of the entire season (with every player on each team being in exactly the same physical condition as they had been initially, and with the information and experience acquired during the season all forgotten), then the same team each time would be the champion. Whether such a rollback were to occur one time or 100 times or 1,000 times or 10,000 times, a deterministic approach would always visualize the same winner. In effect, this is how a deterministic approach *must* visualize the operation of social processes. If the outcomes were not to be always the same under these assumptions, then any actual outcome would have to be treated as a member of a set of different outcomes compatible with the initial conditions rather than a uniquely inevitable product of those conditions.

From the probabilistic perspective adopted here, the actual championship team is not necessarily the best team. That is, if we likewise visualize the counterfactual condition of rolling time back to the beginning of the season in the way as described above, the same team would almost certainly not win in each try. To be sure, the distribution of winners would differ greatly from what would be expected under chance (which itself would produce a somewhat uneven distribution). But if the teams do differ in their potential to win the championship, some teams would almost never be winners, and others would win far more frequently than their share. Hopefully, the team that actually did win the championship is the team that most frequently wins in this series of rollbacks. Indeed, knowing only about the one actual series, we would have grounds to assume that the actual winner is the team most likely to be the most frequent winner in the rollback of 1,000 times. But this point needs to be put very carefully. It only means that the winner in a series of 1,000 rollbacks is the most likely winner if we have only one championship season. Depending on the distribution of teams, which is unknown to us, it could easily be the case that far in excess of 50 percent of the time the winning team is not the best team (as determined in a large series of rollbacks). Thus, what we mean by the "best team" is no longer what one typically means by such a phrase. Unless one team is extraordinary compared to any and all other teams—and this is going to be rare—we do not know which team in a league is the best as customarily defined. Since there are no rollbacks of 1,000 times, the best team is unknown, for the best team is the one with the greatest likelihood of being the winner in a series of such rollbacks. This likelihood might be almost unity or it could be far less than .5.

One could argue that this description of the deterministic perspective is too extreme, in requiring that the winning team would have to repeat in an infinite number of rollbacks. One might claim that this creates a straw man. Suppose, in a thousand trials, the same team wins, say, 995 times. Would that not be good enough? In point of fact, the deterministic perspective is incorrect if there is an expectation of anything less than an infinite stream of rollbacks with the same outcome. However, if the probability of the same team winning is virtually 1.0, then the deterministic viewpoint provides an outstanding approximation of the results obtained under a probabilistic approach. However, do we know this to be the case? On the other hand, do we know this *not* to be the case? We do know that the deterministic approximation of the probabilistic model will vary from situation to situation. Unless we have strong grounds for assuming a close approximation, we can make enormous errors by pursuing a deterministic approach.

A nice example of the role played by chance in such cases is the partially random interaction between the schedule of opponents and the health of the team. For example, an injury to a key player during the regular season might be of considerably less consequence if it occurs during a time when weak opponents are being played. If it occurs during a tough part of the schedule, the team's record for the season could be seriously affected. This is a good example because any variation in skills in avoiding injuries and any variation in skills in scheduling to optimize a team's outcome can only be loosely tied together, and hence chance factors will play a role in the combination of timing of injuries and opponents. But this is only one example; a somewhat more systematic set of observations is in order.

### The Super Bowl

Let us examine the issue in a somewhat more rigorous empirical manner. We will look at the winner of each of the twenty-seven Super Bowl professional football championships in the United States through 1993. In each case, we will start with the winner of each Super Bowl and ask how "certain" or "clean" was the team's progress to that championship. Now this is complicated. First of all, we are dealing with wall-to-wall counterfactuals of a heroic nature. If a winner barely edges out a loser because of the latter's misplay, for example, then we have no idea how the winner would have done if more effort had been required, i.e., if the misplay had not occurred. Secondly, it is reasonable to assume that misplays are not random. In other words, some teams have better players or better training programs or better strategies that reduce their likelihood of misplays and/or increase the likelihood that the team will cause a misplay by its opponent. There are many examples: a superior defense may force the quarterback to throw inappropriately and thereby increase the chances of an interception; a well-executed tackle may be aimed at knocking the ball loose from the carrier thereby creating a fumble. Likewise, coaches may differ in their strategies for certain situations or their ability to detect a weak spot in an opposing player and thereby capitalize on it in a situation which the naïve observer views as simply a bad break. Certainly it is important to minimize the role of chance in influencing outcomes, and certainly it is not entirely separable from such matters. There is reason to believe that good teams make their own breaks. But it is also sound to recognize that to judge a team to be superior because it won a game by making its own "breaks" may well involve circular reasoning. Finally, there is a danger in a game with a close final score of finding the "chance" events which helped the winner, but ignoring those events which worked against them. For all we know a team may win by a narrow margin only because it was the victim of a string of bad breaks.

I shall focus exclusively on the Super Bowl champion teams. I will start with the Super Bowl game itself, asking if the championship game is won by a relatively narrow margin, seven points or less. If not, then we will ask the same question about the playoff game win which allowed the team to reach the Super Bowl (in other words, the semi-finals, which consist of the final two teams in the league). Again, if the margin is more than seven points, we go on to the previous round in the playoffs, and so forth. We then ask if the team got into the playoffs because its record was at least two wins better than the next team.[7] If none of these conditions occurred, we will conclude that the path to victory was not marked by any obvious chance events. Of course this set of criteria incorporates far less information than may genuinely be relevant. On the one hand, a more extensive analysis would also consider more carefully the factors affecting what teams end up playing the eventual winner. There is reason to think that chance factors will also affect which opponents are faced by the eventual Super Bowl winner—and therefore the actual opponents are not necessarily the strongest of the possible opponents. Likewise, we do not deal with how another team from the other division or league might have fared if they had won some playoff which could have led them to the Super Bowl against the winner. There are many events in the season that could be viewed (at least in part) as being accidental or chance occurrences, e.g., a key player (say the quarterback or the leading runner) is injured for the bulk of the season and therefore the team is knocked out of contention. On the other hand, as observed earlier, a close game (as defined here) might have been close only because the winning team had a lot of bad luck. Hence, the opposite conclusion would be justified. There are many circumstantial factors that are unanalyzed as well: for example, calls made by the game officials are sometimes arbitrary or difficult ( for example, pass interference, or the exact point where the ball is placed after a tackle, or the exact point on the sideline where a punt crossed out of the playing field).

The analysis presented below is based on the data reported in David S. Neft, Richard M. Cohen, and Rick Korch (1993). I used their volume not only to trace the scores of all relevant games, but also as a source of descriptions of the Super Bowl and league championship games which they supply (usually, there is no text accompanying the results of the playoff games within the leagues). Keep in mind that the accounts of the games given in this source were presumably not written by the authors with the intent of underwriting my thesis here about a probabilistic approach to broad social events. Their accounts provide, therefore, an independent source of descriptive information about the games.

The following facts are not subjective: essentially half of the eventual winners won their Super Bowl game *and* their league championship *and* any preceding playoffs all by eight or more points, *and* initially entered the playoffs because they had a record two or more wins better than the team just below them (see chart 1). However, in thirteen of the twenty-seven seasons the margin of victory for the ultimate winner in the Super Bowl game was seven or less points in at least one of the post-regular season games: either in the Super Bowl game itself and/or in the league championship and/or in a playoff game leading to the league championship game. In five of these thirteen seasons, the margin of victory was close in more than one of these games. In Super Bowl XXV, for example, the New York Giants won the Super Bowl game by one point, 20–19, after winning their NFC Championship by two points, 15–13. In Super Bowl XVI, San Francisco won the Super Bowl by five points, 26–21, after winning its NFC Championship game by one point, 28–27. The most extreme example thus far is the Miami team that won Super Bowl VII by seven points, 14–7 after winning the AFC Championship 21–17, which in turn was preceded by a conference playoff victory of 20–14.

To be sure, a close score may still mean that the "better" team (as defined here in terms of the unknowable counterfactual outcomes of playing the season over and over again) was the winner. After all, a series of rollbacks in which one team was a fairly consistent winner would, from time-to-time, include some close games. In making such a point, however, keep in mind that the opposite would also hold. To wit, any given game might easily have been won by more than seven points (and hence would not be included here) even if the opponents were fairly evenly matched in a series of rollbacks.

**CHART 1.**  Close Scores for the Team Winning the Super Bowl[a]

| Super Bowl and Season | | Playoff | Conference Championship | Super Bowl | No Post-Season Close Scores |
|---|---|---|---|---|---|
| I | 1966–67 | | 34–27 | | |
| II | 1967–68 | | 21–17 | | |
| III | 1968–69 | | 27–23 | | |
| IV | 1969–70 | | 13–6 | | |
| V | 1970–71 | | | 16–13 | |
| VI | 1971–72 | | | | X |
| VII | 1972–73 | 20–14 | 21–17 | 14–7 | |
| VIII | 1973–74 | | | | X |
| IX | 1974–75 | | | | X |
| X | 1975–76 | | 16–10 | 21–17 | |
| XI | 1976–77 | | | | X |
| XII | 1977–78 | | | | X |
| XIII | 1978–79 | | | 35–31 | |
| XIV | 1979–80 | | | | X |
| XV | 1980–81 | 14–12 | 34–27 | | |
| XVI | 1981–82 | | 28–27 | 26–21 | |
| XVII | 1982–83 | | | | X |
| XVIII | 1983–84 | | | | X |
| XIX | 1984–85 | | | | X |
| XX | 1985–86 | | | | X |
| XXI | 1986–87 | | | | X |
| XXII | 1987–88 | 21–17 | 17–10 | | |
| XXIII | 1988–89 | | | 20–16 | |
| XXIV | 1989–90 | | | | X |
| XXV | 1990–91 | | 15–13 | 20–19 | |
| XXVI | 1991–92 | | | | X |
| XXVII | 1992–93 | | | | X |
| | TOTAL | 3 | 10 | 7 | 14 |

[a] Games won by seven or fewer points.

The descriptions of some of these games support the speculations suggested by the statistical pattern in the twenty-seven seasons thus far. For example, the Super Bowl V game, in which the Baltimore Colts defeated the Dallas Cowboys by 16–13 is described by Neft, Cohen, and Korch as "a comedy of errors" (1993, 207). Consider the close score and the wide range of misplays, such as fumbles and intercepted passes at key points, as well as critical injuries. Some examples (all from page 207):

> Baltimore quarterback Johnny Unitas threw a long pass down the center of the field to wide receiver Eddie Hinton; the ball bounced off Hinton's hands, back up into the air, grazed the fingertips of Dallas cornerback Mel Renfro, and came right down to the surprised John Mackey. Taking the ball around mid-field, Mackey sprinted the rest of the way to the end zone.

Or, consider the other side of the equation: a Colt player fumbled the opening kickoff in the second half

> deep in Baltimore territory. The Cowboys then drove from the 31-yard line to the two-yard line on five plays, with Thomas' hard running the key element. With the ball in the shadows of the goal posts, Thomas took a handoff and fumbled the ball, the Colts recovering on the one-foot line.

With the Cowboys ahead by seven points in the fourth quarter,

> a Morton pass [Morton being the Cowboy quarterback] bounced off the fingers of fullback Walt Garrison into the hands of Colt safety Rick Volk, who returned the ball 17 yards to the Dallas three-yard line. In short order, Tom Nowatzke smashed over for the touchdown, and Jim O'Brien added the tying extra point.

There are many other examples of such factors which, if not entirely random, are hardly to be viewed as non-accidental. Miami, the winner of Super Bowl VII by a 14–7 score, had first won its AFC Championship by 21–17 over Pittsburgh. Early in the first quarter of the AFC game, the Pittsburgh quarterback was knocked dizzy when he fumbled the ball and was out of the game until the final seven minutes (Neft, Cohen, and Korch 1993, 242).

Consider the following events in Super Bowl X, in which Pittsburgh defeated Dallas, 21–17:

> Through the first three quarters Dallas held a 10–7 lead. Then, at 3:32 of the final quarter, Reggie Harrison, a Pittsburgh reserve running back who plays on special teams, blocked a punt by Mitch Hoopes at the

Dallas 9. The ball bounced off Harrison's face hard enough to wind up in the Dallas end zone, good enough for a two-point safety and run the score to 10–9. It was a play which was considered the turning point of the game. . . .

Dallas coach Tom Landry blamed the defeat on the blocked punt by Harrison, which he said changed the momentum of the game around. He may have been right, but Swann's performance [Swann being a wide receiver for Pittsburgh]—which earned him the game's Most Valuable Player award—was momentum enough for the Steelers. Hospitalized only two weeks earlier with a concussion, and dropping passes in practice, the fleet-footed receiver returned to catch four passes for an astonishing total of 161 yards—a Super Bowl record certain to stand for many years. (Neft, Cohen, and Korch 1993, 297)

This last account is a helpful example of the operation of both skill and chance factors. A blocked punt represents a variety of skills, such that the blocker successfully evades those trying to prevent him from getting close enough to the punter, the inability of the punter to get the kick off quickly enough and/or at an initial arc which could not have been blocked. Presumably there was a combination of skills which worked to Pittsburgh's advantage, but the direction of a ball bouncing off the defender's face (hopefully the authors meant to say his face mask) is hard to allocate to a skill. Yet it did take a bounce such as to end up in the Dallas end zone.

Likewise, Swann, a pass receiver who was a key player in Pittsburgh's offense, is a victim of an initial bad break, to wit the concussion suffered two weeks earlier. Apparently, he recovered just in time to play a critical role in the offense. Had he recovered just a tad later (remember Swann was dropping passes in practice), the impact could have been great. Here again the events reflect a mixture of skill, the pass-catching ability of Swann, and of unfortunate chance events (the injury to Swann) coupled with a recovery which occurs just in time for the game (assuming recovery from a concussion is not simply a psychosomatic response to the urgency of the game). Injuries are inherent in the sport. Differences may occur between players in their proneness to injuries and their ability to play under less than optimal conditions. Likewise, differences may exist between teams in the exposure of their players to injuries. For example, teams will differ in the protection they are able to provide their quarterback and this, in turn, obviously affects the chances that such a key player will be tackled frequently and possibly injured.

I could provide many other examples of close games that could have gone either way and in which small "random" events seriously impacted the outcome.

The New York Giants defeated Buffalo by one point in Super Bowl XXV. In the final two minutes, Buffalo missed a 47-yard field goal attempt with a kick that went wide to the right (p. 627). Now a 47-yard field goal would be a real accomplishment, but it is not unheard of. We could view the matter as follows: what are the chances of the Buffalo kicker making such a field goal under the defensive conditions provided by the New York Giants? The actual failure could then be viewed in the context of these odds.[8]

Focusing on the issue at hand, in summary many of these close games suggest victories that might easily have been losses had not some events occurred which may be best described as having a random component. Contests are mixtures of luck and skill, no different than many card games, say, bridge or poker. Presumably random factors will wash out in the long run. But in small-N research there is no way to approximate this condition, and in large-N studies other factors obtrude to prevent us from ascertaining more than that certain events are probabilistically linked with others.

### Social Relevance

It might be thought that while random factors do play a role in sports outcomes, it is otherwise in the case of broad macrosocial events. Is it reasonable to also conceptualize large-scale social processes in terms of probabilistic forces? This is a less risky step than one might otherwise assume.[9] First, there is a non-trivial technical matter: with so many measurement errors in social research, a probabilistic model is in order even if one wants to be a determinist. Second, there is evidence that a probabilistic approach is the best that can be achieved in many areas of the "hard sciences," where the influences on some outcome are best described in terms far weaker than would be required on a deterministic model. See, for example, the discussion in Lieberson (1985, 226–27) of the "amplification effect" described by Weisskopf as operating in a variety of scientific contexts where very minor causal perturbations sometimes have very large effects; or the review of probabilistic issues by Lieberson (1992, particularly pp. 7–9), the discussion by Salmon (1971, 56–57) of "incomplete explanations" of paresis and radioactive decay, as well as the literature cited in these publications. Finally, there are grounds for taking a probabilistic perspective even if one is convinced that a deterministic model is actually operating, and that everything allocated to the residual category of "chance" consists of nothing more than unexplained deterministic process.

The rationale for this latter claim rests on a simple fact: in social as well as in many other complex processes, elucidating all of the factors that actually contributed to a given outcome would be highly counterproductive given the goal of generalizable knowledge. Even if it were possible to do so, the massive effort required would involve assessing the roles of all sorts of factors whose occurrence or failure to occur in a particular instance would be wholly incidental from the viewpoint of seeking to understand a particular type of process. Suppose we take a deterministic view of history such that a given outcome, say a war or the assassination of a prominent political figure, *had* to happen when and how it did. What knowledge would it take to locate and understand every conceivable influence for and against that outcome's occurring in the exact time and place that it did? It would be, for all practical purposes, infinite. We would need to account for everything from the birth of the assassin, the life history leading to a desire to kill the victim, to the decision of the political figure to go to the specific place where he or she was assassinated, to the availability of the weapon used, to the behavior of the Secret Service agents, and so forth. Surely it would be more reasonable to renounce such a goal, recognizing that complex chains of events in their particularity cannot be subsumed under generalizable laws or theory (more on this below). In examining a single case, or when engaged in a small-N study, how confident can we really be that the observed outcome(s) would be repeated over and over again?

The role of random factors is, of course, relevant to a wide variety of social contexts. Consider a simple example: undergraduate admission to Notre Dame or Harvard, or to any other selective institution with more "qualified" applicants than it can accept. If a highly selective college were to reconsider the same set of applicants for admission, using the same admissions personnel, and following the same guidelines, there would presumably be a correlation between the admissions at time 1 and time 2. But it would hardly be a correlation of 1.0 . Some who were admitted in the first review would be denied admission on the second trial; in other cases, the opposite would occur. There would be a huge population denied admission on both trials (a numerical necessity when the number of admissions is much smaller than the number of applicants) and there would be many who would be admitted in both cases because of exceptionally distinguishing characteristics. But the key point is that there would be a certain number of winners who fall out in time 2, and a certain number making it at time 2 who were denied admission at time 1. This is interesting to me for two reasons. First of all, we can see that there is a certain random dimension to the process that is being ignored when we use admission (like, victory) to settle questions about who are the "best applicants" (or "best teams"). Most interesting are the questions that inevitably arise here about the specific features of the structures involved. That is, there is no reason to assume that correlations among trial outcomes would converge on a specific value in all similar systems of this sort. Systems could vary because of the nature of the judges, the nature of the criteria, the distribution of qualities among the applicants, and the ratio between the number to be admitted and the number of applicants. Also, except for sampling errors, we would expect the characteristics of the aggregate of successful applicants to remain approximately constant from trial to trial, even if the same applicants were not always successful. So there is much that could be done here to explore the relative role of chance factors and the ways in which the features of particular structures influence this role.

There is a second direct application. When there is an elimination tournament such that a team is out when it loses one game (as is the case for the Super Bowl playoffs), then we have an irreversible dependent variable (winning or losing). This model has relevance insofar as there are social events that we want to understand which have the same quality, e.g., assassinations or wars or the emergence of social movements. From the point of view of social processes, there is every reason to think that the analogy holds. Just as a team's outcome is a function of skills as well as of chance events, so too social processes involve both complex and recurring structures and highly contingent events—and the parallel is particularly significant when outcomes are not directly reversible, in the sense that the absence or removal of whatever factors are responsible for a particular outcome will not (or not alone) enable the situation to return to its previous state (see the discussion of asymmetrical forms of causation in Lieberson 1985, chapter 4).

An intercepted pass or a fumble is the product of an interaction between the skills of one team's offense, the skills of the other team's defense, and chance events. Recurring social structures operate as do such skills to increase or decrease the probability of certain outcomes. But as in the sports event there are always other factors, best called chance factors, which may impact on social outcomes. The limits within which such factors may operate to affect decisively a particular outcome will vary, as will the importance of what we have characterized as "chance." But this variability, too, is mirrored in the differing roles played by chance vs skill in different card games and in different sports (depending on both the sport and the rules). The ramifications of these analogies for understanding social processes are enormous, as I will attempt to show below.

## Some Proposals

On the basis of our analysis thus far, the remainder of this essay sketches out some proposals for how to think about major social processes. We have started by reviewing some grounds for dissatisfaction with current methodologies used on small-N data sets as well as with the deterministic mode of thinking which underlies them. In a certain sense the description of small-N research also revealed difficulties for large-N research and conceptualization as well. It is just that the difficulties of the latter are compounded when causal inferences are drawn on the basis of small N's.

## A Multi-layered Probabilistic Approach to Social Processes

I start with three assumptions. First, there is no alternative to a probabilistic perspective—even if we believe social causes operate in a deterministic mode. The second assumption is that there will always be too many different causal influences for us to take all of them into account at one and the same time. My third assumption is that not all causal influences operate in the same way. By that I mean that it will frequently be useful to think of causal factors as hierarchically organized such that some set the precondition for others to operate and that the latter, in turn, regulate what further factors come into play. Ultimately, of course, we must reach factors capable of directly influencing the behavior of individuals. We might visualize "layers" of such causal principles (i.e., formulations of relatively stable relationships among causal variables) in the following way. Suppose we have a given principle, $P_2$, and there are certain conditions which determine its operation or non-operation. Then principles which determine these conditions (say, $P_4$, $P_7$, $P_{14}$) are at a higher level than $P_2$. On the other hand, if $P_2$ represents a relationship affecting conditions which in turn influence the operation of some other principle, say $P_1$, then $P_2$ is, in that respect, at a higher level than $P_1$. A fourth assumption is that genuine stability in social processes is rare; in modern societies at least, the conditions affecting the operation of social causes are forever subject to change along with their influence on lower level processes and events.

If these assumptions are correct, then in seeking to establish principles expressing causal influences we shall have to recognize the following:

1) These principles will typically be expressible only in probabilistic terms.

2) There are bounds on the operation of each principle, and that one task will be to specify the conditions under which particular principles operate. Indeed, a principle may even be totally irrelevant in a specific setting at a specific time. The principle is not true or false, per se, if it fails to hold in a given setting. It is false only if it *generally* fails to hold in a setting in which we have reason to believe it should operate (more on this below in discussing Central Place Theory) or, conversely, operates when its bounds predict it should not.

3) If a principle is to be useful, we must be able to state the bounds or conditions under which it will or will not operate. Universal social principles, intended to apply without limits of time and place, are extremely

unlikely (other than those which are biologically determined). Hence, no principle is acceptable unless it is possible to provide a rough approximation of the assumptions underlying its operation. It is impossible for such bounds to be fully tested and they may even prove to be impermanent, but one starts with such limitations and then adjusts as the future unfolds and the principle's application to the present and past is expanded (into, say, different places, settings, combinations of conditions, and so forth).

The various permutations and combinations of principles which can tug against each other in conflicting directions or pull together in tandem towards a given outcome makes for an interesting set of problems, namely figuring out the conditions under which the principles operate both individually and jointly. Under which conditions will one triumph over the other, or are the outcomes simply intermediate between the two?

Such questions, however fascinating, cannot be further explored here. Rather, I will briefly describe several cases that illustrate what I intend in calling for a multi-layered probabilistic approach to the conceptualization of social phenomena.

1) There is good reason for believing that race riots have both underlying causal conditions and immediate precipitants. The immediate precipitants are events which trigger a riot. Examples are: abusive interaction between police and members of a given racial or ethnic group; members of one group who assault someone in another group (particularly, if the latter is a small child, woman, elderly, or otherwise someone for whom there are rules of special care and concern, such as someone who is blind, highly regarded in the community, a member of the clergy, a person risking his or her life for the good of the community). The nature of the action is also relevant for the immediate precipitant, for example, a rape is more provocative than a purse snatching. At any rate, these events occur all the time, but do not always lead to riots. Indeed the vast majority of such incidents do not. When do riots occur? There are certain institutional conditions which appear to affect the likelihood that such painfully common incidents will increase or decrease the chances of a riot (see Lieberson and Silverman 1965). For the purposes at hand, we visualize two principles operating. One is a principle dealing with events which trigger riots, the other—at a higher level—deals with conditions which increase or decrease the chances that such incidents will trigger a riot. We can visualize two separate switches— if both are turned on, the chances of a riot are greater than if either is turned off. If both are turned off, then the chances are extremely low (observe, incidentally, that I am describing this in probabilistic form to begin with—not in flat terms

such as: there will or there will not be a riot). If the precipitating incident switch is on, but the underlying condition switch is not, then the chances of a riot are extremely modest; likewise, if the opposite conditions hold.

2) Another race and ethnic example from my favorite author. In a theory of race and ethnic relations that I developed many years ago, I drew a distinction between two types of ethnic/racial subordination: groups that migrated into such situations from another setting *vs* groups that were conquered or otherwise overrun and thereby became subordinated. I argued that different consequences flowed from these conditions. Of interest here is the claim that separatism and reduced levels of assimilation were far more likely to occur when groups were conquered than when groups entered subordination through migration.[10] I think this proposition holds—separatist movements are far more likely to occur among conquered peoples, and these people are likely to be slower to assimilate. But not all conquered groups in such subordinated settings develop separatist movements. So there are several interesting possibilities: 1) that is the best we can do; we cannot *a priori* separate the two subsets of conquered subordinate groups into those who do or do not push towards separatism; 2) in addition, for those who do push towards separatism, can we offer any principles as to when they will do so? Although the essay does not answer this, it is easy to see that a number of relevant lower levels of principles could be operative. To wit, given the conditions of inter-ethnic contact, what will generate the next stage and when? Likewise, we have the need for higher-level principles as well: to wit, under what conditions will groups migrate into subordination? Under what conditions, will one group conquer another?[11]

3) The third example, based on Central Place Theory as propounded by Lösch (1954), I have discussed in greater detail elsewhere (Lieberson 1992, 8). Basically this is a theory about the distribution of cities of different sizes and their distances from one another. It fails to go very far in helping us understand the spatial distribution of different-sized cities in the United States. The location of New York, for example, is off by about 1,000 to 1,500 miles! It is, by my tastes, an elegant and sensible theory and works beautifully in some specific instances. The key fact is that Lösch has carefully stated the conditions under which his model operates. When these conditions are met, the model seems to work very nicely. When these conditions are not met, then the principles he proposed carry very little weight (although I suspect that over time we would still see movements in the direction of his model). The point is that the Central Place Theory formulated by Lösch includes its own preconditions. We thus are given the higher-order principles that set the limitations on when this theory will work or not work. This is, then, an excellent illustration of the explicit use of multi-layered principles in social theorizing.

4) The first names given to children—a topic I am currently studying—provides another example of the potential fruitfulness of an approach to theorizing which emphasizes the multi-layering of principles. At this point I speculate that certain broad social conditions will impact on the likelihood that different names will be given to children. And, in turn, this makes it possible for lower-level principles of taste to enter into the setting. For example, the feminist movement (which itself is a product of the operation of other principles) influences the kind of names that are acceptable or unacceptable to parents when they name daughters. If certain names decline in popularity because of the feminist movement, then new names will replace them. What new names will they be? Certainly, the feminist movement will influence the choices, but there is now room for other principles of taste to affect the new choices that are made from within the boundaries or limitations set by the feminist movement. Hence the specific choices will reflect a lower-level principle that is activated by the condition caused by the higher-level principle.

5) Finally, let us think through the meaning of the phrases "an accident ready to happen" or "the time is ripe" or "not ripe." Each of these phrases suggests that there is some set of conditions which is necessary for something else to happen: say, a war, or a social movement, or a major institutional change, a political shift, etc. The implication is that some such set of conditions plays a critical role in either impeding or making possible a particular outcome. The outcome has a certain independence in that the "higher-level" conditions do not guarantee its occurrence or non-occurrence, but clearly a point can be reached where lower-level events are almost certain to trigger the outcome simply because the relevant "triggers" are extremely common, and/or because the probability is extremely high that the outcome will occur whenever potential triggering events are encountered. So we can develop a wide variety of conditional linkages between the sets of layers.

## Chain Models

If important aspects of social life are accurately describable in terms of multi-layered processes of the sorts discussed above, and if there is multicausality for just about any given dependent variable, then it seems reasonable to assume that many of our questions will require the unraveling of complex probabilistically linked chains of events rather than simply searching for "the" cause or causes of some phenomenon of interest. This means that many questions about society, which may initially sound quite sensible, in reality have very limited meaning. If we try to identify a causal connection between two broad and significant characteristics of society, say the influence of some "key" economic condition on certain major political characteristics, or perhaps the effect of a social characteristic (say linguistic or ethnic diversity) on some economic characteristic, there is a good chance that a variety of causal pathways exists between two such phenomena, each one conditioned by different predisposing circumstances. Moreover, such pathways themselves may lead to different outcomes when activated. Hence some pathways from, say, key economic conditions to a major political characteristic may move the latter in one direction, other paths in the opposite direction, and yet others serve to maintain the political status quo.

Now obviously we can choose as our independent variable one "so close" to the dependent variable that there will not be much of a chain. But few interesting questions could be answered by adopting that strategy. For example, suppose I want to account for the social characteristics of those who become physicians. Where do I start? If I start very late in the process, say, with the behavioral orientations of those who are already in their second year of medical school, I am unlikely to learn anything of much interest. If I start with people just finishing high school—or even earlier—I am going to find a very complex chain of events leading to differentials in who becomes a physician. There are going to be a lot of points in this complex set of paths to consider, though some, to be sure, will prove to be more important than others.

Many links in the chains may be very weak—weak in the sense that a movement from one point to a distant point may occur through a variety of connections each of which has a low probability of occurring. In other words, it may be that nothing can be said with certainty about what outcomes will occur—particularly if we do more than go back to the equivalent of second-year medical students. This is the sociological equivalent, as it were, of the Super Bowl process. There are a lot of "chance" influences at work. If we start by asking who wins the Super Bowl game, and take as given that the two actual finalists are playing each other, then we are dealing with something closer to the case of second-year medical students. If we go back to the beginning of the season, that is, if we do not take for granted who the finalists are, then we have a much more complex and less certain chain of events. It has to be more likely that the best of the two finalists will win the Super Bowl than that the best team at the start of the season will win it. Indeed, this is definitional, since the chances of the latter can be no higher than the chances of the former (and easily considerably lower).[12]

This view of long chains fits nicely with the position taken by Popper (1964, 115):

> although we may assume that any actual succession of phenomena proceeds according to the laws of nature, it is important to realize that

practically *no sequence of, say, three or more causally connected concrete events proceeds according to any single law of nature.* If the wind shakes a tree and Newton's apple falls to the ground, nobody will deny that these events can be described in terms of causal laws. But there is no single law, such as that of gravity, nor even a single definite set of laws, to describe the actual or concrete succession of causally connected events; apart from gravity, we should have to consider the laws explaining wind pressure; the jerking movements of the branch; the tension in the apple's stalk; the bruise suffered by the apple on impact; all of which is succeeded by chemical processes resulting from the bruise, etc. The idea that any concrete sequence or succession of events (apart from such examples as the movement of a pendulum or a solar system) can be described or explained by any one law, or by any one definite set of laws, is simply mistaken. There are neither laws of succession, nor laws of evolution.

Even if fairly high probabilities operate to link points along an actually observed chain of events, the likelihood of any chain of events repeating itself with any reasonable frequency is minuscule when the chain is long. Under those circumstances a theory concerning a complex set of events predicated on very specific outcomes is not possible. Or, if developed for a single case, has no applicability to the general problem. If we could replay history, which is no more possible than replaying a football season, then the probabilistic view is that we would encounter a rather different history each time. Presumably some outcomes would occur more often than other outcomes, but the distribution of outcomes also would depend on what specific types of events we had chosen to study. Consider the different expectations we might have if we started with, say, manners in 1800 in the United States and then looked at manners in 1900, or slavery in 1800 *vs* 1900, or a country's form of government over an extended period, and so forth. The outcome distribution also might depend on how precisely and concretely we identify an outcome of interest: a specific manner *vs* a theme in manners (see, for example, Elias 1978); a specific feature of government as opposed to the general degree of democracy, etc.

To be sure, one can visualize points on some chains where the path to a given outcome is not preordained, but still where a set of interrelated outcomes all point to a certain final event. Consider the comments in the preceding section about an accident waiting to happen, the time being ripe, etc. Hence, the system of chains is an interesting problem to consider in understanding social events. As is the case with some diseases, in which the body is so weakened that one or another, otherwise minor, illness is likely to prove fatal, so too, we can at least visualize social processes proceeding in such a way that at some point there is a very strong likelihood of a certain outcome, even if the particular path to that outcome has a probability far less than 1.0. Of course, this will always be an empirical question rather than an assumption we can ordinarily make. I think it unlikely, for example, that long complex chains of causal influence in the social realm will have such "absorbing" qualities, except perhaps close to the end of a relevant sequence (which of course is what Popper is in effect claiming in the quotation above).

To diverge briefly from the main point, this analysis also implies something about second guessing. The problem is not that the second-guesser uses 20/20 hindsight to reach the correct solution—rather it is that the original plan may well have been the absolutely best one even though it failed. This is because such efforts are often based on complex probabilistic chains of reasoning. Hence the best shot is simply the one which seeks to activate the most reliable of what could be a set of low-probability chains of causal influence.[13] Incidentally, this illustrates a difference between the probabilistic and deterministic approaches to policy issues. The latter says there is a correct way, which when followed, will produce the desired results. If the desired result does not ensue, then some error was made. A probabilistic view may acknowledge that one action is more likely to produce the desired outcome than any others, while yet recognizing that the outcome is not inevitable; it simply has a higher probability. The failure of a policy to produce a desired outcome thus does not imply that it would be inappropriate to seek to implement it again in the future, so long as it was based on a well-confirmed theory and adequate data. However, such an approach may be politically very difficult unless both decision makers and those likely to be most affected by the decision are prepared to take a probabilistic view of failure.

## Explaining Specific Events

As a general rule, there are limitations to how well we can explain (or understand the cause[s] of) a specific social event. These limitations exist even if the research is based on very large N's. There is the difficulty of applying existing theory (regardless of whether it is derived from small- or large-N research) to help us understand the cause of a specific event. In effect, the explanation of a specific event is a special small-N problem even if the knowledge applied to it draws on large-N research. When we deal with a specific event and its explanation, the total number of all such events is immaterial since we are only concerned with one of them. In small-N research we are always dealing with a small number of events; but if we are trying to explain one specific instance of a type of event, then there are special difficulties regardless of whether there are many such events or only a few—we are still trying to account for only one specific event.

At best, we can come up with a probabilistic statement of the causes of a given event—a probabilistic statement which normally will involve two parameters, and very likely more than two. It will almost inevitably involve a stochastic variable as well as at least one known causal variable because, typically, research will show that the substantive variable fails to fully account for the dependent variable. More often we will have good reason to believe that additional substantive variables are operating (the multivariate case) if for no other reason than our usual inability to experiment or otherwise *situationally* measure or control other influences.

Visualize the simplest case in which there is one independent variable and a stochastic variable ($X_1$ and $U$, respectively), and where both the independent variable and the dependent variable are dichotomous (simple yes/no variables). If there is a relationship such that "yes" on the independent variable is associated with the occurrence of the dependent variable, can we say that the presence of $X_1$ is the cause of Y? No, we cannot, even though this is a common form of thought. Suppose we know that there is a causal linkage between smoking and lung cancer. If someone we know dies of lung cancer and was a heavy smoker, it is hard for us to avoid thinking we know the cause of this particular individual's lung cancer. But if there are non-smokers who die of lung cancer, then we really cannot explain the cause of the specific event—the death from lung cancer of our smoker-friend. Depending on the proportion of all smokers who die of lung cancer in comparison with the proportion of all non-smokers similarly afflicted, I can make some statements about the relative probability of the death's being due to smoking—but no more unless there is additional information. In other words, the excess in the rate of lung cancer deaths for smokers compared to non-smokers presumably tells us about the increased likelihood of fatal cases of lung cancer due to smoking. But it does not tell us about the cause of any particular smoker's lung cancer. In counterfactual conditional terms, unless we know a lot more about other causes of lung cancer death, we can safely say that some percentage of smokers' deaths would not have occurred if they had not smoked, but we cannot say anything about any specific smoker who dies of lung cancer other than to list the probabilities of different causes. Although I can do this very nicely for a population, I cannot really explain any individual case.[14] Notice how different this is from the very clean and clear statement about the probability of lung cancer for those who smoke *vs* the probability for non-smokers.

Suppose persons residing in a given area were subjected to exceptionally large doses of radioactivity (say due to chemical wastes or nuclear testing, etc.). Suppose those now dying from a cancer known to be caused by radioactivity decide to sue those responsible for the radioactivity. Again all we can do is say that a certain percentage of these victims (possibly an extremely large percentage) are dying for this reason. But unless the probability of getting this cancer is otherwise zero, we cannot say why any given person is dying. We can only make a probability statement about the likely cause. The key is not whether all people subjected to radioactivity get the cancer, but rather whether there are people not subjected to this level of radioactivity who get that type of cancer. If only radioactivity can cause cancer, we have to that degree an explanation.[15]

I do not believe these examples concerning illness and disease are at all difficult to appreciate. But I have a hunch that we are likely to initially draw a conclusion unsupported by present knowledge whenever we hear of a smoker who died of lung cancer. The temptation to draw an unwarranted conclusion is even more difficult to resist when we have no way to evaluate the likelihood of alternative outcomes, or information is unavailable concerning the likelihood of the same outcome under different circumstances.[16] But the examples so far are less obtuse than examples involving large-scale social events: wars, macro-level societal changes, institutional developments, forms of social organization, political events, and the like. In the typical research problem in nonexperimental social science one will virtually always have reason to believe that more than one independent variable is operating. Here, if there is also a stochastic factor operating, only a set of probability statements can be made about the cause(s) of a specific event. Thus only in an incredibly unlikely set of circumstances would it be possible to talk about *the unique, necessary and sufficient cause* of a specific event. Suppose we complicate matters by assuming that there are two known causes ($X_1$, $X_2$), both being continuous variables, that the dependent variable, $Y$, is also a continuous variable, and that the relationship between each variable is linear. Then for any individual case, knowing the values of $X_1$ and $X_2$ and $Y$, we could only estimate the likely causal influence of each relevant variable. Here it is even easier to fall into the mistake of thinking that we can find the definite cause of each specific outcome. Unlike the medical examples used above, it is harder to visualize these sorts of cases as being probabilistic in nature and hence to recognize that it may be extremely difficult, and in many cases impossible, to say what really caused a specific event to occur.

## In Conclusion

Returning to the main point, in brief we should be suspicious of pied pipers who purport to tell us why such and such *had* to happen. This applies "in spades" to customary small-N analyses which seek to simulate the sort of results which might be attainable with large-scale data sets.[17] But it also applies to large-scale data analysis as well. The conjectures I have made here about how social processes operate may or may not prove to be wholly adequate; they are certainly open to correction and revision. However, I believe they provide a viewpoint from which it may be possible to pursue our questions in more fruitful and realistic ways, given both the obvious complexity of social phenomena and the limitations of our methods. Above all we should resist seeking a precision that is inappropriate for either our subject matter or the nonexperimental tools with which we must work.

## NOTES

I am indebted to the following for helpful discussions of some of the problems raised in this paper: William Alonso, John Campbell, Zvi Griliches, S. M. Miller, and Debra Minkoff. Angelina C. Wong assisted in the analysis of Super Bowl data.

1. For some problems, 100 cases is more than sufficient; for others a data set ten times that number will be inadequate.

2. Ragin and Becker (1992) provide a general discussion of the various ways in which a case can be defined in social research, and the consequences of doing so.

3. This is oversimplified and hardly does justice to the theory, but it serves to illustrate the way in which variance is forced.

4. Sometimes practitioners of this approach pool two variables together such that a certain level with one variable and a specific level with the other are viewed as combining to cause an outcome, whereas neither alone is a cause. But this is not really two causes, in the sense that either of two different conditions will lead to the same outcome. Rather it is one cause involving two separate conditions.

5. The Boolean solution proposed by Ragin (1987), although appealing in its ability to lay out all of the permutations and combinations of variables in relatively small-N situations, is still ultimately a deterministic approach.

6. The comparison between deterministic and stochastic models used in statistical research is relevant here. See, for example, Hanushek and Jackson (1977, 11–13).

7. As a general rule, it was determined if the eventual Super Bowl winner got into the initial playoff rounds with only a one game edge over the next best team.

8. To be sure there are unmeasured intangibles such as the kicker choking up or, at the other extreme, having a clutch player who does better with such incentives.

9. Keep in mind that the deterministic model is also an unproven assumption.

10. In fairness, the original paper (Lieberson 1961) did not describe these principles in probabilistic form, but I would now restate them in this fashion.

11. Some of these higher-principle questions are at least partially answered in Lieberson (1961).

12. The Super Bowl analysis of chance factors, reported earlier, is therefore relatively conservative since it did not consider the far more complex model.

13. If they are all more or less similarly low in probability, then substantial differences in the probable harmful effects of failure may be a more important basis for deciding than the small differences in the chances of success (a cost-benefit issue).

14. All of this assumes that the usual factors of age, sex, frequency of smoking, occupation, and the like are taken into account.

15. Incidentally, if not all people who smoke die of lung cancer and if not all subjected to radioactivity get leukemia, then there is the added question of why some do and some don't. This is inevitably intriguing, but it is not always answerable and may in fact reflect factors best allocated to randomness as well as to more definite influences (again see Salmon 1971, 56–57).

16. To use a sports analogy again, if the manager orders a given action which backfires, say a runner at a key point in a baseball game is caught attempting to steal a base, or if a football team unsuccessfully goes for a first down rather than punting, the usual assumption is that an error was made. Yet in such cases we do not have anything like the equivalent of data on lung cancer among non-smokers. Hence, we are in no position to say that the failure shows it was a bad call. We have no real grip on relevant counterfactuals, and the relevant sets of probabilities are clearly unknown.

17. I exclude here research which really can build on single instances through the use of exceptionally strong quasi-experimental conditions, such as the work by Snow on cholera (described in Freedman 1991).

# CONTRIBUTORS

**NANCY CARTWRIGHT** is Professor in the Department of Philosophy, Logic and Scientific Method at the London School of Economics, where she is also Co-Director of the Centre for the Philosophy of the Natural and Social Sciences. She has written *How the Laws of Physics Lie* (Oxford University Press, 1983), *Nature's Capacities and Their Measurement* (Oxford University Press, 1989), and (with J. Cat, K. Fleck, and T. Uebel) *Otto Neurath: Philosophy between Science and Politics* (Cambridge University Press, forthcoming).

**CLIFFORD CLOGG** was, until his recent untimely death, Professor of Sociology and Statistics at Pennsylvania State University. His research dealt with statistical methods for categorical data and with methodology for social research more generally. He was the editor of *Sociological Methodology* (1989–1991) and the Coordinating and Applications Editor of the *Journal of the American Statistical Association* (1989–1991). He was also a Fellow of the American Statistical Association and of the AAAS.

**DAVID A. FREEDMAN**, Professor of Statistics at the University of California, Berkeley, is the author of over 100 papers and technical reports as well as five books. He has made numerous contributions to statistical theory and pedagogy, in addition to his well-known work in the area of applied statistics. In the latter context, he has written a number of studies critical of ways in which statistical methodologies have been employed in the social sciences.

**CLARK GLYMOUR**, Alumni Professor of Philosophy at Carnegie Mellon University and Adjunct Professor of Intelligent Systems at the University of Pittsburgh, has contributed to many areas in the philosophy of science. In addition to his early work on scientific methodology, *Theory and Evidence* (Princeton University Press, 1980), he has authored, co-authored, or edited eight other books and written dozens of scholarly and popular essays. Most recently he has been senior scholar in the working group at Carnegie Mellon seeking to exploit artificial intelligence methods to create more powerful techniques for identifying causal models through statistical research. The results of this research program have appeared as *Discovering Causal Structure: Artificial Intelligence, Philosophy of Science and Statistical Modeling* (Academic Press, 1987) and *Causation, Prediction and Search* (Springer-Verlag, 1993).

**ADAMANTIOUS HARITOU** is Professor in the Department of Applied Information at the University of Macedonia (Greece). His research interests include generalized linear models and statistical computing.

**LARRY V. HEDGES** is the Stella M. Rowley Professor of Education and the Social Sciences at the University of Chicago. His research involves the development and application of statistical models in the social sciences. A particular focus of that work concerns the analysis of replicated research studies (so-called meta-analysis) and what implications bodies of replicated research may have for the accumulation of knowledge in the empirical sciences.

**PAUL HUMPHREYS** is Professor of Philosophy at the University of Virginia. He is the author of a number of papers on probabilistic causality and of *The Chances of Explanation*. (Princeton University Press, 1989). His current interests include reductionism and the role that computational methods play in the natural sciences.

**STANLEY LIEBERSON** is a sociologist who has written widely on a variety of subjects, including a book addressing epistemological issues in social research, *Making It Count: The Improvement of Social Research and Theory* (University of California Press, 1985) as well as many papers addressing methodological issues. He is Abbott Lawrence Lowell Professor of Sociology at Harvard.

**VAUGHN R. McKIM,** a philosopher of science at the University of Notre Dame, currently directs the University's Reilly Center for Science, Technology, and Values. His research interests include both methodological and conceptual issues in the human sciences.

**MARY S. MORGAN** is author of *The History of Econometric Ideas* (Cambridge University Press, 1990) and co-editor of *The Probabilistic Revolution: Ideas in the Sciences* (MIT Press, 1987). Her current research project is an investigation of the history and methodology of modeling in economics. She is both Reader in the History of Economics at the London School of Economics and Professor of the History and Philosophy of Economics at the University of Amsterdam.

**RICHARD SCHEINES** received his doctorate in the History and Philosophy of Science from the University of Pittsburgh. Scheines is now a Senior Research

Scientist in the Department of Philosophy at Carnegie Mellon. His research involves the foundations of statistical causal inference and intelligent computerized proof construction in formal logic.

**PETER SPIRTES** is Associate Professor of Philosophy at Carnegie Mellon University. He is one of the co-authors of *Causation, Prediction, and Search* and the *TETRAD II* program.

**STEPHEN P. TURNER** is Distinguished Research Professor of Philosophy at the University of South Florida. His books include *The Search for a Methodology of Social Science: Durkheim, Weber, and the Nineteenth Century Problem of Cause, Probability, and Action* (Boston Studies in the Philosophy of Science, 1986). He has written extensively on the history of social research and the philosophy of social science.

**JAMES WOODWARD** is Professor of Philosophy and Executive Officer for the Humanities at the California Institute of Technology. He is in the early stages of a book on causality and explanation.

# BIBLIOGRAPHY

Aachen, C. H. 1982. *Interpreting and Using Regression*. Beverly Hills, Calif.: Sage.

Agresti, A. 1990. *Categorical Data Analysis*. New York: Wiley.

Aldrich, J. 1994. Correlations Genuine and Spurious in Pearson and Yule. Department of Economics, University of Southhampton, UK. Unpublished.

Andvig, J. Forthcoming. "From Macrodynamics to Macroeconomic Planning: A Basic Shift in Ragnar Frisch's Thinking." *European Economic Review*.

Arminger, G. 1994. "Mean Structures." In *Handbook of Statistical Modeling in the Social Sciences*, ed. G. Arminger, C. C. Clogg, and M. Sobel. New York: Plenum.

Arntzenius, F. 1990. "Physics and Common Causes." *Synthese* 82: 77–96.

———. 1992. The Common Cause Principle. Preprint, Department of Philosophy, University of Southern California.

———. 1993. "The Common Cause Principle." In *PSA 1992*, ed. D. Hull, M. Forbes, and K. Okruhlik. Philosophy of Science Association. 227–237.

Bartels, L. M. 1991. "Instrumental and 'Quasi-instrumental' Variables." *American Journal of Political Science* 35: 777–800.

Bartels, L. M., and H. E. Brady. 1993. "The State of Quantitative Political Methodology." In *Political Science: The State of the Discipline II*, ed. A. W. Finifter. Washington, D.C.: American Political Science Association.

Bartholomew, D. 1987. *Latent Variable Models and Factor Analysis*. Oxford: Oxford University Press.

Becker, B. J. 1992. "Using Results from Replicated Studies to Estimate Linear Models." *Journal of Educational Statistics* 17: 341–362.

———. Forthcoming. "The Generalizability of Empirical Research Results." In *From Psychometrics to Giftedness: Papers in Honor of Julian C. Stanley*, ed. C. Benbow and D. Lubinski. Baltimore: Johns Hopkins University Press.

Becker, B. J., and C. M. Schram. 1994. "Examining Explanatory Models through Research Synthesis." In *The Handbook of Research Synthesis*, ed. H. Cooper and L. V. Hedges. New York: The Russell Sage Foundation. 357–381.

Begg, C. B., and J. A. Berlin. 1988. "Publication Bias: A Problem in Interpreting Medical Data (with discussion)." *Journal of the Royal Statistical Society,* Series A, 151: 419–463.

Beinlich, I., H. Suermondt, R. Chavez, and G. Cooper. 1989. "The ALARM Monitoring System: A Case Study with Two Probabilistic Inference Techniques for Belief Networks." In *Proceedings of the Second European Conference on Artificial Intelligence in Medicine*. London. 247–256.

Bentler, P. 1985. *Theory and Implementation of EQS: A Structural Equations Program.* Los Angeles: BMDP Statistical Software Inc.

Berk, R. A. 1988. "Causal Inference for Sociological Data." In *Handbook of Sociology*, ed. N. J. Smelser. Newbury Park, Calif.: Sage.

Berkson, J. 1950. "Are There Two Regressions?" *Journal of the American Statistical Society* 45: 164–180.

Bertillon, J. 1882. "Mariage." *Dictionnaire Encyclopédique des Sciences Medicales*, 2nd. Series L- P, vol. 27, pp. 7–50. Paris: Asselin.

Beveridge, W. H. 1912. *Unemployment, A Problem of Industry.* London: Longmans, Green.

Blalock, H. M. 1964. *Causal Inferences in Nonexperimental Research.* Chapel Hill, N.C.: University of North Carolina Press.

———. 1979. *Social Statistics.* Rev. 2nd. ed. New York: McGraw-Hill.

———. 1982. *Conceptualization and Measurement in the Social Sciences.* Beverly Hills, Calif.: Sage.

Blau, P. M., and O. Duncan. 1967. *The American Occupational Structure.* New York: Wiley.

Bollen, K. 1989. *Structural Equations with Latent Variables.* New York: Wiley.

Bollen, K., and J. Scott Long. Eds. 1993. *Testing Structural Equation Models.* Newbury Park, Calif.: Sage.

Bowley, A. L. 1901. *Elements of Statistics.* London: P. S. King and Son.

Brunner, E., G. Hughes, and M. Patten. 1927. *American Agricultural Villages.* New York: Doran.

Bryk, A. S., and S. W. Raudenbush. 1992. *Hierarchical Linear Models.* Newbury Park, Calif.: Sage.

Cairns, J. 1978. *Cancer: Science and Society.* San Francisco: W. H. Freeman.

Campbell, D.T. 1957. "Factors Relevant to the Validity of Experiments in Social Settings." *Psychology Bulletin* 54: 297–312.

Campbell, D. T., and J. Stanley. 1966. *Experimental and Quasi-Experimental Designs for Research.* Skokie, Ill.: Rand McNally.

Campbell, D. T., and D. W. Fiske. 1959. "Convergent and Discriminant Validation by the Multitrait-Multimethod Matrix." *Psychological Bulletin* 56: 81–105.

Caramazza, A. 1986. "On Drawing Inferences about the Structure of Normal Cognitive Processes from Patterns of Impaired Performance: The Case for Single Patient Studies." *Brain and Cognition* 5: 41–66.

Carmelli, D., and W. F. Page. 1996. "24-year Mortality in World War II U.S. Male Veteran Twins Discordant for Cigarette Smoking." *International Journal of Epidemiology* 25: 554–59.

Cartwright, N. 1983. *How the Laws of Physics Lie.* New York: Oxford University Press.

———. 1989. *Nature's Capacities and Their Measurement.* Oxford: Clarendon.

———. 1993. "Aristotelian Natures and the Modern Experimental Method." In *Inference, Explanation and Other Philosophical Frustrations*, ed. J. Earman. Berkeley: University of California Press.

———. 1994. "Probabilities and Experiments." LSE Centre for the Philosophy of the Natural and Social Science Working Papers 943. London: CPNSS.

———. 1995. "Causal Structures in Econometric Model." In *The Reliability of Economic Models*, ed. D. Little. Dordrecht: Kluwer.

Clogg, C. C. 1992. "The Impact of Sociological Methodology on Statistical Methodology." *Statistical Science* 7: 183–207.

Clogg, C. C., and G. Arminger. 1993. "On Strategy for Methodological Analysis." In *Sociological Methodology 1993*, ed. P. V. Marsden. Oxford: Basil Blackwell.

Clogg, C. C., and A. Dajani. 1991. "Sources of Uncertainty in Modeling Social Statistics: An Inventory." *Journal of Official Statistics* 7: 7–24.

Clogg, C.C., E. Petkova, and A. Haritou. 1995. "Statistical Methods for Comparing Regression Coefficients between Models." *American Journal of Sociology.*

Clogg, C. C., E. Petkova, and E. S. Shihadeh. 1992. "Statistical Methods for Analyzing Collapsibility in Regression Models." *Journal of Educational Statistics* 17: 51–74.

Conant, J. B. 1951. *On Understanding Science.* New York: New American Library.

Cook, T. D. 1993. "A Quasi-sampling Theory of the Generalization of Causal Relationships." *New Directions in Program Evaluation* 57: 39–82.

Cook, T. D., and D. T. Campbell. 1979. *Quasi–experimentation.* New York: Houghton-Mifflin.

Cooper, G., and E. Herskovitz. 1992. *A Bayesian Method for the Induction of Probabilistic Networks from Data.* Machine Learning.

Cooper, H. M. 1984. *The Integrative Research Review.* Beverly Hills, Calif.: Sage.

Cooper, H. M., and L. V. Hedges. Eds. 1994. *The Handbook of Research Synthesis.* New York: The Russell Sage Foundation.

Cornfield, J., W. Haenszel, E. C. Hammond, A. M. Lilienfeld, M. B. Shimkin, and E. L. Wynder. 1959. "Smoking and Lung Cancer: Recent Evidence and a Discussion of Some Questions." *Journal of the National Cancer Institute* 22: 173–203.

Cox, D. R., and N. Wermuth. 1993. "Linear Dependencies Represented by Chain Graphs (with discussion)." *Statistical Science* 8: 204–283.

Cronbach, L. J. 1982. *Designing Evaluations of Educational and Social Programs.* San Francisco: Jossey-Bass.

Cronbach, L. J., G. C. Gleser, H. Nanda, and N. Rajartram. 1972. *The Dependability of Behavioral Measurements: The Theory of Generalizability of Scores and Profiles.* New York: Wiley.

Daggett, R., and D. Freedman. 1985. "Econometrics and the Law: A Case Study in the Proof of Antitrust Damages." In *Proceedings of the Berkeley Conference in Honor of Jerzy Neyman and Jack Kiefer*, vol. 1, ed. L. LeCam and R. Olshen. Belmont, Calif.: Wadsworth. 126–175.

Darroch, J. N., S. L. Lauritzen, and T. P. Speed. 1980. "Markov Fields and Log-linear Interaction Models for Contingency Tables." *Annals of Statistics* 8: 522–539.

Dawid, A. 1979. "Conditional Independence in Statistical Theory (with discussion)." *Journal of the Royal Statistical Society*, Series B, 41: 1–31.

Desrosières, A. 1993. *La Politique des Grands Nombres.* Paris: Éditions la Découverte.

Duncan, O. D. 1966. "Path Analysis: Sociological Examples." *American Journal of Sociology* 72: 1–16.

———. 1975. *Introduction to Structural Equation Models*. New York: Academic Press.

Duncan, O., D. Featherman, and B. Duncan. 1972. *Socioeconomic Background and Achievement*. New York: Seminar Press.

Durkheim, Émile. 1951. *Suicide*. Trans. J. A. Spaulding and G. Simpson. New York: Free Press. [Original 1897.]

Ehrenberg, A. S. C., and J. A. Bound. 1993. "Predictability and Prediction." *Journal of the Royal Statistical Society,* Series A, 156 (Part 2): 167–206.

Eitelberg, M. J. 1988. *Manpower for Military Occupations*. Office of the Assistant Secretary of Defense (Force Management and Personnel).

Elias, N. 1978. *The History of Manners: The Civilizing Process*. Vol. 1. New York: Pantheon.

Ellis, J., and A. van den Wollenberg. 1993. "Local Homogeneity in Latent Trait Models." *Psychometrika*.

Engle, R. F., D. F. Hendry, and J. F. Richard. 1983. "Exogeneity." *Econometrica* 51: 277–304.

Eyler, J. M. 1979. *Victorian Social Medicine: The Ideas and Methods of William Farr*. Baltimore: Johns Hopkins University Press.

Farr, W. 1880. *Forty-First Annual Report of the Registrar-General of Births, Deaths and Marriages in England*. London: HMSO.

———. 1885. *Vital Statistics: A Memorial Volume of Selections from the Reports and Writings of William Farr*. London: Offices of the Sanitary Institute.

Ferguson, T. S. 1967. *Mathematical Statistics: A Decision Theoretic Approach*. New York: Academic Press.

Fisher, R. 1951. *The Design of Experiments*. Edinburgh: Oliver and Boyd.

———. 1959. *Smoking. The Cancer Controversy*. Edinburgh: Oliver and Boyd.

Floderus, B., R. Cederlof, and L. Friberg. 1988. "Smoking and Mortality: A 21-year Follow-up Based on the Swedish Twin Registry." *International Journal of Epidemiology* 17: 332–340.

Freedman, D. 1983a. "A Note on Screening Regression Equations." *The American Statistician* 37: 152–155.

———. 1983b. "Structural-equation Models: A Case Study." Report, Department of Statistics, University of California, Berkeley.

———. 1987. "As Others See Us: A Case Study in Path Analysis (with discussion)." *Journal of Educational Statistics* 12: 101–128.

———. 1991. "Statistical Models and Shoe Leather." In *Sociological Methodology 1991*, ed. P. Marsden. Washington, D.C.: American Sociological Association.

Freedman, D., and D. Lane. 1981. *Mathematical Methods in Statistics*. New York: W. W. Norton.

Frisch, R. 1938. "Statistical versus Theoretical Relations in Economic Macrodynamics." Mimeo dated July 1938 in *Autonomy of Economic Relations*, ed. R. Frisch, collection of mimeo articles issued by the University of Oslo, 1948.

Frydenberg, M. 1990. "The Chain Graph Markov Property." *Scandinavian Journal of Statistics* 17: 333–353.

Fuller, W. A. 1987. *Measurement Error Models*. New York: Wiley.

Gauss, C. F. 1809. *Theoria Motus Corporum Coelestium*. Hamburg: Perthes et Besser. Reprint, New York: Dover, 1963.

Geiger, D. 1990. Graphoids: A Qualitative Framework for Probabilistic Inference. Doctoral dissertation, Department of Computer Science, University of California, Los Angeles.

Glymour, C. 1983. "Social Science and Social Physics." *Behavioral Science* 28 (no. 2 April): 126–134.

Glymour, C., T. Richardson, and P. Spirtes. 1996. "Note on Cyclic Graphs of Discrete Variables." Techincal Report, Carnegie Mellon University.

Glymour, C., R. Scheines, P. Spirtes, and K. Kelly. 1987. *Discovering Causal Structure. Artificial Intelligence, Philosophy of Science, and Statistical Modeling*. Orlando, Fla.: Academic Press.

Glymour, C., and P. Spirtes. 1993. "Comment." *Statistical Science*.

Glynn, R. J., N. M. Laird, and D. B. Rubin. 1986. "Selection Modeling versus Mixture Modeling with Nonignorable Nonresponse." In *Drawing Inferences from Self Selected Samples*, ed. H. Wainer. New York: Springer. 115–142

Godfrey, L. G. 1988. *Misspecification Tests in Econometrics*. Cambridge: Cambridge University Press.

Goldberger, A. S. 1973. "Structural Equation Models: An Overview." In *Structual Equation Models in the Social Sciences*, ed. A. S. Goldberger and O. D. Duncan. New York: Seminar Press.

Goldberger, A. S., and O. D. Duncan. Eds. 1973. *Structual Equation Models in the Social Sciences*. New York: Seminar Press.

Goodman, L. A. 1973. "Causal Analysis of Data from Panel Studies and Other Kinds of Surveys." *American Journal of Sociology* 78: 1135–1191.

Granger, C. 1969. "Investigating Causal Relations by Econometric Models and Cross-Spectral Methods." In *Rational Expectations and Econometric Practice*, ed. R. Lucas and T. Sargent. London: Allen and Unwin.

———. 1988. "Causality Testing in a Decision Science." In *Causation, Chance and Credence*, vol. I, ed. B. Skyrms and W. Harper. Dordrecht: Kluwer.

———. 1990. "Causal Inference." In *Econometrics*, ed. J. Eatwell, M. Milgate, and P. Newman ( from *The New Palgrave: A Dictionary of Economics*). New York: W. Norton.

Gujarati, D. N. 1988. *Basic Econometrics*. 2nd ed. New York: McGraw-Hill.

Haavelmo, T. 1943. "The Statistical Implications of a System of Simultaneous Equations." *Econometrica* 11: 1–12.

———. 1944. "The Probability Approach to Econometrics." *Econometrica* 12 (supplement): 1–117.

Hacking, I. 1965. *The Logic of Statistical Inference*. Cambridge: Cambridge University Press.

———. 1990. *The Taming of Chance*. Cambridge: Cambridge University Press.

Hakama, M., M. Lehtinen, P. Knekt, A. Aromaa, P. Leinikki, A. Miettinen, J. Paavonen, R. Peto, and L. Teppo. 1993. "Serum Antibodies and Subsequent Cervical Neoplasms: A Prospective Study with 12 Years of Follow-up." *American Journal of Epidemiology* 137: 166–170.

Hanushek, E. A., and J. E. Jackson. 1977. *Statistical Methods for Social Scientists*. New York: Academic Press.

Hardle, W. 1990. *Applied Nonparametric Regression Analysis*. Cambridge: Cambridge University Press.

Hausman, D. 1984. "Causal Priority." *Nous* 18: 261–279.

Hausman, J. A. 1978. "Specification Tests in Econometrics." *Econometrica* 46: 1251–1271.

Heckman, J. J. 1979. "Sample Selection Bias as a Specification Error." *Econometrica* 47: 153–161.

Heckman, J., and V. J. Hotz. 1989. "Choosing among Alternative Nonexperimental Methods for Evaluating the Impact of Social Programs: The Case of Manpower Training Programs." *Journal of the American Statistical Association* 84: 862–874.

Heckman, J., and R. Robb. 1985. "Alternative Methods for Evaluating the Impact of Interventions: An Overview." *Journal of Econometrics* 30: 239–267.

———. 1986. "Alternative Methods for Solving the Problem of Selection Bias in Evaluating the Impact of Treatments on Outcomes." In *Drawing Inferences from Self-Selected Samples*, ed. H. Wainer. New York: Springer. 63–107.

Heckerman, D., D. Geiger, and D. Chickering. 1994. "Learning Bayesian Networks: The Combination of Knowledge and Statistical Data." Technical Report MSR-TR-94-09, Microsoft Research, Advanced Technology Division, Redmond, Wash.

Hedges, L. V. 1984. "Estimation of Effect Size under Nonrandom Sampling: The Effects of Censoring Studies Yielding Statistically Insignificant Mean Differences." *Journal of Educational Statistics* 9: 61–85.

———. 1992. "Modeling Publication Selection Effects in Random Effects Models in Meta-analysis." *Statistical Science* 7: 246–255.

Hedges, L. V., and I. Olkin. 1985. *Statistical Methods for Meta-analysis*. New York: Academic Press.

Hedges, L. V., and T. Waddington. 1993. "From Evidence to Knowledge to Policy: Research Synthesis for Policy Formation." *Review of Educational Research* 63: 345–352.

Heise, D. R. 1975. *Causal Analysis*. New York: Wiley.

Hendry, D. F., and M. S. Morgan. 1995. *The Foundations of Econometric Analysis*. Cambridge: Cambridge University Press.

Hofferth, S. L., and K. A. Moore. 1979. "Early Childbearing and Later Economic Well-being." *American Sociological Review* 44: 784–815.

Holland, P. 1986. "Statistics and Causal Inference (with discussion)." *Journal of the American Statistical Association* 81: 945–960.

———. 1988. "Causal Inference, Path Analysis, and Recursive Structural Equation Models." In *Sociological Methodology 1988*, ed. C. Clogg. Washington, D.C.: American Sociological Association. 449–484.

Hooker, R. H. 1898. "Is the Birth Rate Still Falling?" *Transactions of the Manchester Statistical Society*, 101–126.

———. 1901. "On the Correlation of the Marriage-rate with Trade." *Journal of the Royal Statistical Society* 64: 485–492.

———. 1905. "On the Correlation of Successive Observations Illustrated by Corn Prices." *Journal of the Royal Statistical Society* 68: 696–703.

———. 1907. "Correlation of Weather and Crops." *Journal of the Royal Statistical Society* 70: 1–41.

Hoover, K. 1988. *The New Classical Economics*. Oxford: Basil Blackwell.

Humphreys, P. W. 1985. "Quantitative Probabilistic Causality and Structural Scientific Realism." In *PSA 1984* , vol. 2, ed. P. D. Asquith and P. Kitcher. East Lansing, Mich.: Philosophy of Science Association.

———. 1989. *The Chances of Explanation.* Princeton, N.J.: Princeton University Press.

Humphreys, P., and D. Freedman. 1996. "The Grand Leap." *British Journal for the Philosophy of Science* 47: 113–123.

Hunter, J. E., and F. L. Schmidt. 1990. *Methods of Meta-analysis: Correcting Error and Bias in Research Findings.* Newbury Park, Calif.: Sage.

International Agency for Research on Cancer. 1986. *Tobacco Smoking. Monographs on the Evaluation of the Carcinogenic Risk of Chemicals to Humans.* Vol. 38. Lyon, France: IARC.

Joreskog, K. G. 1971. "Simultaneous Factor Analysis in Several Populations." *Psychometrika* 36: 409–426.

Joreskog, K., and D. Sorbom. 1984. *LISREL VI User's Guide.* Mooresville, Ind.: Scientific Software.

Kadane, J., and T. Seidenfeld. 1990. "Randomization in a Bayesian Perspective." *Journal of Statistical Planning and Inference* (North–Holland) 25: 329–345.

———. 1992. "Statistical Issues in the Analysis of Data Gathered in the New Designs." In *Toward a More Ethical Clinical Trial*, ed. J. Kadane. New York: John Wiley & Sons.

Kaprio, J., and M. Koskenvuo. 1989. "Twins, Smoking and Mortality: A 12-year Prospective Study of Smoking-discordant Twin Pairs." *Social Science and Medicine* 29: 1083–89.

Kendall, M. 1948. *The Advanced Theory of Statistics.* London: Charles Griffin.

Keynes, J. M. 1939. "Professor Tinbergen's Method." *The Economic Journal* 49: 558–570.

———. 1940. "Comment on Tinbergen's Response." *The Economic Journal* 50: 154–156.

Kiiveri, H., and T. Speed. 1982. "Structural Analysis of Multivariate Data: A Review." In *Sociological Methodology 1982*, ed. S. Leinhardt. San Francisco: Jossey Bass.

Kish, L. 1965. *Survey Sampling.* New York: Wiley.

Klein, J. 1994. "The Method of Diagrams and the Black Arts of Inductive Economics." In *Measurement, Quantification and Economic Analysis*, ed. I. Rima. London: Routledge.

———. Forthcoming. *Statistical Visions of Time: A History of Time Series Analysis From 1662–1938.* Cambridge: Cambridge University Press.

Koster, J. 1996. "Marcov Properties of Recursive Causal Models." *Annals of Statistics.*

Krüger, L., G. Gigerenzer, and M. S. Morgan. Eds. 1987. *The Probabilistic Revolution*, Vol.II: *Ideas in the Sciences.* Cambridge, Mass.: MIT Press.

Kruskal, W. 1989. "Hooker and Yule on Relative Importance: A Statistical Detective Story." *International Statistical Review* 57: 83–88.

Kruskal, W., and F. Mosteller. 1979a. "Representative Sampling, I: Non-scientific Literature." *International Statistical Review* 47: 13–24.

———. 1979b. "Representative Sampling, II: Scientific Literature, Excluding Statistics." *International Statistical Review* 47: 111–127.

———. 1979c. "Representative Sampling, III: The Current Statistical Literature." *International Statistical Review* 47: 245–265.

Lachin, J. M., N. Tygstrup, and E. Juhl. Eds. 1982. *The Randomized Clinical Trial and Therapeutic Decisions*. New York: Marcel Dekker.

LaLonde, R. 1986. "Evaluating the Econometric Evaluations of Training Programs with Experimental Data." *American Economic Review* 76: 604–620.

LaLonde, R., and R. Maynard. 1987. "How Precise Are Evaluations of Employment and Training Programs: Evidence from a Field Experiment." *Evaluation Review* 11: 428–451.

Lauritzen, S., B. Thiesson, B., and D. Spiegelhalter. 1993. "Diagnostic Systems Created by Model Selection Methods—A Case Study." Fourth International Workshop on Artificial Intelligence and Statistics.

Lauritzen, S., and N. Wermuth. 1989. "Graphical Models for Associations between Variables, Some of Which Are Qualitative and Some Quantitative." *Annals of Statistics* 17: 31–37.

Leamer, E. E. 1985. "Vector Autoregressions for Causal Inference." In *Understanding Monetary Regimes*, ed. K. Brunner and A. Meltzer. Supplement to the *Journal of Monetary Economics*. Amsterdam: North-Holland.

Legendre, A. M. 1805. *Nouvelles Méthodes pour la Détermination des Orbites des Comètes*. Paris: Courcier. Reprint, New York: Dover, 1959.

Levasseur, E. 1885. "La Statistique Graphique." In *Jubilee Volume of the Statistical Society*. London: Edward Stanford. 218–250.

Lieberson, S. 1961. "A Societal Theory of Race and Ethnic Relations." *American Sociological Review* 26: 902–910.

———. 1985. *Making It Count: The Improvement of Social Research and Theory*. Berkeley: University of California Press.

———. 1991. "Small N's and Big Conclusions: An Examination of the Reasoning in Comparative Studies Based on a Small Number of Cases." *Social Forces* 70: 307–320. Reprinted (slightly revised) in *What is a Case?* ed. C. C. Ragin and H. S. Becker. New York: Cambridge University Press, 1992. 105–118.

———. 1992. "Einstein, Renoir, and Greeley: Some Thoughts about Evidence in Sociology." *American Sociological Review* 57: 1–15.

Lieberson, S., and A. Silverman. 1965. "The Precipitants and Underlying Conditions of Race Riots." *American Sociological Review* 30: 887–898.

Lindley, D., and R. Novick. 1981. "The Role of Exchangeability in Inference." *Annals of Statistics* 9: 45–58.

Liu, T. C. 1960. "Under-identification, Structural Estimation, and Forecasting." *Econometrica* 28: 855–865.

Longstaff, G. B. 1891. *Studies in Statistics*. London: Edward Stanford.

Lösch, A. 1954. *The Economics of Location*. New Haven, Conn.: Yale University Press.

Lucas, R. E., Jr. 1976. "Econometric Policy Evaluation: A Critique." In *The Phillips Curve and Labor Markets*, ed. K. Brunner and A. Meltzer. Vol. 1 of the Carnegie-Rochester Conferences on Public Policy. Supplementary series to the *Journal of Monetary Economics*. Amsterdam: North-Holland. 19–64.

Mackenzie, D. A. 1981. *Statistics in Britain 1865–1930*. Edinburgh: Edinburgh University Press.

Maddala, G. S. 1992. *Introduction to Econometrics*. 2nd ed. McGraw-Hill.

Manski, C. F. 1991 "Regression." *Journal of Economic Literature* 29: 34–50.

————. 1993. "Identification Problems in the Social Sciences." In *Sociological Methodology 1993*, ed. P. V. Marsden. Oxford: Basil Blackwell. 1–56.

Marshall, A. 1885. "On the Graphic Method of Statistics." In *Jubilee Volume of the Statistical Society*. London: Edward Stanford. 251–261.

McCullagh, P. M., and J. A. Nelder. 1989. *Generalized Linear Models*. 2nd. ed. London: Chapman and Hall.

Meehl, P. 1954. *Clinical versus Statistical Prediction: A Theoretical Analysis and a Review of the Evidence*. Minneapolis: University of Minnesota Press.

Mill, J. S. 1965. *A System of Logic*. 8th ed. London: Longmans, Green.

Montgomery, D. C., and E. A. Peck. 1982. *Introduction to Linear Regression Analysis*. New York: Wiley.

Moore, H. L. 1911. *Laws of Wages*. Reprint, New York: Augustus M. Kelley, 1967.

Moore, K. A., and S. L. Hofferth. 1980. "Factors Affecting Early Family Formation: A Path Model." *Population and Environment* 3: 73–98.

Morselli, H. 1882. *Suicide: A Comparative Study*. New York: D. Appleton.

Morgan, M. S. 1990. *The History of Econometric Ideas*. Cambridge: Cambridge University Press.

Mosteller, F., and J. W. Tukey. 1977. *Data Analysis and Regression*. Reading, Penn.: Addison-Wesley.

Muñoz, N., F. X. Bosch, K.V. Shah, and A. Meheus. Eds. 1992. *The Epidemiology of Human Papilloma Virus and Cervical Cancer*. Lyon: International Agency for Research on Cancer; New York: distributed in the USA by Oxford University Press.

Needleman, H., S. Geiger, and R. Frank. 1985. "Lead and IQ Scores: A Reanalysis." *Science* 227: 701–704.

Neft, D. S., R. M. Cohen, and R. Korch. 1993. *The Sports Encyclopedia: Pro Football*, 11th ed. *The Modern Era, 1960–1992*. New York: St. Martin's Press.

Nemeth, E. 1981. *Otto Neurath und der Wiener Kreis: Wissenshaftlichkeit als revolutionärer*. Frankfurt a.M. and New York: politischer Anspruch.

Neter, J., W. Wasserman, and M. H. Kutner. 1989. *Applied Linear Regression Models*. 2nd ed. Homewood, Ill.: Irwin.

Nichols, E. 1986. "Skocpol and Revolution: Comparative Analysis vs. Historical Conjuncture." *Comparative Social Research* 9: 163–186.

Norton, P. N. 1902. *Statistical Studies in the New York Money Market*. New York: Macmillan.

Ogle, W. 1890. "On Marriage Rates and Marriage Ages, with Special Reference to the Growth of Population." *Journal of the Royal Statistical Society* 53: 253–280.

Papineau, D. 1985. "Causal Asymmetry." *British Journal of Philosophy of Science* 36: 273–289.

Pearl, J. 1986. "Fusion, Propagation and Structuring in Belief Networks." *Artificial Intelligence* 29: 241–288.

————. 1988. *Probabilistic Reasoning in Intelligent Systems*. San Mateo, Calif.: Morgan Kaufmann.

————. 1993. "Comment: Graphical Models, Causality and Intervention." *Statistical Science* 8: 266–273.

———. 1994a. "On the Statistical Interpretation of Structural Equations." Technical Report, Computer Science Department, University of California, Los Angeles.

———. 1994b. "On the Identification of Nonparametric Structural Equations." Technical Report, Computer Science Department, University of California, Los Angeles.

———. 1995. "Causal Diagrams for Empirical Research." *Biometrika* 82: 669–710, with discussion.

Pearl, J., and R. Dechter. 1996. "Identifying Independencies in Causal Graphs with Feedback." Technical Report, Computer Studies Department, University of California, Los Angeles.

Pearl, J., R. Dechter, and T. Verma. 1991. "Knowledge Discovery vs. Data Compression." Proceedings KDD-91 Workshop. Anaheim, Calif. 207–211.

Pearl, J., D. Geiger, and T. Verma. 1989. "The Logic of Influence Diagrams." In *Influence Diagrams, Belief Nets and Decision Analysis*, ed. R. M. Oliver and J. Q. Smith. New York: John Wiley & Sons. 67–87.

Pearl, J., and T. Verma. 1991. "A Theory of Inferred Causation." In *Principles of Knowledge, Representation, and Reasoning: Proceedings of the Second International Conference*, ed. J. A. Allen, R. Fikes, and E. Sandewall. San Mateo, Calif.: Morgan Kaufmann. 441–452.

Pearl, J., and N. Wermuth. 1993. "When Can Association Graphs Admit a Causal Explanation?" In *Proceedings of the Fourth International Workshop on Artificial Intelligence and Statistics*, 141–150. Reprint in *Artificial Intelligence and Statistics*, ed. P. Cheeseman and W. Oldford. Berlin: Springer-Verlag, 1994.

Pearson, K. 1897. "Mathematical Contributions to the Theory of Evolution—On a Form of Spurious Correlation Which May Arise When Indices Are Used in the Measurement of Organs." *Proceedings of the Royal Society* 60: 489–498.

Persons, W. M. 1910. "The Correlation of Economic Statistics." *American Statistical Association* 12: 287–232.

———. 1919. "An Index of General Business Conditions." *Review of Economic Statistics* 1: 111–205.

Peto, R., and H. zur Hausen. Eds. 1986. *Viral Etiology of Cervical Cancer.* Cold Spring Harbor Laboratory. Banbury Report No. 21.

Playfair, W. 1796. *A Real Statement of the Finances and Resources of Great Britain.* London: Whittingham.

Popper, K. R. 1964. *The Poverty of Historicism.* New York: Harper.

Porter, T. M. 1986. *The Rise of Statistical Thinking 1820–1900.* Princeton, N.J.: Princeton University Press.

Pratt, J., and R. Schlaifer. 1984. "On the Nature and Discovery of Structure." *Journal of the American Statistical Association* 79: 9–21.

Pratt, J., and R. Schlaifer. 1988. "On the Interpretation and Observation of Laws." *Journal of Econometrics* 39: 23–52.

Raftery, A. 1995. "Bayesian Model Selection in Sociology." *Sociological Methodology.*

Ragin, C. C. 1987. *The Comparative Method: Moving Beyond Qualitative and Quantitative Strategies.* Berkeley: University of California Press.

Ragin, C. C., and H. S. Becker. Eds. 1992.*What is a Case?* New York: Cambridge University Press.

Rao, C. R. 1973. *Linear Statistical Inference and Its Applications.* 2nd ed. New York: Wiley.

Reichenbach, H. 1956. *The Direction of Time.* Berkeley, Calif.: University of California Press.

Reiss, I., A. Banwart, and H. Forman. 1975. "Premarital Contraceptive Usage: A Study and Some Theoretical Explorations." *Journal of Marriage and the Family* 37: 619–630.

Rindfuss, R. R., L. Bumpass, and C. St. John. 1980. "Education and Fertility: Implications for the Roles Women Occupy." *American Sociological Review* 45: 431–447.

———. 1984. "Education and the Timing of Motherhood: Disentangling Causation." *Journal of Marriage and the Family* 46: 981–984.

Robinson, W. S. 1950. "Ecological Correlations and the Behavior of Individuals." *American Sociological Review* 15: 351–357.

Rodgers, R., and C. Maranto. 1989. "Causal Models of Publishing Productivity in Psychology." *Journal of Applied Psychology* 74: 636–649.

Rosenbaum, P. R., and D. B. Rubin. 1983. "The Central Role of the Propensity Score in Observational Studies for Causal Effects." *Biometrika* 70: 41–50.

Rosenberg, M. 1968. *The Logic of Survey Analysis.* New York: Basic Books.

Rosenthal, R. 1984. *Meta-analytic Procedures for Social Research.* Beverly Hills, Calif.: Sage.

Rubin, D. B. 1973. "Matching to Remove Bias in Observational Studies." *Biometrics* 29: 159–183. (Also erratum 30: 728.)

———. 1974. "Estimating Causal Effects of Treatments in Randomized and Nonrandomized Studies." *Journal of Educational Psychology* 66: 688–701.

———. 1977. "Assignment to Treatment Group on the Basis of a Covariate." *Journal of Educational Statistics* 2: 1–26.

———. 1978. "Bayesian Inference for Causal Effects: The Role of Randomization." *Annals of Statistics* 6: 34–58.

———. 1987. *Multiple Inputation for Nonresponse in Surveys.* New York: Wiley.

Russell, B. 1968. *Mysticism and Logic.* Garden City, N.Y.: Doubleday Anchor. [Original 1917.]

Salmon, W. C. 1971. *Statistical Explanation and Statistical Relevance.* Pittsburgh: University of Pittsburgh Press.

———. 1984. *Scientific Explanation and the Causal Structure of the World.* Princeton, N.J.: Princeton University Press.

Scheines, R. 1996. "Estimating Underidentified Structural Equation Models." Comstat, 1996 Proceedings.

Scheines, R., P. Spirtes, C. Glymour, and C. Meek. 1994. *TETRAD II: User's Manual*, Hillsdale, N.J.: Lawrence Erlbaum.

Scheines, R., P. Spirtes, C. Glymour, and S. Sorensen. 1990. "Causes of Success and Satisfaction among Naval Recruiters." Report to the Navy Personnel Research Development Center, San Diego, Calif.

Schmidt, F. L., and J. Hunter. 1977. "Development of a General Solution to the Problem of Validity Generalization." *Journal of Applied Psychology* 62: 529–540.

Seneta, E. 1987. "Discussion." *Journal of Educational Statistics* 12: 198–201.

Shaffer, C. 1993. "Selecting a Classification Method by Cross-Validation." Preliminary Papers of the Fourth International Workshop on Artificial Intelligence and Statistics.

Sherman, K. J., J. R. Daling, and J. Chu et al. 1991. "Genital Warts, Other Sexually Transmitted Diseases, and Vulvar Cancer." *Epidemiology* 2: 257–262.

Silberman, G., J. A. Droitcour, and E. W. Scullin. 1992. *Cross Design Synthesis: A New Strategy for Medical Effectiveness Research*. Washington, D. C.: U. S. General Accounting Office. Report No. GAO/PEMD–92–18.

Simon, H. A. 1954. "Spurious Correlation: A Causal Interpretation." *Journal of the American Statistical Association* 49: 467–479.

———. 1980. "The Meaning of Causal Ordering." In *Qualitative and Quantitative Social Research*, ed. R. K. Merton, J. S. Coleman, and P. H. Rossi. New York: Free Press. 65–81.

Singer, B., and M. M. Marini. 1987. "Advancing Social Research: An Essay Based on Stanley Lieberson's Making It Count." In *Sociological Methodology, 1987*, ed. C. C. Clogg. Washington, D.C.: American Sociological Association. 373–391.

Skyrms, B. 1980. *Causal Necessity: A Pragmatic Investigation of the Necessity of Laws*. New Haven, Conn.: Yale University Press.

Smart, J. J. C. 1963. *Philosophy and Scientific Realism*. London: Routledge and Kegan Paul.

Smelser, N. J., and D. R. Gerstein. 1986. *Behavioral and Social Science: Fifty Years of Discovery*. Washington, D.C.: National Academy Press.

Smith, H. L. 1990. "Specification Problems in Experimental and Nonexperimental Social Research." In *Sociological Methodology, 1990*, ed. C. C. Clogg. Vol. 20. Oxford: Basil Blackwell.

Snow, J. 1936. "On the Mode of Communication of Cholera." 2nd expanded edition of pamphlet originally published in 1849. In *Snow on Cholera.*, ed. W. F. Frost Cambridge: Harvard University Press. iii–139.

Sobel, M. E. 1994. "Causal Inference in the Social and Behavior Sciences." In *Handbook of Statistical Modeling in the Social Sciences*, ed. G. Arminger, C. C. Clogg, and M. E. Sobel. New York: Plenum.

Sober, E. 1988. "The Principle of the Common Cause." In *Probability and Causality: Essays in Honor of Wesley C. Salmon*, ed. J. Fetzer. Dordrecht: Reidel.

Sosa, E. 1980. "Varieties of Causation." *Grazer Philosophische Studien* 11: 93–103. Reprinted in *Causation*, ed. M. Tooley and E. Sosa. Oxford: Oxford University Press, 1993. 234–242.

Speed, T. P., and H. T. Kiiveri. 1986. "Gaussian Markov Distributions over Finite Graphs." *Annals of Statistics* 14: 138–50.

Spirtes, P. 1992. "Building Causal Graphs from Statistical Data in the Presence of Latent Variables." In *Proceedings of the IX International Congress on Logic, Methodology, and the Philosophy of Science*, ed. B. Skyrms. Uppsala, Sweden.

———. 1993. "Directed Cyclic Graphs, Conditional Independence, and Non-Recursive Linear Structural Equation Models." CMU Philosophy, Methodology, and Logic Technical Report CMU-PHIL-35.

———. 1994. "Conditional Independence in Directed Cyclic Graphical Models for Feedback." Technical Report CMU-PHIL-54, Department of Philosophy, Carnegie Mellon University, Pittsburgh, Penn.

Spirtes, P., and C. Glymour. 1993. "Inference, Intervention, and Prediction." Preliminary Papers of the Fourth International Workshop on Artificial Intelligence and Statistics.

Spirtes, P., C. Glymour, and R. Scheines. 1991. "An Algorithm for Fast Recovery of Sparse Causal Graphs." *Social Science Computer Review* 9: 62–72.

Spirtes, P., C. Glymour, and R. Scheines. 1993. *Causation, Prediction, and Search*. Lecture Notes in Statistics 81. New York: Springer-Verlag.

Spirtes, P., and C. Meek. 1995. "Learning Bayesian Networks from Data." In *Proceedings of the First Conference on Knowledge Discovery and Data Mining*.

Spirtes, P., R. Scheines, C. Glymour, and C. Meek. 1993. "*TETRAD II*. Documentation for Version 2.2." Technical Report, Department of Philosophy, Carnegie Mellon University, Pittsburgh, Penn.

Spirtes, P., R. Scheines, C. Meek, and C. Glymour. 1994. *Tetrad II: Tools for Discovery*. Hillsdale, N.J.: Lawrence Erlbaum.

Spirtes, P., and T. Verma. 1992. "Equivalence of Causal Models with Latent Variables. Philosophy, Methodology and Logic." Technical Report 33, Carnegie-Mellon University.

Stigler, S. M. 1986. *The History of Statistics: The Measurement of Uncertainty before 1900*. Cambridge, Mass: Belknap Press.

Suppes, P. 1970. *A Probabilistic Theory of Causality. Acta Philosophica Fennica, XXIV.* Amsterdam: North-Holland.

Thomas, D. S. 1925. Some Social Aspects of the Business Cycle. Doctoral dissertation, London University.

Thurstone, L. 1935. *The Vectors of Mind*. Chicago: University of Chicago Press.

Timberlake, M., and K. Williams. 1984. "Dependence, Political Exclusion and Government Repression: Some Cross National Evidence." *American Sociological Review* 49: 141–146.

Tinbergen, J. 1939. *Statistical Testing of Business Cycle Theories, Vol I: A Method and its Application to Investment Activity*. Geneva: League of Nations.

———. 1940. "Reply to Keynes." *Economic Journal* 50: 141–154.

Tufte, E. 1974. *Data Analysis for Politics and Policy*. Englewood Cliffs, N.J.: Prentice-Hall.

Turner, S., and R. A. Factor. 1994. *Max Weber: The Lawyer as Social Thinker*. London: Routledge.

Verma, T., and J. Pearl. 1990a. "Causal Networks: Semantics and Expressiveness." In *Uncertainty in AI 4*, ed. R. Shachter, T. S. Levitt, and L. N. Kanal. Elsevier Science Publishers. 69–76.

———. 1990b. "Equivalence and Synthesis of Causal Models." In *Proceedings of the Sixth Conference on Uncertainty in AI*. Mountain View, Calif.: Association for Uncertainty in AI. 220–227.

———. 1990c. "On Equivalence of Causal Models." Technical Report, R-150, Department of Computer Science, University of California at Los Angeles, April.

Wainer, H. Ed. 1986. *Drawing Inferences from Self Selected Samples*. New York: Springer.

———. 1989. "Eelworms, Bullet Holes, and Geraldine Ferraro: Some Problems with Statistical Adjustment and Some Solutions." *Journal of Educational Statistics* 14: 121–140.

Wedelin, D. 1993. Discovering Causal Structure from Data. Preprint, Department of Computer Science, Chalmers University of Technology, Sweden.

Welsh, J. R., S. K. Kucinkas, and L. T. Curran. 1990. *Armed Services Vocational Battery (ASVAB): Integrative Review of Validity Studies*. Air Force Human Resources Laboratory Report AFHRL-TR-90-22.

Wermuth, N., and S. Lauritzen. 1983. "Graphical and Recursive Models for Contingency Tables." *Biometrika* 72: 537–552.

———. 1990. "On Substantive Research Hypotheses, Conditional Independence Graphs and Graphical Chain Models." *Journal of the Royal Statistical Society*, Series B, 52: 21–50.

White, H. 1980. "A Heteroscedasticity-Consistent Covariance Matrix Estimator and a Direct Test for Heteroscedasticity." *Econometrica* 48: 817–838.

———. 1982. "Maximum Likelihood Estimation of Misspecified Models." *Econometrica* 50: 1–25.

Whittaker, J. 1990. *Graphical Models in Applied Multivariate Statistics*. New York: Wiley.

Wold, H. O. A. 1949. "Statistical Estimation of Economic Relationships." Read to the International Statistical Conference, Washington, D.C. Proceeding published as a Supplement to *Econometrica* 17: 1–21.

Woodward, J. 1995. "Causation and Explanation in Econometrics." In *On the Reliability of Economic Models*, ed. D. Little. Boston: Kluwer. 9–61.

Yule, G. U. 1895 and 1896. "On the Correlation of Total Pauperism with Proportion of Out-Relief." *Economic Journal* 5: 603–611 and 6: 614–623.

———. 1896. "Notes on the History of Pauperism in England and Wales from 1850, Treated by the Method of Frequency Curves." *Journal of the Royal Statistical Society* 59: 318–349.

———. 1897. "On the Theory of Correlation." *Journal of the Royal Statistical Society* 60: 812–854.

———. 1899. "An Investigation into the Causes of Changes in Pauperism in England, Chiefly during the Last Two Intercensal Decades." *Journal of the Royal Statistical Society* 62: 249–295.

———. 1903. "Notes on the Theory of Association of Attributes in Statistics." *Biometrika* 2: 121–134.

———. 1906. "On the Changes in the Marriage and Birth-Rates in England and Wales during the Past Half Century; with an Inquiry as to Their Probable Causes." *Journal of the Royal Statistical Society* 69: 87–130.

———. 1909. "The Application of the Method of Correlation to Social and Economic Statistics." *Journal of the Royal Statistical Society* 72: 721–729.

Yule, G. U., and M. G. Kendall. 1937. *An Introduction to the Theory of Statistics*. 11 ed. rev. London: C. Griffin.

# INDEX